Nursing Care in the Genomic Era: A Case-Based Approach

Jean F. Jenkins, PhD, RN, FAAN
Senior Clinical Advisor
National Human Genome Research Institute
National Institutes of Health
Bethesda, Maryland

Dale Halsey Lea, MPH, RN, CGC, APNG, FAAN
Assistant Director, Southern Maine Genetics Services
Foundation for Blood Research
Scarborough, Maine

Including personal stories about how genetic discoveries—old and new—have shaped hopes and dreams

JONES AND BARTLETT PUBLISHERS
Sudbury, Massachusetts
BOSTON TORONTO LONDON SINGAPORE

World Headquarters
Jones and Bartlett Publishers
40 Tall Pine Drive
Sudbury, MA 01776
978-443-5000
info@jbpub.com
www.jbpub.com

Jones and Bartlett Publishers Canada
2406 Nikanna Road
Mississauga, ON L5C 2W6
CANADA

Jones and Bartlett Publishers International
Barb House, Barb Mews
London W6 7PA
UK

Library of Congress Cataloging-in-Publication Data
Jenkins, Jean F.
Nursing care in the genomic era : a case based approach / Jean Jenkins, Dale Lea.
p. ; cm.
Includes bibliographical references and index.
ISBN 0-7637-3325-3 (pbk.)
1. Genetic counseling. 2. Nursing. 3. Genomics. 4. Medical genetics.
[DNLM: 1. Nursing Care—trends—Case Reports. 2. Genetics, Medical—trends—Case Reports. 3. Genomics—trends—Case Reports. 4. Nurse-Patient Relations—Case Reports. WY 100 J52n 2004]
I. Lea, Dale Halsey. II. Title.
RB155.7.J46 2004
616'.042—dc22

2004019591

Production Credits
Acquisitions Editor: Kevin Sullivan
Production Manager: Amy Rose
Editorial Assistant: Amy Sibley
Production Assistant: Kate Hennessy
Associate Marketing Manager: Emily Ekle
Manufacturing and Inventory Coordinator: Amy Bacus
Composition: ATLIS Graphics and Design
Cover Design: Kristin E. Ohlin
Printing and Binding: Malloy, Inc.
Cover Printing: Malloy, Inc.

Cover Art: Reproduced with kind permission of Wessex Regional Genetics Laboratory. The quilt is currently displayed in the Post-Graduate Medical Centre at Salisbury District Hospital. The quilt was designed by Janet McCallum and made by the Sarum Quilters, Salisbury, England.

Printed in the United States of America
08 07 06 05 10 9 8 7 6 5 4 3 2

For Rebecca Lynn Jenkins, the light of my life.

Jean (Grammy)

For Tom, Allie, Halsey, and Annie, with thanks for your love, support, and patience and for serving as a special reminder that "The fruit of love is service." (Mother Teresa)

Dale

And for all those who shared their stories with us, you have so eloquently taught us that, "We must be willing to let go of the life we have planned so as to have the life that is waiting for us." (E.M. Forster) Thank you!

CONTENTS

FOREWORD Francis Collins MD, PhD & Alan Guttmacher MD xiii

INTRODUCTION *Why is genomics important for nurses to utilize in their practice?* xv

CHAPTER 1 *Connecting Genomics to Biology* 1

Introduction: Genetics Old and New 1
Elucidating the Structure and Function of the Human Genome 6
Human Genome Discoveries 7
Chromosomes, Genes, and DNA: Basic Human Genetics Terminology 7
DNA and RNA: How Proteins Are Made 9
Genetic Variation and Basic Patterns of Inheritance 12
New Genomics and Biology 16
Nursing Implications: The Importance of Family History in Assessment, Planning, Diagnosis, and Treatment 16
Summary 20
Chapter Activities 20

CHAPTER 2 *Connecting Genomics to Practice* 23

Family History as a Tool 23
Variation in the Human Genome 25
Family History Tools 27
Ethical Issues and Family History 52
Genetics and Genomics in Practice 54
Resources for Public Information 55
Summary 55
Chapter Activities 56

CHAPTER 3 *Connecting Genomics to Health Benefits* 59

Introduction 59
Genetic Approaches to Prevention and Early Detection 60

Genetic Testing 61
Disease and Drug Response 63
Therapeutic Approaches 63
Health Modifiers 64
Conveying Genetic Risk Information 65
Monitoring for Residual Disease and Medical Surveillance 65
Health Care Resources and Systems 66
Summary 66
Chapter Activities 67

CHAPTER 4 *Connecting Genomics to Health Benefits: Genetic Testing* **69**

Introduction 69
Genetic Testing: What it is and How it is Used Clinically 75
Genetic Testing Defined 75
Types of Genetic Testing 78
Clinical Genetic Tests 79
Research Genetic Tests 80
Oversight of Genetic Testing 80
Information Provided by Genetic Testing 80
Clinical Uses of Genetic Testing: Newborn Screening, Carrier Testing, Prenatal Screening and Diagnosis 84
Newborn Screening 84
Family Education and Support 90
Carrier Testing 93
Prenatal Screening and Diagnosis 95
Potential Risk of Genetic Testing 103
Cultural and Economic Issues and Genetic Testing 104
Summary 106
Chapter Activities 107

CHAPTER 5 *Connecting Genomics to Health Outcomes: Pharmacogenetics and Pharmacogenomics* **111**

Introduction 111
Pharmacogenetics and Pharmacogenomics 116
Genetic Mechanisms and Variable Drug Response 118
Pharmacokinetics 119

Pharmacodynamics 120
Clinical Applications of Pharmacogenetics 120
Gene-Environment Interactions and Clinical Interventions 129
Nursing Implications 130
Nursing Roles 131
Ethical and Societal Issues 136
Summary 141
Chapter Activities 142

CHAPTER 6 *Putting the Pieces Together: How I Learned of the Genetic Condition in My Family* **147**

Introduction 147
Indications for Making a Genetics Referral: Putting the Pieces Together 153
A Nursing Framework for Making a Genetics Referral 154
Nursing Assessments and Genetic Referrals 156
Finding Peer Support 164
Summary 165
Chapter Activities 166
COMMENTARY: "Breaking the News" Talking with Parents about Their Child's Birth Defect or Genetic Condition, Physician, Alan Guttmacher, MD 167

CHAPTER 7 *Finding What I Need* **175**

Introduction 175
How Genetics Professionals Became a Part of My Life 182
Genetics Counseling and Evaluation: What Is It and Who Are the Providers? 183
Genetics Specialists: Who They are and What They Do 188
Nursing and Genetic Education and Counseling 191
Future Models for Genetic Counseling and Evaluation 194
Summary 200

Chapter Activities 201
COMMENTARY: Without a Bigger Team, Our "Genomics" Cup Will Runneth Over, Genetic Counselor, Don Hadley, MS, OGC 203

CHAPTER 8 *Transitions and New Understandings: How Genetics Has Transformed My Life* **209**

Introduction: The Impact of Genetic Information 209
Transitions and New Understandings: Genetics and My Personal Life 217
Before Testing: How I Coped 218
To Test or Not To Test 218
How My Test Result Affected Me 227
More Decisions: Sharing My Genetic Information 229
New Roles and Relationships: Genetics and My Family 230
Nursing Implications 232
Summary 237
Chapter Activities 237
COMMENTARY: The Role of the Psychologist in Genetics Clinical Care, Psychologist, Andrea Farkas Patenaude, PhD 241

CHAPTER 9 *Connecting Genomics to Society* **247**

Introduction 247
Genomic Data and Race, Ethnicity, and Culture 248
Genomics and the Concept of "Self" 253
Genomics and the Family: Traits and Behaviors 255
Genomics and Community Identity 256
Genomics, Race, Ethnicity, and Research 257
Ethical Boundaries and Use of Genomics 258
Translation of Research to Policy: Integration of Genetics and Genomics 259
Implications for Nursing: Transcultural Nursing 262
Chapter Activities 264

CHAPTER 10 *Connecting Genomics to Society: Spirituality and Religious Traditions* **267**

Introduction 272

Genomics and the Concept of Normal: What is Normal? 274
Genomics and the Concept of Soul: When Does Life Begin? 276
Genomics and Identity: What Does It Mean to Be Human? 277
Religious Boundaries and Use of Genomics: What Can We Do? What Should We Do? 277
Gene Therapy: Is It All in the Genes? 278
Cloning 280
Stem Cell Research 281
Translation of Religion to Policy: Integration of Genetics and Genomics 282
Implications for Nursing: Spirituality in Care 283
How Then Should We Face the Future? 284
Chapter Activities 285

CHAPTER 11 *My Hopes and Fears: How I Feel about Genetic Research* **287**

Introduction 287
Genetic Research 292
Genetc Testing Research 293
Genetic Mapping Research 294
Treatment Research: Clinical Trials 296
Pharmacogenetic Research 298
Ethical and Social Considerations with Genetic Testing and Treatment Research 301
Genetics Research and Racially or Ethnically Diverse Populations 311
Genetic Testing and Community Research 312
Summary: Implications for Nurses 313
Chapter Activities 315

CHAPTER 12 *Braving New Frontiers: What Is My Future with My Genetic Condition?* **317**

Introduction 317
Profound Change 317
Power 319
Possibilities 319
Policy 320

Patience 320
Persuaded 321
Professionals 321
Public Perception 322
Public Health 322
Preventive 323
Predictive 323
Perplexities 324
Performance 324
Promise 325
Priorities 328
Passionate 329
Chapter Activities 329
COMMENTARY: Kevin Lewis, Chairman of the Board, Colon Cancer Alliance 330

CHAPTER 13 *Integration of Genomics to Improve Health Care* **333**

Introduction 333
Research Advances and Integration of Genomics into Health Care 334
HapMap 335
Genomics and Future Applications 335
Genomics and Nursing Practice: Expanding Roles for Nurses 335
Educating Nurses to Face the Challenge 351
Research Directions 353
Moving Forward 354
Chapter Activities 354
COMMENTARY: Summary of the Contribution of the Credentialing Process to Assurance of Quality Nursing Care That Integrates Genetics/Genomics, Rita Black Monsen, DSN, MPH, RN 358

CHAPTER 14 *Genomics Resources for Nurses* **363**

The Nursing Workforce 363
Nurses and Genomics Resources 364
Professional Nursing Societies: Genetics Resources for Nurses 364

New Information Technologies and Genomics Resources 380
Chapter Activities 380

ADDENDUM NCHPEG Competencies **383**
Purpose 384
Implementation 385
Recommendations 386
References 389
Acknowledgments 389
NCHPEG Member Organizations 390

INDEX **395**

FOREWORD

With the recent completion of the human genome sequence, and thus the successful conclusion of the Human Genome Project, we stand at the dawn of the genome era. And, as this book eloquently details, it is an era of unparalleled possibility in biology and of unequalled promise for our patients.

The possibility and promise come from the ability of genomics to help us understand the pathophysiology of human disease—and human health—in much greater detail than ever before. Genomics-based approaches and resources will play a key role in redefining our categorization of disease, from one based on symptomatology to one focused on biologic causation. And, by focusing us on the biological pathways that lead to disease, it will provide new approaches to prevention and therapy, even for diseases in which genetic factors play a relatively small role in causation or progression. As an example, for too long we have labeled everyone who wheezes as having the same condition, "asthma," and thus approached them with one standard that fits some patients, but not others. If, instead, we recognize the specific combination of genetic and environmental mechanisms that lead to the particular individual's complaint of wheezing, and target those specific mechanisms with our therapies and preventive strategies, we will be able to promote benefit to our patients much more frequently.

Genomics will not only inform us on risk factors for disease, it will also point to reasons for good health. Why do some people who have many risk factors for a particular disease live long and healthy lives without ever developing the disease? For some, it is not merely good luck, but genetic variants that help them avoid the disease. For instance, individuals with variants in the gene *CCR5* prove relatively resistant to HIV and AIDS. And, by understanding the mechanism that underlies this resistance, we may be able to develop new strategies for lowering the risk of HIV and AIDS in the many individuals who lack this genetic variant. As we use genomics to identify such genetic factors and to learn how they help prevent disease in the few, we will achieve a new understanding of biology that we can harness to benefit the many.

But, whatever the breakthroughs in biological understanding that genomics provides, the key to benefiting our patients will be genomics-based health care. And, it will take a broad coalition of health professionals to determine how to integrate genomics into health care most effectively. Even when genetic testing becomes more widely useful and available, how do we optimally utilize it to educate patients, encourage them to alter their health behaviors, and, most importantly, improve health outcomes? As pharmacogenomics testing allows us to predict much more reliably who will benefit from a particular medication, who will show no effect, and who will suffer toxic effects, how do we explain such testing to patients in a way that helps, rather than confuses? As health professionals with a tradition of patient-oriented care, nurses must play a key role in both developing and applying these and many other aspects of genomic health care. Indeed, unless nursing plays a key role in genomics, the genome era will fail to live up to its promise. And, unless genomics plays a key role in nursing, the nursing profession will fail to live up to its promise.

This text is designed with care and skill to provide nursing students, faculty, and nurses in practice with up-to-date and accessible information on this powerful new approach to understanding, preventing, and treating disease. Welcome to the genome era.

Francis S. Collins, MD, PhD
Director, National Human Genome Research Institute, NIH

Alan E. Guttmacher, MD
Deputy Director; Senior Clinical Advisor to the Director; Director Office of Policy, Communications and Education (OPCE); National Human Genome Research Institute, NIH

INTRODUCTION

How many ideas have there been in history that were unthinkable ten years before they appeared?
Fyodor Dostoyevsky

Translating new human genome research discoveries into clinical practice is happening every day. This translation includes new ways to diagnose genetic conditions; in the past diagnosis relied on clinical symptoms and physical presentation. Genetic testing can now be used to make a diagnosis, to identify a carrier of a genetic condition such as cystic fibrosis, to provide presymptomatic testing of conditions such as Huntington disease, and to provide susceptibility testing for breast, ovarian, and colon cancers. New genetic technologies are providing new understandings of the genetic components of common diseases such as diabetes, heart disease, and Alzheimer's disease and are paving the way for individualized prevention and early intervention strategies.

Nurses are at the interface of this translation and will increasingly care for individuals and families who have a genetic condition or a genetic component to their health and disease. Some individuals will have lived with the knowledge or diagnosis of their condition for a long time, while others are just learning of the genetic contribution to their own and their family's health. Nurses will be gathering family history and other clinical information, offering genetic testing, and administering genetic therapeutics. Listening to and witnessing an individual's and his or her family's story of the genetic condition or issues of concern is an important part of these activities. Stories give meaning to and shed light on individual family values and health beliefs and convey the impact of a particular genetic condition in a family. Listening to an individual's story helps the nurse understand how genetics has been or is being integrated into daily living. It isn't just a story—it's someone's life.

The word "story" comes from the Greek language and means knowing, knowledge, and wisdom (Yoder-Wise & Kowalski, 2003). Stories told by individuals living with genetic conditions, both old and new, are a powerful way for nurses to connect with the patients' values, information, and beliefs. In the context of genetic health, patients' stories can help nurses gain knowledge of the importance of family history, the role of genetics in clinical care, and the values and health beliefs each individual brings to their situation.

For this book, we have gathered stories from persons willing to share their individual and family experiences with a variety of genetic conditions ranging from those that are well established, to those that are more newly recognized to have a genetic component. Each of the stories highlights a particular condition and the related biological, personal, and psychosocial issues. Each of the chapters is built on the stories and discusses the knowledge, skills, and attitudes needed by nurses and all health professionals today as recommended by the National Coalition for Health Professional Education in Genetics (NCHPEG) *(http://www.nchpeg.org)*.

We begin by connecting genetics–genomics to biology. Building on this foundation, we present the importance of the family history in risk assessment and intervention, emerging genetic testing and therapeutics, genetic counseling, and the implications of a genetic condition for an individual and family. Ethical and social issues are woven throughout the chapters, particularly with regard to the individual's concerns about privacy and confidentiality of genetic information and concerns regarding discrimination and stigmatization. Personal accounts of individual and family attitudes and thoughts regarding genomic research and future directions are also presented.

Understanding these issues and how individuals and families regard themselves in light of genomic health care is just a beginning step for nurses in being able to make a difference when providing nursing care in the genomic era. Nursing involvement in research related to how individuals and families come to understand their genetic condition, share information within and outside of the family, and redefine themselves in relationship to new genetic information is needed. Nurses, because of their unique and holistic approach to caring for individuals and families, have a wealth of knowledge, resources, and research ideas to utilize to further enhance and improve clinical care in the genomic age.

Jean Jenkins, PhD, RN, FAAN

Dale Halsey Lea, MPH, RN, CGC, APNG, FAAN

Reference

Yoder-Wise, P. S. & Kowalski, K. (2003). The power of storytelling. *Nursing Outlook, 51,* (1), 37–42.

CHAPTER 1
Connecting Genomics to Biology

A solid foundation in the basics of genetics and of the way in which genomics relates to biology increases comfort and flexibility during the counseling and treatment process. We have utilized the competencies set forth by the National Coalition for Health Professional Education in Genetics for all health professionals to design the book content. Information provided will help establish a foundation of knowledge on which to build future learning about how to translate genomics into clinical care.

Introduction: Genetics Old and New

The state of the science at a particular point in time has tremendous relevance to how and when people learn about a genetic condition in themselves or a family member. For instance, although people have recognized the inheritance of traits and diseases in families for centuries, knowledge of genes and deoxyribonucleic acid (DNA)—the underlying blueprint for health and disease—was not gained until recently (Collins et al., 2003). Watson and Crick's famous discovery of the double helix of DNA did not occur until the 1950s. The first genetic test, chromosome analysis, the diagnostic test for Down syndrome, was not even in clinical use until the 1960s. Genetics was generally housed in a research laboratory and was not considered a medical specialty service until the late 1970s when the federal government began funding public health genetics programs. Genetic diseases were considered rare, and there were limited resources and support available to families. Furthermore, health care providers including nurses have not had genetic information or genetic technology routinely included in their training until recently.

This brief historical perspective illustrates the old way in which genetics was used in clinical practice. Times are changing! As a result of the "new genetics" (Collins et al., 2003), the potential to improve the health of society is extraordinarily high. This potential for improved patient care will be realized only if health professionals become aware of the opportunities brought about by genome research and applications to practice.

Included in this chapter, Judy's story about her family history of cystic fibrosis (CF) provides a wonderful introduction to genetics—old and new—and to the ongoing translation of genetic knowledge and technologies into clinical care. Through her story, we learn that CF was in the 1940s just becoming recognized in clinical practice as a specific genetic condition inherited when both parents are carriers of a genetic change (mutation) in the DNA that, when passed on to the next generation, results in disease. The sweat test, used to diagnose CF, was not developed until the late 1970s. Only within the past 15 years have the gene changes (mutations) that cause CF been discovered, making carrier testing, diagnosis, and prenatal diagnosis more reliably available. There are now more than 1000 different mutations that are known to cause CF, some conferring only mild symptoms in persons not previously identified with CF, but who are now recognized as having CF of a lesser clinical variation (Jorde et al., 2000).

More than 10,000 identified genetic disorders known to be inherited in predictable patterns in families have been identified and classified (*http://www.omim.org*). Through research, more is being learned about the genetic contribution to common and complex diseases such as cancer, diabetes, and heart disease. Nurses in all practice settings will therefore be addressing patients' genetic health concerns. Provision of modern health care will require nurses to become knowledgeable about genetic science, genetic evaluation, and treatment services. This knowledge will enhance the ability of nurses to support individuals and families in the process of recognizing the genetic aspects of their health and well-being. The support begins with having an awareness that how and when individuals learn of a genetic condition in their family shape their personal experience and abilities to integrate new genetic information into their daily lives.

In this chapter, through Judy's story, we explore the transformation of old genetics into new genomics and the impact of new genomic technologies on this process. Nursing competencies—the knowledge and skills recommended by the National Coalition of Health Care Professionals—needed to understand this transformation and to support individuals and families living through and with this experience are highlighted.

Judy's Story

Judy tells of having lived with the diagnosis of CF in her family for a long time before genetics became a recognized medical specialty and newer genetic diagnostic criteria and testing became available.

I am the sole survivor of three siblings born to parents who were both carriers of cystic fibrosis. I really don't remember much about life before CF. My brother was born 13 months before me and lived for 2 weeks. He died because "his intestines were all stuck together." My sister was born shortly after my fourth birthday. From what I gather, she had symptoms that would now be classic for CF, but in 1949, there wasn't anything that was known about it. My parents took her to a major children's hospital, and the doctors there told them she would never be right, just to take her home, put a blanket over the bassinet, and let her die—something that was not an option for my parents. They knew something was wrong with her, but they didn't know what.

The family pediatrician, a kindly man with generalist knowledge, was the one who first figured out that my sister and brother had CF. He had never seen a case of CF but had read about it in a journal. He suggested my parents take my sister to the major children's hospital in our area, which they did. It was there they met Dr. L., one of the early and greatest pioneers in the field. He told them what CF was and what the prognosis would be: but he also told them that as long as there is life, there is hope. He was the most incredibly wonderful person! My brother had been treated and died in that hospital. For some reason, the case number for his autopsy was ingrained on my father's brain, and he mentioned this to Dr. L. The autopsy report from January 1944 was retrieved, and 5 years after the fact a diagnosis of CF, with death complications from meconium ileus, was made. I was only 4 years old at the time. Most of what I remember is from hearing the story over and over.

I have lots of memories . . . I remember the fact that, as long as possible, I was kept out of school so that I wouldn't be bringing home infections that could be fatal to my sister. I never attended nursery school or kindergarten, and I never had lots of playmates in the house . . . going to first grade was a difficult period for me. My mother lived in fear of something bad happening, and I think I knew this all the time. I spent a great deal of the period of my life between ages 4 and 10, when my sister died, sitting in hospital lobbies and waiting rooms because I was never allowed in the inpatient unit.

Dr. L. and his staff were great. This was in the days before the kind of health insurance we have today . . . most of it was self-pay. My mother often said they "put over $100,000 into the ground" when they buried Linda, my sister. Dr. L. would often use personal funds to pay for medication for those patients who couldn't afford it. I remember a couple of things about him. One was that, 3 days before my sister died, he came to the house in the evening . . . traveling about 2 hours to get there. It was March and it was cold. He was in the house for a couple of hours, and my parents found out when he went to leave that his wife had been sitting out in the car. He had come over to see Linda, and to tell my parents that the end was near and that she needed to be hospitalized. Now, this was a famous MD from a major academic teaching center . . . even back in 1955, folks like this did not make house calls.

Cystic fibrosis has been a pervasive issue in the background for me. I believe that it was a contributing factor to my choice of nursing as a profession. I know it delayed this decision. I had wanted to be a nurse for as long as I could remember, but my folks thought it was out of a misguided sense of guilt or because I had known only illness as I grew up. As a result, I was persuaded to try something else. I ended up going to a liberal arts school and majoring in English and American literature. After I graduated, with no clear sense of what to do with this magnificent education, I returned to school the following September to study nursing. I never regretted my liberal arts education where I learned a great deal, but clearly the family situation was an influence.

I had a great deal of difficulty conceiving, and often thought that maybe this was God's way of telling me that I shouldn't have children. When I became pregnant, I did worry a great deal about having a child with CF—more because of what it would have done to my mother than with the thought of dealing with a sick child. My mother always felt incredible guilt about the CF and died blaming herself for it. When my son was born, she was with us. He was breastfed; and the first time she changed a diaper and saw a breast stool, she freaked out, because it was loose and bulky like my sister's CF stools. I quickly had to remind her that CF stools were foul smelling and that this wasn't.

I selected my pediatrician because his practice partner was the local expert on CF. We were told that we had to wait until our son was 6 weeks old (remember, he was born in 1977) before he could be tested for CF, and when we called to schedule the test it was another 3 weeks before we could get in for testing. There was never a question that he would be tested. The day before the test was scheduled, I just happened to get a call from Dr. L., who was in San Francisco on vacation. He congratulated me on having had a baby and asked about CF status. I said he was being tested the next day, and Dr. L. got very angry that I had been made to wait that long, saying that if it were his practice, the baby would have been tested when the first meconium stool was passed. Thankfully, my son's test was negative.

That summer, Dr. L. offered to test us with an experimental test he was doing, using toenail clippings. He tested my mother, my husband, and me. Of course, my mother's test was positive. My husband and I tested negative, but I never have followed up with further testing. We are not going to have any more children, and I never wanted to be tested while my mother was alive because I know I couldn't have kept the results from her. I also knew that, if the test were positive, she'd feel incredible guilt; and I could not have done this to her.

My son knows about my sister and brother. He is now 25 and at present has no plans to marry or have children. When the time comes, he'll have lots of genetic testing choices because we are Jewish and there is the worry about Tay-Sachs and other conditions. I'll provide an opinion if asked, but the choice to be tested will be his.

Here is another piece of interesting information about my family and our communication. My mother died in 1997, 42 years after burying her daughter. As my father and I drove to the funeral parlor to make arrangements for her funeral, he told me something I had never known—and we had had many talks over the years.

I didn't realize that there were still secrets. My mother had three bouts with tuberculosis when I was little, one shortly after I was born, and a second the year after Linda, my sister, was born. The third bout was the year before Linda died. My father had not shared this information with anyone, but had lived with sole knowledge of it for 42 years. He probably had not told me because he never wanted my mother to know that her illness had been a contributing factor to Linda's death. I told him that day that the usual cause of death for CF patients was a superimposed bacterial infection, and that in Linda's case it just happened that the bacterium was tuberculosis. He said that he knew that CF was the gun and tuberculosis was the bullet, but he had never been able to share his feelings with my mother because it would have driven her over the edge. I felt so bad for this man, my father, who had lived with this for so many years. It was clear that he couldn't wait beyond the day of my mother's death to share it with another human being.

I do worry about the utilization of genetic information. I know that, back 50+ years ago, I was shunned because people were afraid that whatever was in my family was contagious. It's hard to sort out why I felt so different growing up. Part of it was my mother's incredible fear that something would happen to her remaining child, and another part was my father's wanting me to be perfect. I always felt that I had to meet the family expectations for all three of us.

I think that knowledge can be both powerful and dangerous. I believe the decision to know or not know rests with the individual, and that all of us have to make our own decisions. While I have always wanted to know my carrier status, I chose not to find out because of the potential implications to my mother. And that was fine. I do not believe that there is such a thing as completely nondirective counseling. I think that, no matter how hard we try, our personal biases come through. Most of us are in the business of patient education and counseling because we believe that information is power. For some folks, it is a terrible burden, like the autopsy results were for my father.

I would like to see genetics become integrated into the mainstream, so that it's just one of many tools in the armamentarium, rather than being seen as something unique. After all, the most powerful genetic test is the family history, and we've been using that for years. As long as things are special or different, they get special attention and scrutiny. I think genetics is a core component of health care and needs to be fully integrated into practice. I think we should talk of "patient education and counseling" for genetic conditions rather than "genetic counseling." As nurses, for example, we provide patients with counseling and education about a variety of human conditions. We need to have the genetics education firmly integrated into professional education so that those of us who provide care can use this as one more tool. I think the specialists should be available for referral in complex cases, the same as the nutritionist or social worker is available for those with specialized needs, but that the generalist nurse should be able to provide the bulk of the care.

I'd like to see the integration be seamless so that the health care needs of the patient would be at the center, rather than the condition or medical specialty! That way the focus of care could be on the human response and it would be driven by the patient . . . sounds a whole lot like nursing to me!

Judy

Elucidating the Structure and Function of the Human Genome

As Judy suggests, to be able to provide patient education and counseling related to genetic topics, and to focus care on the patient, nurses will need to integrate the science of human genetics into daily practice. Such integration requires a firm understanding of basic human genetics concepts and terminology, patterns of biological inheritance and variation both within families and within populations, and the importance of family history in assessing predisposition to disease. The following section of Chapter 1 provides a general overview of these concepts. A listing of genetics resources that provide a more in-depth overview of human genetics and genomics is found in **Table 1-1.**

Table 1-1 Genomics to Biology: Resources for Nurses

National Library of Medicine Web Site
http://ghr.nlm.nih.gov/ghr/page/understandGenetics;jsessionid=8B67FFB15A99653742 54A7E459 893707 *Help Me Understand Genetics.* Presents basic information about genetics in understandable language and provides links to other online resources.
National Institutes of Health
National Institute of General Medical Sciences
NIH Publication No. 01-5021 December 2001 *http://www.nigms.nih.gov/news/science_ed/genepop/faq.html* *Genes and Populations.* Does everybody have the same genes? Presents information on basic genetics, genes, and genetic research.
Gene Clinics
http://www.geneclinics.org A publicly funded medical genetics information resource developed for physicians, other health care providers, and researchers.

Human Genome Discoveries

Since the initiation of the Human Genome Project in 1990—an international research endeavor whose goal is to uncover the mysteries of the human genome and translate this knowledge into new prevention and treatment modalities—the pace of genetic discoveries and their application to clinical practice has accelerated. A new model of health and disease has evolved that includes consideration of the contribution of genetics to the cause of disease, predisposition, inheritance, and treatment options. Guttmacher and Collins (2002) propose a broader definition of genetics and medicine—genomic medicine, which includes the study of the functions and interactions of all genes in the human genome, including their interactions with environmental factors.

Chromosomes, Genes, and DNA: Basic Human Genetics Terminology

The totality of the approximately 30,000 genes that humans have in their genetic makeup is called a genome. Genes are contained in chromosomes, and chromosomes are located within the cell nucleus (**Figure 1-1A**). There are 46 chromosomes in all human cells of the body, and these occur in pairs with the exception of reproductive cells (oocytes and sperm), which contain 23 chromosomes. Genes on each chromosome pair are homologous to each other, meaning they have the same position and order. Twenty-two pairs of chromosomes, called autosomes, are the same in women and men. The twenty-third pair is referred to as the sex chromosomes. A woman has two X chromosomes, while a man has one X chromosome and one Y chromosome (**Figure 1-1B**).

The makeup of each gene in the human genome, as well as factors determining gene expression, is specified in the DNA of the 46 chromosomes. Each individual gene is made up of a segment of DNA. DNA is a chemical that contains genetic instructions for making proteins needed for proper body functioning. Genes are responsible for making specific proteins, and control the rate at which proteins are made. In the case of CF, the gene for CF encodes a protein called "cystic fibrosis transmembrane regulator" (CFTR). CFTR is involved in regulating the transport of sodium ions across epithelial cell membranes. The fact that CFTR is involved in sodium and chloride transport gives us an understanding of the multiple effects of gene changes (mutations) at the CF gene locus (Nussbaum, McInnes, & Willard, 2001).

The structure of DNA carries the chemical information that allows the precise transmission of genetic information from one cell to its daughter cells and from one generation to the next. The specific structure of DNA, as identified by Watson and Crick in 1953, is a double helix. The double helix resembles a right-handed spiral

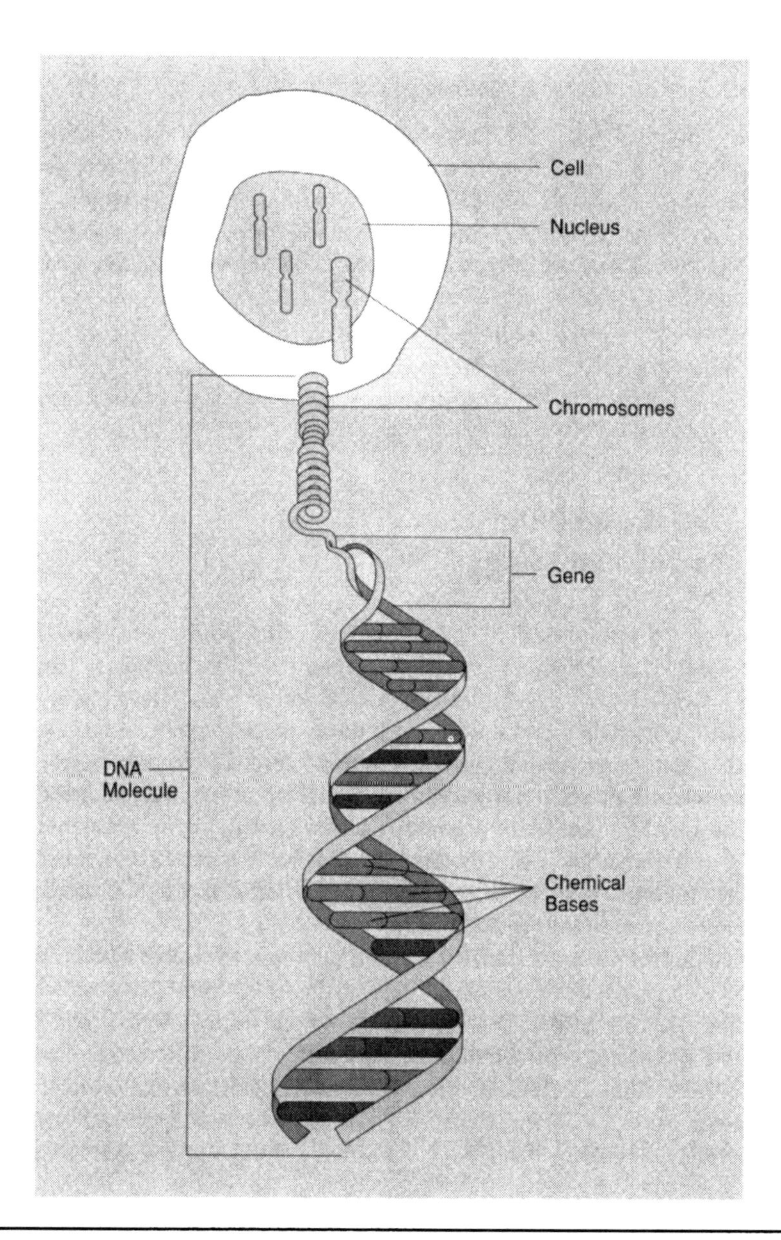

FIGURE 1-1A DNA, which carries the instructions that allow cells to make proteins, is made up of four chemical bases. Tightly coiled strands of DNA are packaged in units called chromosomes, housed in the cell's nucleus. Working subunits of DNA are known as genes.

Understanding Gene Testing
Published by the National Institutes of Health, National Cancer Institute

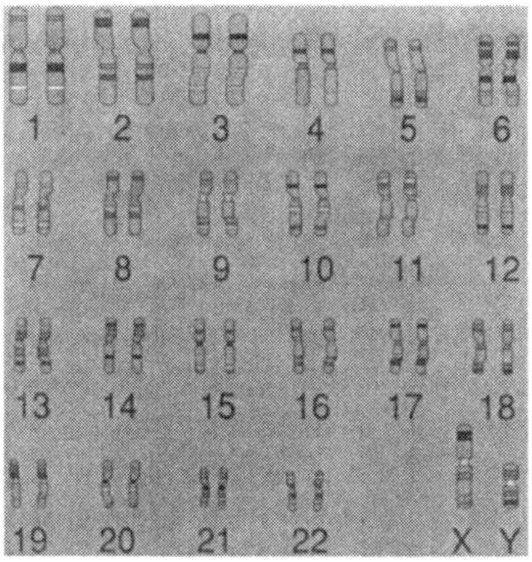

Figure 1-1B Each human cell contains 23 pairs of chromosomes, which can be distinguished by size and by unique banding patterns. This set is from a male, since it contains a Y chromosome. Females have two X chromosomes.

Understanding Gene Testing
Published by the National Institutes of Health, National Cancer Institute

staircase in which two strands (polynucleotide chains) run in opposite directions, held together by the hydrogen bonds between pairs of bases: Adenine (A), Thymine (T), Guanine (G), and Cytosine (C). DNA is composed of a sugar (deoxyribose), a phosphate group, and one of the four nitrogenous bases. A sugar group combined with a phosphate group and one of the four bases is called a nucleotide. The bases on each strand of DNA are paired in a specific manner. Adenine on one strand is always paired with thymine on the other strand, and guanine is always paired with cytosine. The pattern of pairing is related to the formation of hydrogen bonds. Adenine and thymine form two hydrogen bonds. Cytosine and guanine form three hydrogen bonds. The formation of hydrogen bonds between bases is known as base pairing, and the bases are said to be complimentary (Figure 1-1A).

DNA and RNA: How Proteins Are Made

Genetic information is contained in DNA in the chromosomes within the cell nucleus. The making of proteins, however, during which the genetic information encoded in the DNA is used, takes place outside of the cell nucleus in the cytoplasm. The link between these two types of information—the DNA code of genes and the amino acids needed to build proteins—is RNA (ribonucleic acid). Although RNA has a similar chemical structure to DNA, each nucleotide of RNA has a ribose sugar component instead of deoxyribose, and a uracil (U) replaces thymine (T) base. RNA also differs from DNA in that it is single-stranded, while DNA consists of a double helix.

The relationships among DNA, RNA, and protein synthesis are clearly delineated. DNA directs the creation and sequence of RNA. RNA directs the creation and sequence of polypeptides, and specific proteins are involved in the synthesis and metabolism of DNA and RNA. Human proteins are made up of building blocks called amino acids. Specific sequences of nucleotides in DNA determine the sequence of amino acids in the gene product. There are 22 amino acids and each is created by a set of three bases called a triplet or codon. Each triplet makes only one amino acid, yet amino acids can be specified by more than one triplet. The following triplets, for example, may specify the amino acid, phenylalanine: UUU, UUG, UUC, and UUA. The relationship of specific triplets to particular amino acids is referred to as the genetic code.

Genetic information is, therefore, stored in DNA by means of the genetic code, in which the sequence of adjacent bases ultimately determines the sequence of amino acids. DNA contains coding regions, called exons, and noncoding regions (introns). Most of DNA is noncoding. The DNA molecule also has regulatory regions for starting and stopping transcription and translation. There are specialized sequences of DNA related to tissue-specific expression.

The process of protein synthesis begins when RNA is made from the DNA template through a process known as transcription. The RNA carries the coded information in a form called messenger RNA (mRNA). Messenger RNA transports the coded information to the cytoplasm, where the RNA sequence is translated (decoded) to determine the sequence of amino acids in the protein being made. The process of translation takes place on ribosomes, small structures in the cytoplasm with binding sites for all of the interacting molecules, including mRNA, involved in protein synthesis. Ribosomes are made up of many different structural proteins in association with a specialized type of RNA known as ribosomal RNA (rRNA). A third type of RNA, transfer RNA (tRNA), involved in translation and the assembly of amino acids provides the molecular link between the coded base sequence of mRNA and the amino acid sequence of the protein. This flow of information from transcription of DNA to RNA to the translation of RNA to a protein product is referred to as the "central dogma" of molecular biology (Nussbaum, McInnes, & Willard, 2001). (**Figure 1-1C**).

Each human cell contains hundreds of structures called mitochondria that are essential for energy metabolism. Each mitochondrion contains about 10 copies of small, circular chromosomes. These chromosomes have double-stranded helices of DNA called mtDNA. Human mtDNA has only exons, and both strands of DNA

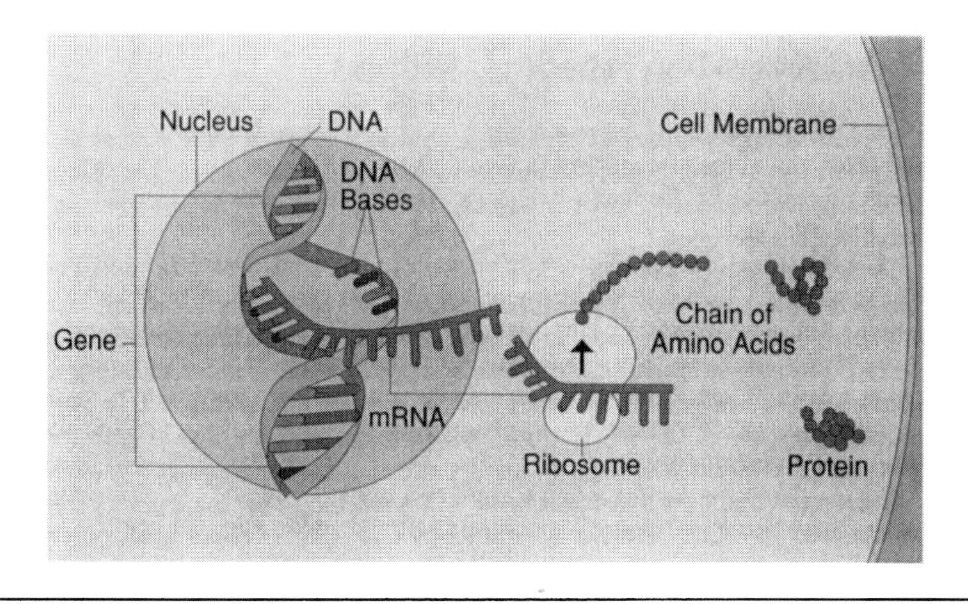

FIGURE 1-1C For a cell to make a protein, the information from a gene is copied, base by base, from DNA into new strands of messenger RNA (mRNA). Then mRNA travels out of the nucleus into the cytoplasm, to cell organelles called ribosomes. There, mRNA directs the assembly of amino acids that fold into a completed protein molecule.

Understanding Gene Testing
Published by the National Institutes of Health, National Cancer Institute

are transcribed and translated. Mitochondria are maternally inherited. The mitochondria in each cell come from the cytoplasm of the ovum at the time of fertilization. Sperm do not pass on mitochondria or mtDNA at the time of fertilization.

The regulated expression of the 30,000 genes contained in the 46 human chromosomes occurs as a result of complex interrelationships within a cell. Proper gene dosage, gene structure, transcription, mRNA stability, translation, protein processing, and degradation are all involved. Any change in any one of these processes may have an impact on a person's health. For some genes, changes in the level of expression can have serious clinical consequences, reflecting the critical nature of those gene products in a particular biological pathway. For other genes, changes in the level of functional gene product due to inherited variation in a particular gene or to changes resulting from nongenetic environmental factors such as diet have relatively little impact. The nature of inherited variation in the structure and function of genes and chromosomes, and the influence of this variation on the expression of particular traits, lies at the heart of genetic influences and health.

Genetic Variation and Basic Patterns of Inheritance

A remarkable degree of genetic variation is seen in humans in traits such as height, skin color, and blood pressure. Disease states such as CF are included in that spectrum of genetic variation, and this aspect of genetic variation is one focus of genomic health care.

Genetic variation occurs as a result of a process referred to as mutation—a change in DNA sequence. Some mutations result in genetic disease; others have no physical effects. Mutations occur in all body cells and can be either somatic or germ line cells (oocytes or sperm). Over time mutations in somatic cells can lead to conditions, such as cancer or Alzheimer's disease, and are not transmitted by parents to future generations. Germ line mutations, on the other hand, can be transmitted from one generation to the next.

Single-gene disorders occur as a result of a mutation in one or both copies of a gene located on an autosome, sex chromosome, or mitochondrial gene. A person who has two identical copies is homozygous, and a person who has two different copies of a gene is heterozygous. Since X and Y chromosomes do not have matching genes, males normally have unpaired genes for all X-linked and Y-linked genes. Males are said to be hemizygous for genes on X and Y chromosomes.

Single-gene conditions usually exhibit obvious and characteristic inheritance patterns in families, and are referred to as Mendelian, after the monk, Gregor Mendel, who originally described the rules of inheritance in garden peas. Traditional Mendelian or single-gene patterns are autosomal dominant, autosomal recessive, and X-linked (**Figures 1-D, E, F, G, H, and I**). Although individually rare, as a group single-gene disorders account for a significant proportion of disease and death. Single-gene conditions affect 2% of the population at some point in the entire life span. For example, the incidence of serious single-gene disorders is estimated to be from 6 to 8% in hospitalized children (Nussbaum, McInnes, & Willard, 2001).

Chromosome disorders are present when the defect is due to an excess or a deficiency of genes contained in whole chromosomes or chromosome segments. Down syndrome is an example of a chromosomal condition in which an extra copy of chromosome number 21 produces specific clinical features, including mental retardation, even though no individual gene on chromosome 21 is abnormal.

Multifactorial inheritance refers to the complex combination of genetic and environmental factors and is responsible for a number of developmental disorders resulting in congenital malformations such as cleft lip and/or palate and spina bifida, and for many common disorders in adult life such as heart disease, cancer, and diabetes. In these disorders, the disease is the result of a combination of small variations in genes that together can produce or predispose to a serious defect, often in concert with environmental influences. Multifactorial disorders tend to recur in

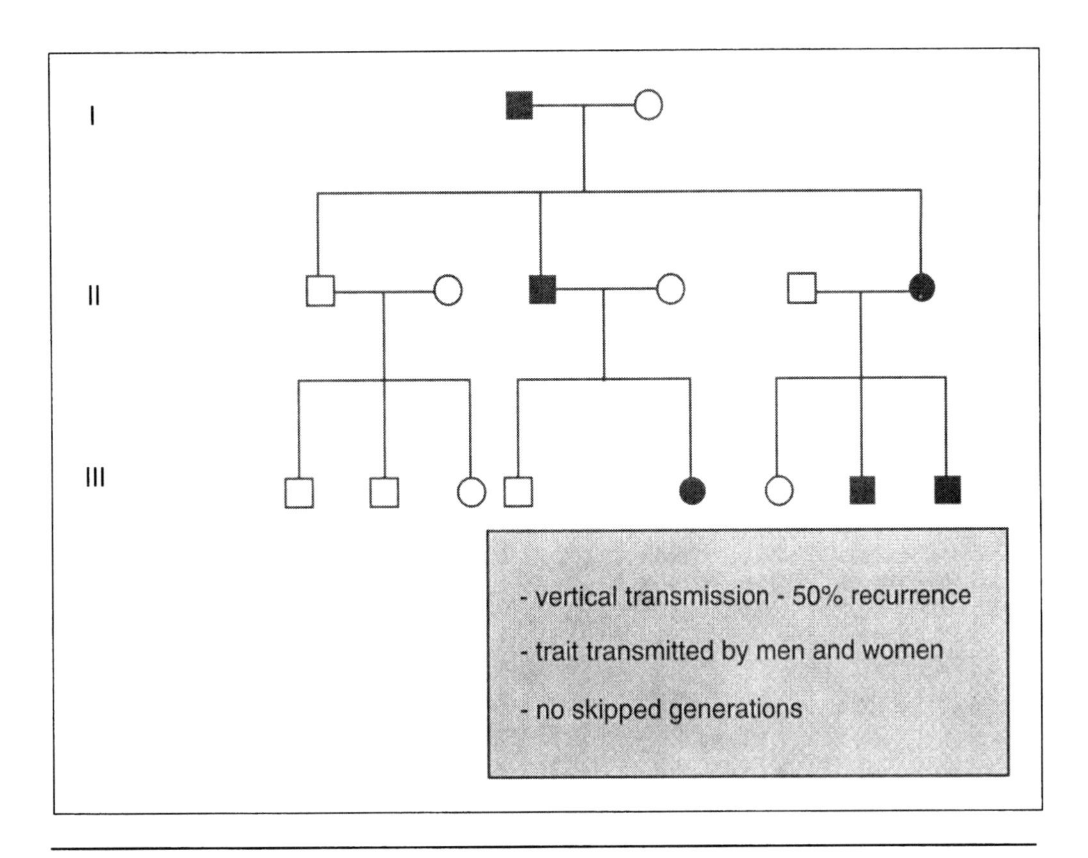

FIGURE 1-D Autosomal Dominant Inheritance

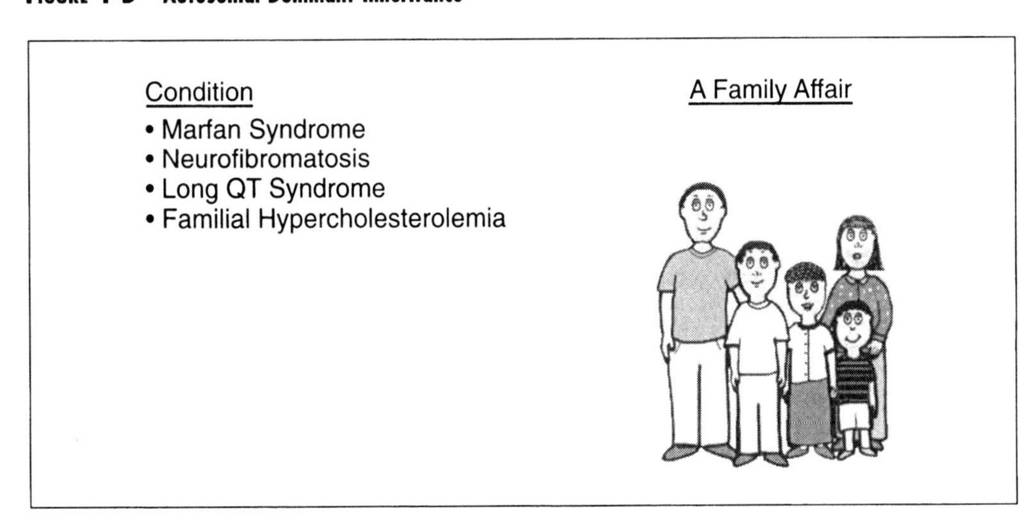

FIGURE 1-E Examples of Autosomal Dominant

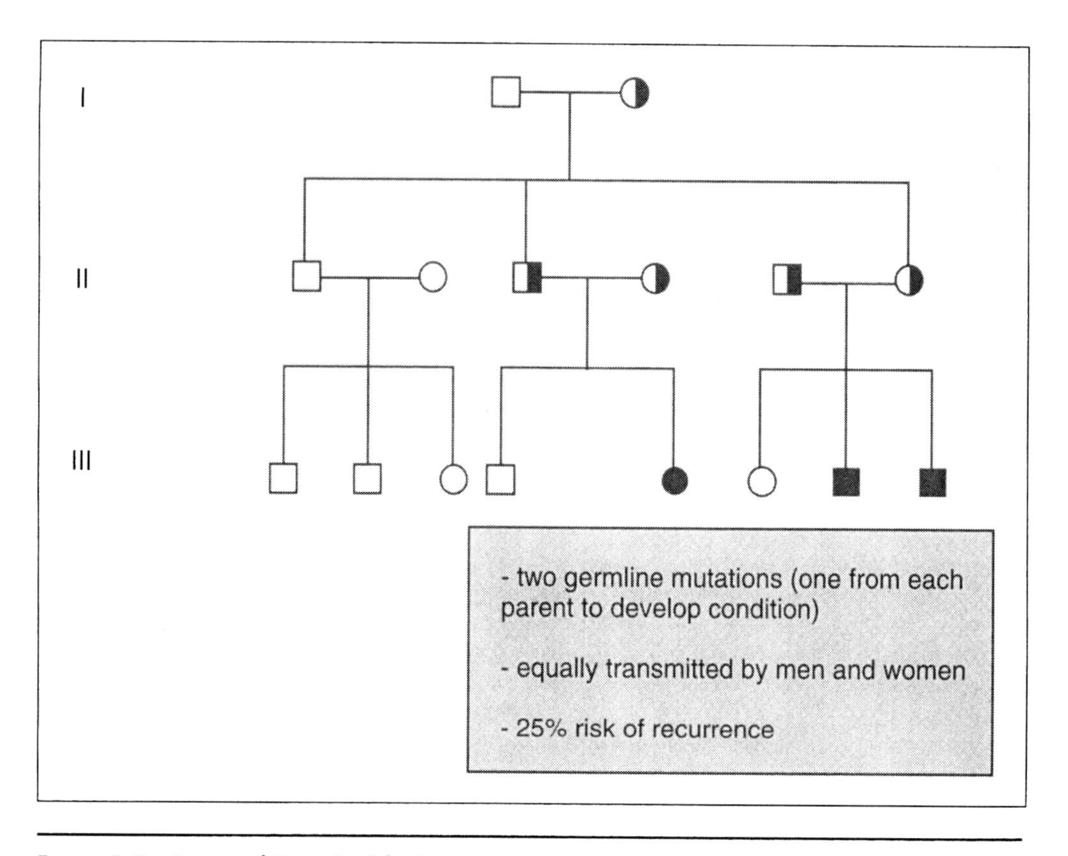

FIGURE 1-F Autosomal Recessive Inheritance

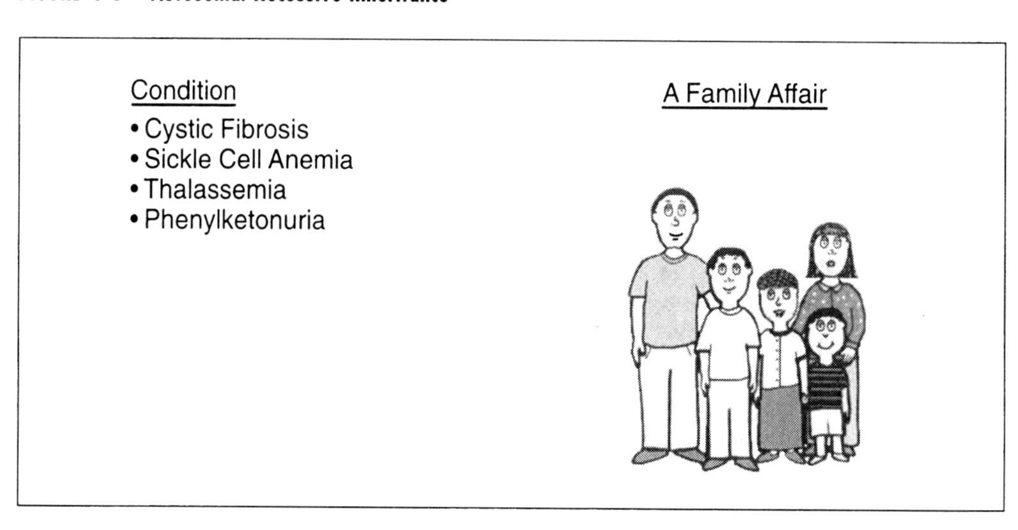

FIGURE 1-G Examples of Autosomal Recessive

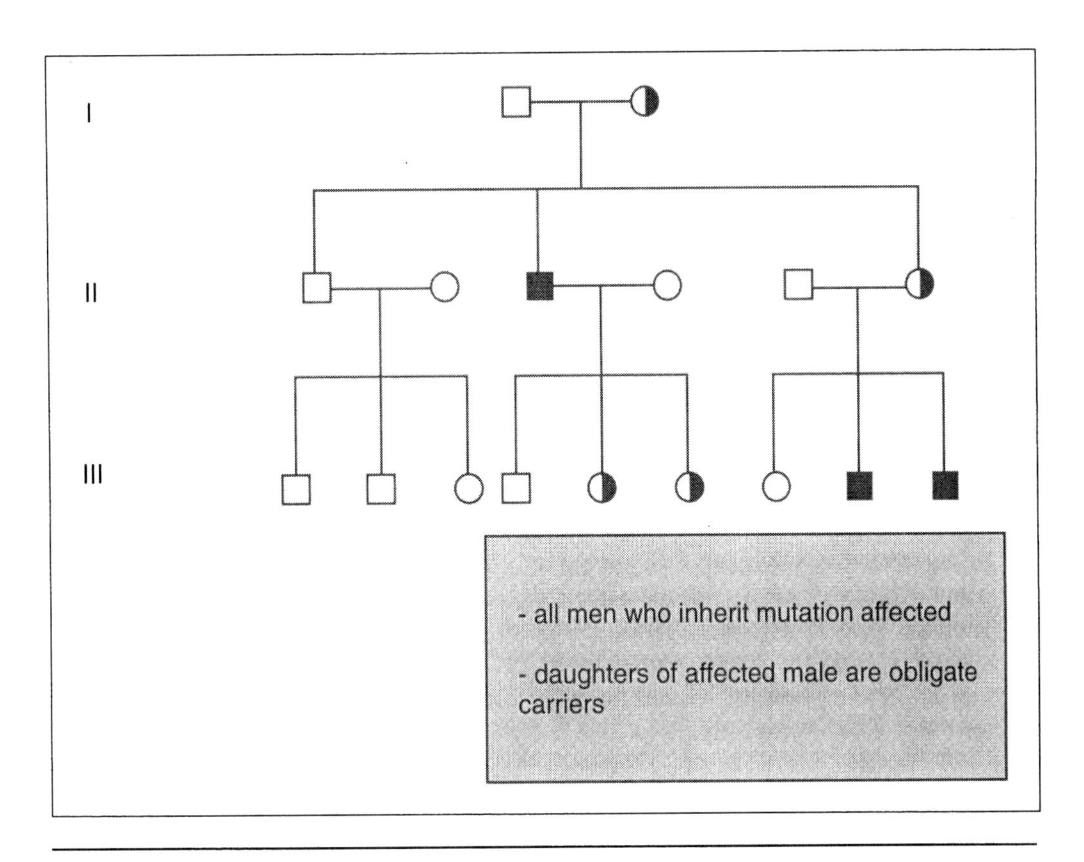

FIGURE 1-H X-Linked Recessive Inheritance

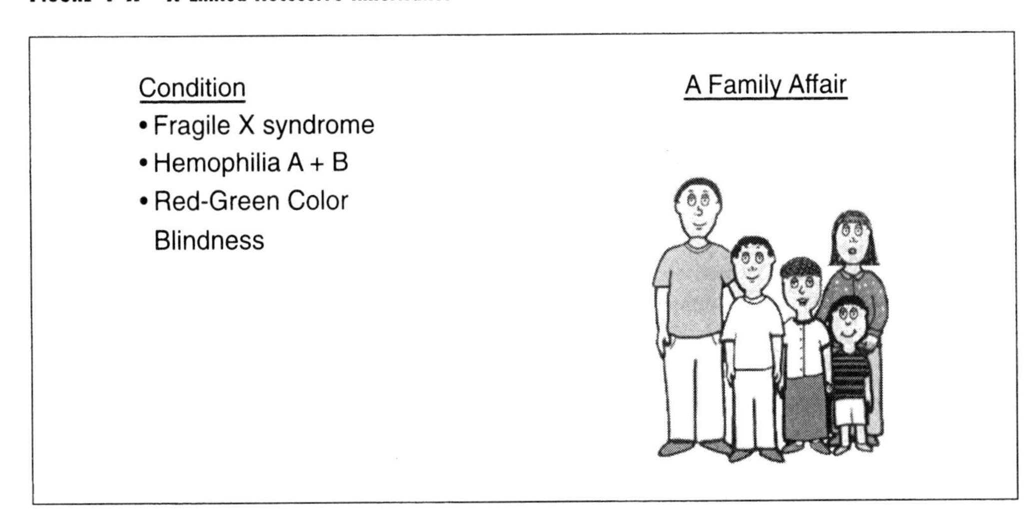

FIGURE 1-I Examples of X-Linked Recessive

families, but do not show the characteristic inheritance pattern seen in single-gene disorders. As more is learned about the contributions of genes interacting with the environment and each other, the differentiation between single-gene disorders and multifactorial disorders becomes less clear.

New Genomics and Biology

The more scientists learn about the human genome and other animal and plant genomes, the more they have come to understand the complexity of DNA's structure and how little is really known about its function. We know that, for example, only from 1 to 2% of DNA bases actually code and make proteins. A full understanding of all protein-coding DNA sequences remains to be established. In addition, scientists believe that an equal amount of the noncoding component of the genome is also functionally important, possibly containing most of the regulatory information that controls the expression of a human's 30,000 genes and other functional elements such as the sequence determinants of chromosome structure and function. Human genome research has revealed that more than half of the genome consists of highly repetitive sequences of DNA, and little is known about the function of these sequences. We also know little about the remaining noncoding, nonrepetitive DNA sequences. And this is just the beginning. Genes and gene products function together in complex, interconnected pathways that, together, contribute to the workings of cells, tissues, and organs. Scientists continue to work toward characterizing, cataloging, and understanding the entire set of functional elements encoded in the human genome, as well as understanding the basis of the heritable variation in the human genome as it relates to new understanding of health and disease, prevention, and treatment strategies (Collins et al., 2003).

Nursing Implications: The Importance of Family History in Assessment, Planning, Diagnosis, and Treatment

Judy's story of her family history of CF highlights the transformation, over time, of old genetics to new genomics, and provides a framework for the following chapters in this book. Judy describes the various stages of understanding the disease process and the cause of CF from a lack of diagnosis in her brother, to attempts at carrier testing using toenail clippings, to Judy's understanding of the genetic testing options available to her son. Today, nurses must have a broad base of genetic knowledge and an awareness of genetic variation and the role of family history so that they can provide appropriate support, education, and counseling to patients and families learning of and living with genetic conditions. In Judy's story, knowledge of the inheritance and variability of CF and how to construct a family tree would be a first step to helping future individuals and families recognize that a genetic condition or genetic component to their health exists. This knowledge also helps nurses to use genetic testing appropriately for diagnosis and for learning about options for treatment. (**Box 1-1**)

Nurses need to have an awareness of how and when individuals learn of the genetic condition present in their family so they can better understand how this knowledge has been incorporated into a person's and a family's life experience. Judy describes the impact of her sister's illness from CF on her life as a child and the expectations she felt from her parents as the only "well child." She talks about the way her family history helped to shape her choice of nursing as a profession, about how her family did not communicate genetic information, and how family secrets were kept until after her mother's death. Nurses can apply their unique philosophy of holism and family-centered approach to genomic patient care to assess and support families who are living and coping with genetic conditions.

Judy's story also provides a glimpse of the spectrum of testing and reproductive choices now available for many genetic conditions, including CF, and the range of individual responses. Judy tells of her decision not to pursue carrier testing for CF because of the potential implications to her mother if she learned that Judy was a carrier. She talks about her son, who will need to make his own decisions about

Box 1-1 Cystic Fibrosis

General Information: Cystic fibrosis (CF) is a relatively common condition occurring in 1 in 2500 births in the non-Hispanic, Caucasian population. It is inherited in an autosomal recessive manner, meaning that both parents of a person with CF are carriers. CF was first identified as a clinical entity in 1938.

Clinical Symptoms: Individuals who have CF have a decreased ability to transport chloride ions across cellular membranes. As a result, the exocrine glands producing sweat, mucus, and intestinal secretions do not function properly. Secretions are thick and can impair delivery of pancreatic enzymes from the pancreas to the digestive tract. Thick mucous accumulates in the lungs, leading to serious and damaging infection. About 85% of persons who have CF have both respiratory and digestive problems, while 15% have only respiratory infections. Males with CF are usually not fertile, and females with CF may experience reduced fertility. CF has highly variable clinical expression, with some individuals experiencing relatively little respiratory difficulty and a nearly normal life span. Improved treatment has improved survival rate of CF patients substantially during the past three decades. The average survival age is now about 30 years.

Box 1-1 continued

Genetic Testing for CF: The CF gene is located on chromosome 7, and over 1000 mutations in the CF gene have been identified. One particular CF mutation (ΔF508) accounts for about 70% of mutations in the Caucasian population. This mutation, along with several other relatively common ones, is tested for in genetic diagnosis of CF. Identification of specific CF mutations in a patient can help in predicting the course of the disease. For example, patients who have two copies of the ΔF508 CF mutation generally have the most severe presentation and nearly always have pancreatic insufficiency. In contrast, individuals with the R117H mutation have a milder clinical presentation and are less likely to have pancreatic insufficiency.

DNA testing can detect about 85 to 90% of Caucasian, about 97% of Ashkenazi Jewish, and about 75% of African-American carrier individuals. The estimated detection rates are lower for individuals of Hispanic and Asian-American heritage. Prenatal diagnostic testing for CF includes amniocentesis or chorionic villus sampling followed by DNA testing when the specific CF gene mutations are known in carrier parents. In the second trimester, high-resolution targeted ultrasound can identify echogenic bowel (characteristic of meconium ileus), which may be indicative of CF in the developing baby. Preimplantation diagnosis using in vitro fertilization techniques is also available to CF carrier couples.

The American College of Obstetrics and Gynecology (ACOG, 2001) currently recommends that CF carrier testing be offered to (1) individuals with a family history of CF, (2) reproductive partners of individuals who have CF, and (3) couples in whom one or both partners are Caucasian and who are planning a pregnancy or seeking prenatal care. It is also recommended that screening should be offered to those at higher risk of having children with CF (Caucasians, including Ashkenazi Jews) and in whom the testing is most sensitive in identifying carriers of a CF mutation. ACOG further recommends that carrier testing be made available to couples in other racial and ethnic groups who are at lower risk and in whom the test may be less sensitive. Screening for CF is offered when couples seek preconception counseling or infertility care, or during the first and early second trimesters of pregnancy.

Some states are now testing for CF as part of their newborn screening panel because a research study showed some benefit to early detection and treatment: improved nutritional status, decreased morbidity, and less deterioration in lung function.

How CF Is Inherited in Families: CF is inherited in an autosomal recessive manner. A parent who is a CF carrier has one CF gene mutation on one of the chromosome number 7 pair and the normal version of the CF gene on the other chromosome number 7. When both parents are CF carriers, they have, with each pregnancy, a 25% chance (1 in 4) to have a baby affected with CF; a 50% chance (1 in 2) to have a baby who will be a CF carrier like themselves; and a 25% chance (1 in 4) to have a baby who will inherit the normal version of the CF gene from both parents and who will be neither a carrier nor affected with CF.

Box 1-1 continued

Genetic counseling for certain CF mutations (such as the R117H, I148T, and D1152H) can be complicated because these mutations can be found in both asymptomatic and affected individuals who have two CFTR mutations. The R117H mutation is modified by a normal variant in another part of the gene. The I148T and D1152H mutations probably have similar modifiers.

Treatment: At present, treatment is directed at controlling pulmonary infection and improving nutrition. This includes supplementing pancreatic enzymes. In the 1990s, DNA-based treatments became available to help with breaking up mucus in the lungs. Bilateral lung transplantation has become a treatment option for some individuals with end-stage disease. Individuals with CF make up approximately 15% of all lung transplant recipients. However, this procedure is available only to a small number of patients and can cause physical, financial, and emotional burdens for CF patients and their families.

In the future, the hope is to design a pharmacological regimen that could directly correct the genetic and biochemical defect. Although there has been some progress in developing gene therapy treatments for CF, there are as yet significant obstacles to successful treatment using this approach.

Patient and Family Resources

Cystic Fibrosis Foundation, *http://www.cff.org*
Phone: 1-800-344-4823

Reference

American College of Obstetrics and Gynecology and American College of Medical Genetics (2001). *Preconception and prenatal carrier screening for cystic fibrosis: Clinical and laboratory guidelines.* Washington, D.C.: Author.

Adapted with Permission: Lea and Smith, 2003. *The genetics resource guide: A handy reference for public health nurses.* Scarborough, ME: Foundation for Blood Research. Written under a grant from the State of Maine Department of Human Services Genetics Program. Grant #BH-01-166B and C.

whether to have carrier testing or prenatal testing should he have children. We learn of Judy's concerns about the use of genetic information, and how she felt shunned because people thought the illness in her family was contagious. Now, with the evolution of genetic testing for CF and the availability of genetic information for individuals, she expresses concerns about societal use of genetic information and issues of privacy and confidentiality. Nurses will need knowledge, education, and skills to help educate individuals and families about the availability of genetic testing, to provide appropriate information about the potential risks and benefits of genetic testing and therapies, and to provide clients with an adequate informed consent process with relation to genetic testing and therapies. These competencies—highlighted in the chapters ahead—will contribute to the "seamless" integration of genetics into nursing care with a focus on the patient and his or her response to the genetic condition, testing, or therapy.

Summary

Judy's story about her family history of CF provides an example of how a more traditional genetic condition such as CF has been transformed through modern genomic discoveries and technologies to become a part of new genetics. We are just beginning to understand the complexities and variability of what were once thought of as "single-gene" conditions like CF, and the underlying biological pathways that create different clinical presentations. This new information is leading to more specific ways to treat individuals with CF. The vision for the future is to translate genome discoveries into health benefits for all populations. Included in this vision are research strategies to identify genes and pathways with a role in health and disease, and to discover how they interact with environmental factors. Scientists foresee expansion of genome research activities to include the development and evaluation of genome-based diagnostic methods for predicting susceptibility to disease, assessing drug responses, and providing accurate molecular classification of disease (Collins et al., 2003).

As more is learned through technology advances about complex and multigenetic conditions such as diabetes, heart disease, asthma, and Alzheimer's, the genetic contribution will be considered along with lifestyle, environment, and other yet-to-be-identified factors that influence health and disease. To do this, we will need to create interdisciplinary resources such as genomics, bioinformatics, and behavioral science that can interpret this information and permit its application to clinical practice. Nurses will have an integral role in this process in assisting individuals and families to understand their health care options and to make informed decisions.

Chapter Activities

1. Judy talks about her son's testing choices when he decides to have a family. She mentions that since the family ancestry is Ashkenazi Jewish, he will need to consider Tay-Sachs and other conditions. Use the Gene Tests Web site *(http://www.genetests.org)* and the Genetic Alliance Web site *(http://www.*

geneticalliance.org) to identify what other testing choices her son will have due to his ancestry.

2. Prepare a plan of care for Mary, a 23-year-old woman who has come for her first clinic visit at 6 weeks of pregnancy. She has a 2-year-old daughter in good health at home. Her husband is 25 and in good health. When reviewing the family history, she tells you that her brother has a daughter with CF. Construct Mary's family tree. What could you tell Mary about the inheritance of CF and her chance to be a CF carrier? What testing would be appropriate to discuss with Mary?

References

Collins, F. S., Green, E. D., Guttmacher, A., & Guyer, M. S. (April, 2003). A vision for the future of genomics research: A blueprint for the genomic era. *Nature, 422*(24), 835–847.

Guttmacher, A. E., & Collins, F. S. (2002). Genomic medicine: A primer. *New England Journal of Medicine, 347,* 1512–1520.

Jorde, L. B., Carey, J. C., Bamshad, M. J., & White, R. L. (2000). *Medical Genetics* (2nd ed.). St. Louis: Mosby, Inc.

Lea, D. H., & Smith, R. (2003). *The genetics resource guide: A handy reference for public health nurses.* Scarborough, ME: Foundation for Blood Research. Written under a grant from the State of Maine Department of Human Services Genetics Program. Grant #BH-01-166B and C.

Nussbaum, R. L., McInnes, R. R., & Willard, H. F. (2001). *Thompson & Thompson. Genetics in Medicine* (6th ed.). Philadelphia: W. B. Saunders Company.

CHAPTER 2
Connecting Genomics to Practice

Why focus on the need for health care professionals to gather family history information now? There is insufficient time during appointments; there is no reimbursement for that service; individuals do not really, accurately know the significant details anyway. All these excuses and more impede utilization of one of the best tools available to assess and determine potential risks for health problems.

Family History as a Tool

Common chronic diseases, such as diabetes, cardiovascular disease, and cancer affect large numbers of people. A positive family history of any of these conditions has long been recognized as a risk factor for many common illnesses (Yoon et al., 2002). Shared environments and behaviors also influence health risk and need to be assessed. Soliciting a three-generation family health history can often highlight potential areas for further investigation and intervention. Additionally, a smaller percentage of the population with a hereditary contribution to their risk for illness (i.e., hereditary cancer syndromes that occur in 5 to 10% of those with cancer) may be directed to resources if appropriately assessed and diagnosed. For instance, Agnes reports the following:

> I learned about my family's cancer risk when I was 10 years old. My grandmother was in the hospital for months fighting cancer. At my age it was hard to know how serious this was, but I later learned that she had survived colon, uterine, and ovarian cancers. When I was 14, my father woke up from an emergency appendectomy with a large section of his colon removed. The doctors had found cancer in his colon and removed the cancerous sections. My father selected to take postsurgical chemotherapy, uncommon at the time.

There are many families who have "lived" with the threat of cancer throughout their entire lives and would benefit from referral for genetic services. Such is the case with the following example.

Gary's Story

Gary came to the attention of the cancer genetics office because his sister called to inquire about a genetic testing research study. Linda reported that she did not have cancer, but that her brother had been diagnosed with colon cancer at the age of 42. Her father was still alive at the age of 74, and her mother had died of colon cancer at age 60. Linda had a sister who had died from ovarian cancer at age 32 and several maternal relatives with cancers, including uterine, leukemia, colon, kidney, or cervical. She stated that at a family reunion they were discussing how there must be

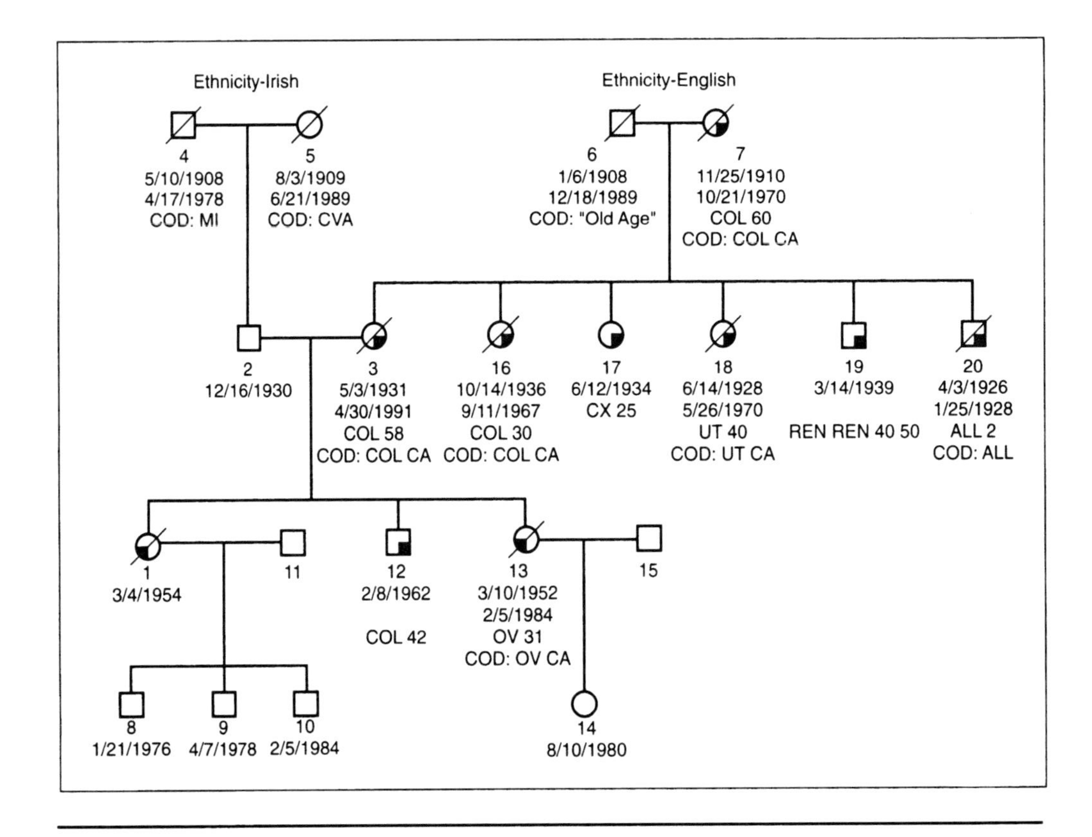

Figure 2-1 Gary's Pedigree

a reason why so many in their family got cancer and that it seemed to be occurring at younger ages with each generation. She had gotten motivated to collect more information about the family's health history and then to seek out hereditary cancer studies (see *http://www.cancer.gov*). Linda discussed with the nurse the family pedigree (see **Figure 2-1**) and the possibility that her family might have a cancer genetics syndrome, hereditary non-polyposis colorectal cancer (HNPCC). Linda expressed the interest of her family in learning more about this cancer syndrome and options for HNPCC testing.

Variation in the Human Genome

All of us have genetic changes within our deoxyribonucleic acid (DNA) that may put us at risk for something. Some of those genetic changes do not create any problems at all. Other changes may increase the potential for the development of a disease or a lifetime susceptibility to that disease. For instance, a genetic change in genes identified as MSH2 or MLH1 increases lifetime cancer risks for individuals in HNPCC families to 70–82% for colon cancer as compared to the 2% general population risk; and in women a 42–60% risk of endometrial cancer (AMA, 2001). Families identified as having a significant family history (**Box 2-1**) may be eligible for genetic testing. Nurses with sufficient understanding that these "red flags" for HNPCC within a family history require further investigation can facilitate referral, screening, and preventive interventions (**Box 2-2**) (Tranin, Masny, & Jenkins, 2003). These "red flags" exist for other complex diseases as well and, if recognized by the nurse, this awareness will improve access to options for clinical care. Agnes reported the following experience with health care practitioners and availability of services:

My experience with health practitioners has varied widely. My first experience was trying to convince my family doctor to refer me to a gastroenterologist (GI) for cancer surveillance. Our family knew we had a problem with hereditary cancer, so we began screening colonoscopies in our early 30s. My doctor didn't know what I was talking about, but felt my insistence meant the referral was important. My GI doctor recorded my family history and agreed with my father that I should begin screening colonoscopies every 3 to 5 years, beginning immediately. At the time, I was less than 10 years younger than my father was when his cancer was diagnosed. This is the guideline for starting screening colonoscopies when a patient meets the Amsterdam criteria, a family history method for determining HNPCC. My first two colonoscopies were performed by this physician, who is very good, but he seemed very much like a restrictor of care rather than a counselor.

Box 2-1 Amsterdam I and Amsterdam II Criteria

1. Histologically verified colorectal cancer in three or more relatives, one of whom is a first-degree relative of the other two;
2. Colorectal cancer involving at least two successive generations;
3. At least one of the relatives with colorectal cancer must have been diagnosed before age 50; and
4. Familial adenomatous polyposis ruled out as a cause for each colorectal cancer.

Amsterdam II Criteria include all the above but also identify at-increased-risk individuals/families that have at least three relatives with a cancer that has been associated with HNPCC (colorectal, endometrial, stomach, ovary, ureters or renal-pelvis, brain, small bowel, hepatobiliary tract, or skin). (Lynch & de la Chapelle, 2003)

When I began participating in clinical studies, I became associated with the best nurses, genetic counselors, and physicians that I have encountered. I learned more from them about my condition than what was available anywhere in the medical community. As I left the clinical studies and moved back to my regular doctors, I found that I knew more about my condition than my doctors. I am now much more aggressive about screening, and I have come to use my doctors as technicians and insurance gatekeepers.

Gary reported a similar experience. No health care provider had recommended that perhaps Gary's family would benefit from genetic testing, but Linda was motivated to track down new options in clinical care. Genetic testing for HNPCC is first offered to individuals who have already experienced cancer. If a mutation is identified, the option of genetic testing is then further available for family members (both those affected or unaffected with cancer). Since Linda was unaffected, she was instructed to contact her brother and determine his interest in pursuing genetic testing. Gary contacted the office a week later to learn more.

When I came down with cancer, I was not surprised. It has been part of my life for many years. Several people in my family have come down with it. Several have died as a result of their individual types of cancer. I was diagnosed with colon cancer and half of my colon was removed. My oncologist told me at the time that, had it been ten years ago, I most likely would not be here today. I believed him. Medical technology has improved greatly.

Box 2-2 Clinical Features of Colon Cancer (CRC)

Sporadic CRC	HNPCC
Average age of onset 63	Average age of onset 45
Often on left side of colon	Often on right side of colon (60–80%)
	Pathology is more often poorly differentiated or undifferentiated
	Improved prognosis
	Excess of extracolonic cancers
	Mismatch repair system defects (MSI) (90%)

(Lynch & de la Chapelle, 2003)

Gary was interested in learning more about the clinical implications of genetic testing for himself, but even more so for his family. Most of his family was interested in knowing about genetic testing, except for his aunt who had expressed that she did not want to know anything about this kind of information. He was concerned that more harm than good could occur for some members of his family because of these dynamics of denial. But he also recognized that earlier screening might have been beneficial for him, and he wanted to offer such an option to others in his family. He decided to participate in the hereditary cancer study and began the process by providing a more comprehensive family history.

Family History Tools

Genetics and genomics in practice necessitate the utilization of tools that may first have to be developed or modified to sufficiently meet the needs of busy practitioners and consumers (Yoon, Scheuner, & Khoury, 2003). A family history tool can be either basic or quite comprehensive. It may be a pen-and-pencil version or available on computer. The information can be gathered and maintained by the individual and his or her family, or provided to the family physician's office to be kept on file. Bennett (1999) provides a useful guide to the available types of family history tools and recommendations for use. Both sample family and medical history forms are found in **Tables 2-1 and 2-2.**

The family history is one part of a genetic assessment that may often take several visits to complete. A questionnaire may be a useful method for having the individual track down details prior to the consultation or clinic visit. This process can facilitate accuracy of details through family communication about their experiences with illness. Once collected, this family health history can then be updated as

Table 2-1 Clinical Cancer Genetics Program Family History Questionnaire

Instructions:
1. When completing this questionnaire, please be sure to turn each page over and complete the back side.
2. If there is not enough space to list all relatives, please include that information on a separate sheet of paper.

For any questions completing this questionnaire, please call ______________________

Name: (Last) ______________________ (First) ______________________ (Maiden) ______________________ **Date of Birth:** ______________________ (MM/DD/YYYY) **Phone No: Home:** __________ **Work:** __________ **When is the best time to contact you?** ☐ Morning (☐ Home ☐ Work) ☐ Afternoon (☐ Home ☐ Work) ☐ Evening (☐ Home ☐ Work)	**What best describes your family's ethnic heritage?** (ex. Irish, Jewish) **Mother's Family** ______________________ **Father's Family** ______________________ **Are you adopted?** ☐ Y ☐ N ☐ Unknown **Are you a twin?** ☐ Y ☐ N If Yes: ☐ Identical ☐ Same sex, unknown if identical ☐ Unknown

Please complete the following table for your parents and children even if they have not had cancer. If they have not had cancer, please write "NONE" in the Cancer Type column. If they have had more than one type of cancer, please list all types of cancer and age at each diagnosis.

YOUR PARENTS AND CHILDREN

Relative and Name	Date of Birth	Cancer Type(s)	Age(s) Diagnosis	Date of Death
Father				
Mother				
Child #1 Sex M F				
Child #2 Sex M F				
Child #3 Sex M F				
Child #4 Sex M F				
Child #5 Sex M F				

TABLE 2-1 continued

Please complete the following table for your brothers and sisters even if they have not had cancer. If they have not had cancer, please write "NONE" in the Cancer Type column. If they have had more than one type of cancer, please list all types of cancer and age at each diagnosis.

YOUR BROTHERS AND SISTERS

Relative and Name	Date of Birth	Cancer Type(s)	Age(s) Diagnosis	Date of Death
Brother #1 __ Full __ Half (same mother) __ Half (same father)				
Brother #2 __ Full __ Half (same mother) __ Half (same father)				
Brother #3 __ Full __ Half (same mother) __ Half (same father)				
Brother #4 __ Full __ Half (same mother) __ Half (same father)				
Sister #1 __ Full __ Half (same mother) __ Half (same father)				
Sister #2 __ Full __ Half (same mother) __ Half (same father)				
Sister #3 __ Full __ Half (same mother) __ Half (same father)				
Sister #4 __ Full __ Half (same mother) __ Half (same father)				

Please complete the following table for your nieces and nephews (children of your brothers and sisters) even if they have not had cancer. If they have not had cancer, please write "NONE" in the Cancer Type column. If they have had more than one type of cancer, please list all types of cancer and age at each diagnosis.

NIECES AND NEPHEWS

Relative and Name of Parent	Date of Birth	Cancer Type(s)	Age(s) Diagnosis	Date of Death
Niece/Nephew #1 Name of Parent				
Niece/Nephew #2 Name of Parent				
Niece/Nephew #3 Name of Parent				
Niece/Nephew #4 Name of Parent				
Niece/Nephew #5 Name of Parent				
Niece/Nephew #6 Name of Parent				
Niece/Nephew #7 Name of Parent				
Niece/Nephew #8 Name of Parent				
Niece/Nephew #9 Name of Parent				
Niece/Nephew #10 Name of Parent				

TABLE 2-1 continued

Please complete the following table for your grandparents, aunts, and uncles on your **FATHER's** side even if they have not had cancer. If they have not had cancer, please write "NONE" in the Cancer Type column. If they have had more than one type of cancer, please list all types of cancer and age at each diagnosis.

GRANDPARENTS, AUNTS, and UNCLES ON YOUR FATHER'S SIDE

Your Father's Side				
Relative	Date of Birth	Cancer Type(s)	Age(s) Diagnosis	Date of Death
Grandfather				
Grandmother				
Uncle #1				
Uncle #2				
Uncle #3				
Uncle #4				
Uncle #5				
Aunt #1				
Aunt #2				
Aunt #3				
Aunt #4				
Aunt #5				

Please complete the following table for your grandparents, aunts, and uncles on your **MOTHER's** side even if they have not had cancer. If they have not had cancer, please write "NONE" in the Cancer Type column. If they have had more than one type of cancer, please list all types of cancer and age at each diagnosis.

GRANDPARENTS, AUNTS, and UNCLES ON YOUR MOTHER'S SIDE

Your Mother's Side				
Relative	Date of Birth	Cancer Type(s)	Age(s) Diagnosis	Date of Death
Grandfather				
Grandmother				
Uncle #1				
Uncle #2				
Uncle #3				
Uncle #4				
Uncle #5				
Aunt #1				
Aunt #2				
Aunt #3				
Aunt #4				
Aunt #5				

Table 2-1 continued

Please complete the following table for your cousins (children of your aunts and uncles) on your **FATHER'S** side even if they have not had cancer. If they have not had cancer, please write "NONE" in the Cancer Type column. If they have had more than one type of cancer, please list all types of cancer and age at each diagnosis.

COUSINS ON YOUR FATHER'S SIDE

Relative and Name of Parent	Date of Birth	Cancer Type(s)	Age(s) Diagnosis	Date of Death
Cousin #1 Sex M F Name of Parent				
Cousin #2 Sex M F Name of Parent				
Cousin #3 Sex M F Name of Parent				
Cousin #4 Sex M F Name of Parent				
Cousin #5 Sex M F Name of Parent				
Cousin #6 Sex M F Name of Parent				
Cousin #7 Sex M F Name of Parent				
Cousin #8 Sex M F Name of Parent				

Please complete the following table for your cousins (children of your aunts and uncles) on your **MOTHER'S** side even if they have not had cancer. If they have not had cancer, please write "NONE" in the Cancer Type column. If they have had more than one type of cancer, please list all types of cancer and age at each diagnosis.

COUSINS ON YOUR MOTHER'S SIDE

Relative and Name of Parent	Date of Birth	Cancer Type(s)	Age(s) Diagnosis	Date of Death
Cousin #1 Sex M F Name of Parent				
Cousin #2 Sex M F Name of Parent				
Cousin #3 Sex M F Name of Parent				
Cousin #4 Sex M F Name of Parent				
Cousin #5 Sex M F Name of Parent				
Cousin #6 Sex M F Name of Parent				
Cousin #7 Sex M F Name of Parent				
Cousin #8 Sex M F Name of Parent				

TABLE 2-1 continued

Please complete the following table for other relatives we have not asked about on your **FATHER's** side even if they have not had cancer. If they have not had cancer, please write "NONE" in the Cancer Type column. If they have had more than one type of cancer, please list all types of cancer and age at each diagnosis.

OTHER RELATIVES ON YOUR FATHER'S SIDE

Relative	Date of Birth	Cancer Type(s)	Age(s) Diagnosis	Date of Death
Other Relative #1 Sex M F				
Other Relative #2 Sex M F				
Other Relative #3 Sex M F				
Other Relative #4 Sex M F				

Please complete the following table for other relatives we have not asked about on your **MOTHER's** side even if they have not had cancer. If they have not had cancer, please write "NONE" in the Cancer Type column. If they have had more than one type of cancer, please list all types of cancer and age at each diagnosis.

OTHER RELATIVES ON YOUR MOTHER'S SIDE

Relative	Date of Birth	Cancer Type(s)	Age(s) Diagnosis	Date of Death
Other Relative #1 Sex M F				
Other Relative #2 Sex M F				
Other Relative #3 Sex M F				
Other Relative #4 Sex M F				

Thank you for completing this questionnaire!

(Reprinted with permission from the Cancer Genetics Branch, National Cancer Institute, Bethesda, MD)

TABLE 2-2 Personal & Health History Questionnaire

Part One: PERSONAL INFORMATION

In this first section, we are gathering information to register you in the clinic. Please try to fill out this form as completely as possible. Thank you.

Today's date: ______________________

1. Name: ____
2. Address:
   Street ____
   City ______________________ State ________ ZIP Code ________
3. Telephone number: Home (____) ______________ Work (____) ______________
4. Birthdate: Month/Day/Year ____/____/____ Sex: ☐ Male ☐ Female
5. Marital Status: ☐ Single ☐ Married ☐ Separated
   ☐ Divorced ☐ Widowed ☐ Significant other
6. Country of birth: ______________________
7. Race: (*Check all that apply*)
   ☐ White (not Hispanic) ☐ African-American ☐ Native American
   ☐ Asian ☐ South Asian/Indian ☐ Hispanic
   ☐ Other (please specify): ______________________
8. Ethnic background (countries of origin):
   a. Paternal: ______________________
   b. Maternal: ______________________
9. Religion: Jewish: ☐ Ashkenazi (descended from Central and Eastern Europe) or
   ☐ Sephardic (descended from Spain and Middle East)
   ☐ Catholic ☐ Protestant ☐ Islam ☐ Other: ______________
10. What is/was your usual occupation? ______________________
11. Active Duty Status:
    ☐ Yes, Active Duty ☐ No, Dependent
    ☐ No, Retired
12. Branch of Service? ______________________ ☐ NA
13. Highest level of education completed:
    ☐ Less than High School Diploma ☐ College Degree
    ☐ High School Diploma ☐ Some Graduate School
    ☐ Vocational School ☐ Graduate Degree
    ☐ Some College ☐ Post-Graduate (Doctoral)

Table 2-2 continued

14. Please give us the name of a friend or relative who could always locate you:

    Name: ______________________ Relationship to you ____________

    Phone Number Home (____)______________ Work (____)______________

    Address ____

    ____

Part Two: HEALTH HISTORY

In this section we are asking you to tell us about your health history. All of your responses will be kept confidential. If you need more space, please continue on the back of each page.

1. Have you ever been diagnosed with any form of cancer? ☐ Yes ☐ No
   [Including any form of skin cancer]

If **Yes,** fill in the grid below.

CANCER TYPE		TREATMENT				
Type of Cancer	Date of Diagnosis Month/Year	Surgery	Radiation Therapy	Chemotherapy	Hormone Therapy	Other Treatment
Example:						
Left Breast Cancer	8/1997	Mastec-tomy	☐ Yes ☑ No	☑ Yes ☐ No	☑ Yes ☐ No	☐ Yes ☑ No
Specify Other Treatment:						
			☐ Yes ☐ No	☐ Yes ☐ No	☐ Yes ☐ No	☐ Yes ☐ No
Specify Other Treatment:						
			☐ Yes ☐ No	☐ Yes ☐ No	☐ Yes ☐ No	☐ Yes ☐ No
Specify Other Treatment:						
			☐ Yes ☐ No	☐ Yes ☐ No	☐ Yes ☐ No	☐ Yes ☐ No

TABLE 2-2 continued

Specify Other Treatment:						
			☐ Yes ☐ No	☐ Yes ☐ No	☐ Yes ☐ No	☐ Yes ☐ No
Specify Other Treatment:						
			☐ Yes ☐ No	☐ Yes ☐ No	☐ Yes ☐ No	☐ Yes ☐ No
Specify Other Treatment:						
			☐ Yes ☐ No	☐ Yes ☐ No	☐ Yes ☐ No	☐ Yes ☐ No

2. Do you have any other serious **medical conditions?** ☐ Yes ☐ No
If **Yes,** fill in the grid below.

Medical Condition	Age at Diagnosis	Treatment	Status
Example:			
Diabetes	48	Insulin	☑ Current ☐ In the past
			☐ Current ☐ In the past
			☐ Current ☐ In the past
			☐ Current ☐ In the past

TABLE 2-2 continued

3. Do you currently take any **medications?** ☐ Yes ☐ No

[Include all prescription drugs and hormones, as well as over-the-counter drugs like antacids, pain relievers, vitamins, and minerals.]

If **Yes,** fill in the grid below.

Drug	How much [e.g., # mg]	How often [e.g., daily, 3 times/day]	When did you start taking it?	For what reason?
Example:				
Lipitor	20 mg	1x/day	8/2000	High Cholesterol

4. Do you have any **allergies** to drugs? ☐ Yes ☐ No

If **Yes,** please list: ______________________________

5. Have you ever had a **breast biopsy?** ☐ Yes ☐ No

   Did any of the biopsies show **atypical hyperplasia**? ☐ Yes ☐ No ☐ Don't Know

   Did any of the biopsies show **lobular carcinoma insitu**? ☐ Yes ☐ No ☐ Don't Know

6. Have you ever had **other surgery** besides those listed above? ☐ Yes ☐ No

If **Yes,** fill in the grid below.

Type of Surgery	Date of Procedure Month/Year	Reason for Surgery
Example:		
Tonsillectomy	8/75	Enlarged tonsils

Table 2-2 continued

MEN please go to question 17

7. At what age did your menstrual periods begin? __________
   Are you still menstruating? ☐ Yes ☐ No
   What is/was the pattern of your menstrual period?
   ☐ Always Regular ☐ Usually Regular ☐ Never Regular
   [By regular we mean the start of your period was predictable within 5 days]

   Date of last menstrual period __________/___________
   Month/Year

8. What is the reason your menstrual period stopped?
   ☐ Natural Menopause ☐ Medication Stopped Period
   ☐ Removal of Ovaries ☐ Radiation Stopped Period
   ☐ Removal of Uterus ☐ Other (specify) ___________________

9. Have you ever been pregnant? ☐ Yes ☐ No Age at your first live birth? _________

If **Yes,** fill in the grid below.

Pregnancy Outcome						Breast Fed?	
Month/Year at End of Pregnancy	# of Live Births	# of Stillborn	Miscarriage ✓	Induced Abortion ✓	Ectopic Pregnancy ✓	Yes/No	Months of Breast Feeding
Example: Twins born in August 1991							
8/91	2(twins)					☑ Yes ☐ No	3 months
Example: Miscarriage in March 1993							
3/93			✓			☐ Yes ☐ No	
						☐ Yes ☐ No	
						☐ Yes ☐ No	
						☐ Yes ☐ No	
						☐ Yes ☐ No	
						☐ Yes ☐ No	

TABLE 2-2 continued

10. Have you ever taken medication to help you become pregnant? ☐ Yes ☐ No
If **Yes,** fill in the grid below.

Name of Medication [Ex. Clomid, Pergonal]	Age Started	Length of Time Used for All Cycles
Example:		
Clomid	28	12 months

11. Have you ever taken oral or injectable birth control? ☐ Yes ☐ No
If **Yes,** fill in the grid below.

Name of Medication [Ex. Orthonovum]	Age Started	Length of Time Used
Example:		
Orthonovum	18	2 years

12. A mammogram is an X-ray taken only of your breasts.
Approximately how many mammograms have you had in your lifetime? _____ ☐ None
Age at first mammogram _____
Date of your most recent mammogram ________/________
Month/Year
Have you ever had an abnormal mammogram: ☐ Yes ☐ No
Date of your most recent breast exam by a health professional ________/________
Month/Year

TABLE 2-2 continued

13. A PAP smear is a test done on the cervix at the time of a pelvic (internal) exam.
    Date of most recent PAP smear _______/_______ ☐ Never had a PAP smear
    Month/Year
    Have you ever had an abnormal PAP smear? ☐ Yes ☐ No
14. Have you ever examined your own breast for lumps (breast self-exam)? ☐ Yes ☐ No
    If **Yes,** how many times have you examined your breasts in the last 12 months? ______
15. How confident are you that your self-exam could detect a new lump in your breast?

[Please circle a number below]

Not at all confident 1 2 3 4 5 6 7 Extremely confident

16. Have you ever taken hormones for any other reason? ☐ Yes ☐ No

If **Yes,** fill in the grid below.

Name of Medication [Ex. Premarin]	Age Started	Length of Time Used	Reason Used
Example:			
Premarin	55	1 year	Hormone Replacement

TABLE 2-2 continued

17. Have you ever had any other cancer screening tests done?
[For example: colonoscopy, transvaginal ultrasound, PSA (prostate specific antigen)]

Type of Screening Test	How Often?	Date of Last Test (Month/Year)
Example:		
CA125	☐ Every 6 months ☑ Every 12 months ☐ Other (specify) ____________	12/2002
	☐ Every 6 months ☐ Every 12 months ☐ Other (specify) ____________	
	☐ Every 6 months ☐ Every 12 months ☐ Other (specify) ____________	
	☐ Every 6 months ☐ Every 12 months ☐ Other (specify) ____________	
	☐ Every 6 months ☐ Every 12 months ☐ Other (specify) ____________	

18. Current height: ______ ft. ______ in. Current weight: ______ lbs.
Approximate weight at age 18: ______ lbs. Birth weight: ______ lbs. ______ oz.
[Not counting pregnancy or illness]

19. To the best of your knowledge, were you born with any physical abnormalities?
☐ Yes ☐ No
If **Yes,** please describe: ____

20. Have you ever smoked cigarettes? ☐ Yes ☐ No
If **Yes,** how old were you when you started? ______________
Do you currently smoke? ☐ Yes ☐ No
If **Yes,** what is the average number of cigarettes you smoke per day? ______________
If **No,** at what age did you stop? ______________
What was the average number of cigarettes you smoked per day? ______________

TABLE 2-2 continued

21. Do you currently consume alcohol? ☐ Yes ☐ No
If **Yes,** fill in the grid below.

Type of Alcohol [Ex. Wine, Beer]	Age	None	Less than or equal to 5 glasses* per week	6–10 glasses* per week	More than 10 glasses* per week
	At ~ 20 years old				
	In the past year				

*drinks

Part Three: WHAT BRINGS YOU TO THE PROGRAM

In this section we are asking you to tell us about what brings you to the program and your thoughts about your risk of developing cancer.

1. What is your primary reason for coming to the Cancer Genetics Program? ____________

____

____

2. Are there any questions or concerns you want to be assured are answered with your visits?

____

____

If you have already been affected with breast cancer, please skip to question 6.

Table 2-2 continued

3. What do you think your chance is of developing breast cancer in your lifetime?
   Please choose a number between 0% (no chance of breast cancer) and 100% (definitely will get breast cancer).
   __________%

4. How often do you worry about developing breast cancer?
   Please circle a number below

1	2	3	4	5	6	7
Not at all						All the time

5. How much does worrying about developing breast cancer interfere with your everyday life?
   Please circle a number below

1	2	3	4	5	6	7
Not at all						All the time

If you are a woman who has already been affected with ovarian cancer, you have completed the questionnaire. Thank you.

6. What do you think your chance is of developing ovarian cancer in your lifetime?
   Please choose a number between 0% (no chance of ovarian cancer) and 100% (definitely will get ovarian cancer).
   __________%

7. How often do you worry about developing ovarian cancer?
   Please circle a number below

1	2	3	4	5	6	7
Not at all						All the time

8. How much does worrying about developing ovarian cancer interfere with your everyday life?
   Please circle a number below

1	2	3	4	5	6	7
Not at all						All the time

Thank you for completing this questionnaire.

(Reprinted with permission from the Cancer Genetics Branch, National Cancer Institute, Bethesda, MD)

needed. Consideration of cultural beliefs, psychosocial state, and cognitive level should also occur during the patient visit (Barse, 2003).

Pedigree Construction

Pedigree construction includes soliciting details about the family history, and recording and interpreting the information. There are many methods of drawing pedigrees once the information is collected. A pedigree is a symbolic representation of family members, indicating specific details about each individual. General family history screening should occur for every family member, with details about members for at least three generations, and include both maternal and paternal sides of the biological family, if possible. This initial pedigree screening can highlight areas for further targeted queries such as questions specific to assessment of certain diseases (i.e., cancer). A pedigree targeting the possibility of a hereditary cancer syndrome will gather even more details about the incidence of cancer, the type and location of the cancer, pathological features, age at diagnosis, and outcome (see **Figure 2-1**). Other pedigrees targeting reproductive issues may focus on the incidence of reproductive events and detail outcomes such as births, reproductive loss, infertility, or birth anomalies.

Standardized Methodologies

The Pedigree Standardization Task Force of the National Society of Genetic Counselors developed a system of pedigree nomenclature that is recommended for use in practice. Samples of standardized pedigree symbols are provided in **Figure 2-2.** Additional examples are available on the National Coalition for Health Professional Education in Genetics (NCHPEG) Web site *(http://www.nchpeg.org)*, under newsletter.

Why Ask About Ethnicity?

Information about maternal and paternal ethnicity should also be included in the collection of family history. Some genetic mutations are found with higher incidence in certain ethnic groups or to be unique to particular communities (Fitzgerald et al., 1996). The sensitivity of certain screening panels may be influenced by ethnicity of the population being screened (i.e., cystic fibrosis) (American College of Medical Genetics, 2002; Grody et al., 2002). Certain genetic conditions occur with increased frequency among specific ethnic groups (Cardeiro, 2003). Further research that assesses the effects of ethnicity on incidence, frequency, and interpretation of genetic variation will guide the importance of including these specific data for clinical utilization. More information about family history and ethnicity can be found at *http://www.nchpeg.org*, under newsletter. Ethnicity also influences health beliefs and may contribute to the family's decision of whether or not to access genetic services. Cost of health care that integrates genetic services is often nonreim-

Examples of Commonly Used, Standardized Pedigree Symbols, Definitions, and Abbreviations

Instructions:
—Key should contain all information relevant to interpretation of pedigree (e.g., define shading)
—For clinical (nonpublished) pedigrees, include:
a) family names/initials, when appropriate
b) name and title of person recording pedigree
c) historian (person relaying family history information)
d) date of intake/update
—Recommended order of information placed below symbol (below to lower right, if necessary):
a) age/date of birth or age at death
b) evaluation
c) pedigree number (e.g., I-1, I-2, I-3)

	Male	Female	Sex Unknown	Comments
1. Individual	b. 1925	30y	4 mo	Assign gender by phenotype.
2. Affected individual				Key/legend used to define shading or other fill (e.g., hatches, dots, etc.).
				With ≥2 conditions, the individual's symbol should be partitioned accordingly, each segment shaded with a different fill and defined in legend.
3. Multiple individuals, number known	5	5	5	Number of siblings written inside symbol. (Affected individuals should not be grouped.)
4. Multiple individuals, number unknown	n	n	n	"n" used in place of "?" mark.
5a. Deceased individual	d. 35 y	d. 4 mo		Use of cross (†) may be confused with symbol for evaluated positive (+). If known, write "d." with age at death below symbol.
5b. Stillbirth (SB)	SB 28 wk	SB 30 wk	SB 34 wk	Birth of a dead child with gestational age noted.
6. Pregnancy (P)	P LMP: 7/1/94	P 20 wk	P	Gestational age and karyotype (if known) below symbol. Light shading can be used for affected and defined in key/legend.
7a. Proband	P	P	P	First affected family member coming to medical attention.
7b. Consultant				Individual(s) seeking genetic counseling/testing.

Instructions:
—Symbols are smaller than standard ones and individual's line is shorter. (Even if sex is known, triangles are preferred to a small square/circle; symbol may be mistaken for symbols 1, 2, and 5a/5b of Figure 1, particularly on hand drawn pedigrees.)
—If gender and gestational age known, write below symbol in that order.

	Male	Female	Sex Unknown	Comments
1. Spontaneous abortion (SAB)	male	female	ECT	If ectopic pregnancy, write ECT below symbol.
2. Affected SAB	male	female	16wk	If gestational age known, write below symbol. Key/legend used to define shading.
3. Termination of pregnancy (TOP)	male	female		Other abbreviations (e.g., TAB, VTOP, Ab) not used for sake of consistency.
4. Affected TOP	male	female		Key/legend used to define shading.

FIGURE 2-2 Standardized Pedigree Symbols

Reprinted from Bennett, R. et al. (1995). Recommendations for standardized human pedigree nomenclature. *Journal of Genetic Counseling, 4*(4), 276–279.

Table 2-3 Indications for Genetic Testing (Policy Statement)

- Personal or family history is suggestive of a hereditary cancer condition
- Test can be adequately interpreted
- Test results will aid in diagnosis or influence care management
- Testing should only be offered with pre- and post-test education and counseling
- Special consideration needs to be given to testing of children (ASCO, 2003)

bursable by insurance. Access to health care is influenced by these multiple factors, each of which needs to be assessed by the clinical team.

Interpretation

HNPCC is an example of a cancer syndrome that can be clinically diagnosed based on information gathered through the family history. Accurate interpretation of the gathered information requires knowledge of which conditions warrant attention and the identification of gaps where additional information is needed. Indications for genetic testing for cancer susceptibility have been proposed by professional societies (i.e., American Society of Clinical Oncology (ASCO), 2003). See **Table 2-3** for more details. Risk estimates can be determined by those knowledgeable about patterns of inheritance. Interpretation of risk for a genetic predisposition to disease is often provided by genetic counselors or geneticists. However, the initial step toward knowing when to make a referral to a specialist is the responsibility of all nurses (NCHPEG, 2001).

Clinical Validity and Utility

Surveillance for cancer The challenge for persons with a hereditary predisposition to cancer lies not only in identifying the mutation, but also in determining and following the recommended surveillance tests and frequency (American Gastrological Association, 2001). Agnes says,

> I am fortunate to have a genetic condition (HNPCC) which leads to a disease that is easily detectable and offers successful treatment if a strict screening regimen is followed. Consequently, I have had annual colonoscopies since discovering my condition (mutation positive). That is my major change in health care. It has not had a major impact on my life. On the other hand, I feel a tremendous responsibility to help others get similar life-saving information, and I have volunteered with a patient group fighting to reduce the deaths associated with colon

cancer. I am more concerned that others learn the value of genetic information than I fear damaging consequences to my life.

One of the proposed benefits of genetic testing is that it allows families with an identified predisposition to a hereditary cancer to determine who is mutation negative and thus does not require enhanced surveillance (i.e., screening tests at an earlier age and more frequent colonoscopies). One study reports that such individuals may still carry lifelong concerns despite a negative mutation result that puts them at normal population risk for cancer (Bleiker et al., 2003). Mutation noncarriers were pleased not to have to undergo colonoscopies, but one-third hesitated to stop surveillance.

Efficacy of surveillance Not all individuals who have a genetic predisposition to an illness benefit from the possibility of earlier disease detection. For instance, women with HNPCC are known to be at increased risk for endometrial cancer and ovarian cancer. Efficacy of surveillance is limited by the inadequacies of tests available to detect these cancers at an early stage of development. Additionally, availability to any surveillance test may be limited by lack of insurance coverage (costs), discomfort associated with tests, or fear of test outcomes. Genetic counseling and testing for HNPCC can significantly influence the utilization of screening tests (Hadley et al., 2004). Burke, Pinsky, and Press (2001) proposed a framework for categorizing genetic tests according to their clinical validity (accuracy with which the test predicts a particular outcome) and the availability of effective treatment for the condition at risk (clinical utility). Nurses can assist patients in having sufficient understanding of the benefits, risks, and limitations of genetic testing; surveillance options; and available interventions for those identified as at-risk for inherited susceptibility to future disease.

Implications of Modifiers to Health Outcomes

Prospects for prevention Gary was reflecting on his experiences. He said,

Why me? This is what most people ask when they come down with cancer. The next questions are "how long do I have?" "what will happen to my family?" and "who will take care of the kids?". All of these are valid questions, but they all have a negative feeling about them. This is what we need to change. In this day and age there is much that can be done to help maintain the lives of those that have come down with the big "C." Attitudes about the disease need to change. It can be beaten.

Agnes also felt that she could do something about her risks associated with the potential development of cancer.

I would like to see a broader understanding of how genes lead to disease and a standard test that discovers a broad range of genetic conditions. When these tests are more developed, I will take advantage of them. In the short term there are some very promising cell-based DNA tests that discover early cancer markers. I will start using these tests in conjunction with my colonoscopies this year.

Both of these individuals sought out the options for potential prevention of future colon cancer. Gary, who was free of disease, and Agnes, unaffected but at-risk, both decided to participate in a clinical study assessing the effectiveness and toxicity of a potential preventive drug intervention. Pharmacologic treatment based on molecular profiling holds great promise for design of targeted interventions (Chemoprevention Working Group, 1999). The vision for the future of genomics research identifies that improved understanding of genes and pathways has the potential to enhance the development of new therapeutic approaches to disease (Collins et al., 2003).

Other modifiers to health Gary expressed thoughts of other factors that influenced his health outcomes. Behavioral, social, and environmental factors (lifestyle, socioeconomic factors, pollutants, etc.) modify or influence genetics in the manifestation of disease. He says:

Another thing we have learned is that teaching both the patient and their family about the disease can be a big factor in the recovery process. The more you know about the risks, the effects of the disease, and the medications, the easier it is to handle some of the situations that come up. Being able to answer questions helps. Letting the family and the patient know ahead of time what to expect also helps. Letting the family share in the healing process gives them a sense of togetherness that sometimes is needed. Sharing their feelings with each other, talking out their questions and concerns with each other is also a good tool.

Communication about genetic information within families is a focus of ongoing research (Peterson et al., 2003; Plantinga et al., 2003).

Ethical Issues and Family History

Disclosure, Confidentiality, and Families

The most commonly expressed concern of persons sharing genetic information, such as a family history or genetic test results, is that somehow that information will be used against them or their family (Clayton, 2003). Distressing changes in thoughts, beliefs, view of self, and family relationships have been reported as potential psychosocial risks related to learning genetic information (Broadstock et al., 2000). The potential for discrimination based on genetic information for insurance coverage or employment decisions is a fear often expressed by individuals considering accessing genetic services (Lerman & Shields, 2004). There is support for establishing a federal law that prohibits discrimination on the basis of genetic information, but currently only state laws exist *(http://www.genome.gov; http://www.ncsl.org/programs/health/genetics.htm)*.

Recording Information

The greatest threat to confidentiality and the client's privacy may be from unintentional disclosures from the medical record. The Health Insurance Portability and Accountability Act defines the federal privacy rule for all personal and private information (including genetic information) *(http://privacyruleandresearch.nih.gov)*. State and federal laws need to be considered when setting guidelines for documenting genetic information that can be inappropriately used. The method and site of any documentation of information collected about the individual and his or her family history needs to be determined prior to offering genetic services. This information must then be described to the consumer so that an informed decision about sharing family details and service utilization can be made after considering all risks and benefits.

Duty to Warn At-Risk Relatives

Once genetic information is gathered and successfully identifies the potential for a future risk of illness, a challenge exists to the common practice of patient–provider confidentiality. The tools of genomic medicine often reveal information not only about the health risks of the individual being seen, but also about other family members who may or may not be aware of the health concerns.

Gary's aunt did not want to pursue genetic testing. She did not feel that, at 75, such information would be of benefit to her. She also asked that other family members not discuss this information with her children. She expressed fear that her children would blame her for their health problems. Although Gary encouraged his aunt to discuss this issue with her physician, she would not do so. Gary discussed his growing concern about this branch of his family with his own physician.

Challenges to the normal expectation of confidentiality of shared health information necessitate open discussions between patient and care provider. Discussion of topics such as planned communication among family members, ways to facilitate such discussion, and expectations of all involved can avert some problems or initiate preparation for how to handle them. A court case that contested confidentiality of the individual versus the duty to warn an at-risk family member ruled that the physician should have notified the daughter directly of her 50% risk of having the disorder (*Safer v. Pack,* 1996). The complexity of integrating genomic tools into practice requires that personal, clinical, and legislative responses be addressed (Leung, 2000).

Genetics and Genomics in Practice

Family History for Preventive Medicine and Public Health

The Centers for Disease Control and Prevention (CDC) Office of Genomics and Disease Prevention recognized that family history could be a valuable tool for improving public health. A multidisciplinary working group implemented in 2002 a Family History Public Health Initiative to begin to evaluate the effectiveness of family history for assessing risk of common diseases *(http://www.cdc.gov/genomics)*. The goal is to influence population health outcomes through risk stratification, thereby leading to early detection and prevention options for commonly occurring diseases.

Prototype tools A prototype family history tool will be piloted in different clinical and public health settings. Information collected through family history surveys may be useful for primary care providers in determining risk levels for at least six conditions: heart disease, stroke, diabetes, breast cancer, ovarian cancer, and colon cancer. A resource manual, when completed, will provide primary care providers with information that facilitates risk classification. The risk will be calculated as average, moderate, or high, and shown in algorithms based on data yet to be analyzed (Yoon, Scheuner, & Khoury, 2003). Once all the tools are developed, the CDC plans to implement a public health campaign. It is also recognized that primary care provider education on how to use such risk identification for appropriate patient referral, coordination, and collaboration with other interdisciplinary team members of health professionals will be necessary.

Utility of Family History in Effecting Behavior Change

The assumption that an awareness of risk associated with family history will significantly effect behavioral change may be insufficient to ensure the benefit of improved disease prevention (Audrain-McGovern, Hughes, & Patterson, 2003). Shared environments and behaviors also influence health risk and may be difficult to change. Significant research attention is needed, along with implementing genetics and genomics tools into practice, to assess outcomes of all interventions. Knowledge of risk alone may be insufficient to effect behavior change. Nurses must actively study the issues, barriers, and facilitators of translating genomic information to improved health care.

Resources for Public Information

American Medical Association—Family History Tools	*http://www.ama-assn.org/ama/pub/category/2380.html*	Tools for gathering family history
Centers for Disease Control and Prevention (CDC)	*http://www.cdc.gov/genomics/default.htm*	Information about the importance of family history in public health
Genetic Alliance	*http://www.geneticalliance.org/DIS*	Family history guidelines
Harvard Center for Cancer Prevention—Your Cancer Risk	*http://www.yourcancerrisk.harvard/edu*	Personalized estimation of cancer risk and tips for prevention
Generational Health	*http://www.generationalhealth.com/*	Tool to help trace a family's medical history and provide information on common diseases
Genetics and Rare Diseases Information Center	*http://www.genome.gov/10000409*	Information service for the general public, including patients and their families, health care professionals, and biomedical researchers
Genetics Home Reference—National Library of Medicine	*http://ghr.nlm.nih.gov/*	Consumer information about genetic conditions and the genes responsible for those conditions

Summary

Details of an individual's family history provide valuable information for the health care provider to use when assessing potential risks for health problems. Limitations to the effectiveness of such information include collection methodology, accuracy of data provided, health care provider time and knowledge, and appropriate interpretation of the gathered information. Individuals are often challenged to remember specifics about actual diagnosis of illness in their families. Some family members express fear of sharing such information and may inaccurately convey details of an illness. Outcomes of the potential for health risks are influenced by many factors, such as behavioral, social, and environmental influences and also need to be considered when interpreting health risks. Health care providers must be able to recognize problem areas indicating need for further investigation or referral.

Chapter Activities

1. A family history is not always easy to obtain and keep current. The accuracy of interpretation of the information is dependent on the accuracy of the data provided. Take time to collect information about your own family history. Identify some of the challenges that you (and therefore your patients) experienced.
2. Go to the Web site *http://www.generationalhealth.com/* and enter your family history information. Determine if there appear to be any indications of potential health issues for your family. Identify what your next steps will be to clarify missing information, talking with other relatives about what you found, or seeking out resources, if needed.

References

American College of Medical Genetics. (2002). Technical standards and guidelines for CFTR mutation testing. *Genetics in Medicine, 3,* 1–25.

American Gastrological Association. (2001). Medical position statement: Hereditary colorectal cancer and genetic testing. *Gastroenterology, 121,* 195–197.

American Medical Association. (2001). Identifying and managing risk for Hereditary Nonpolyposis Colorectal Cancer and Endometrial Cancer (HNPCC). Chicago, IL: AMA.

American Society of Clinical Oncology. (2003). American Society of Clinical Oncology Policy Statement Update: Genetic testing for cancer susceptibility. *Journal of Clinical Oncology, 21*(12), 1–10.

Audrain-McGovern, J., Hughes, C., & Patterson, F. (2003). Effecting behavior change. Awareness of family history. *American Journal of Preventive Medicine, 24*(2), 183–189.

Barse, P. (2003). How to perform a genetic assessment. In: Tranin, A., Masny, A., & Jenkins, J. *Genetics in oncology practice. Cancer risk assessment.* PA: Oncology Nursing Society Press, pp. 57–76.

Bennett, R. (1999). *The practical guide to the genetic family history.* New York: Wiley-Liss.

Bennet, R., et al. (1995). Recommendations for standardized human pedigree nomenclature. *Journal of Genetic Counseling, 4*(4), 276–279.

Bleiker, E., Menko, F., Taal, B., Kluijt, I., Wever, L., Gerritsma, M., Vasen, H., & Aaronson, N. (2003). Experience of discharge from colonoscopy of mutation negative HNPCC family members. *Journal of Medical Genetics, 40,* e55 *(http://www.jmedgenet.com/cgi/content/full/40/5/e55).*

Broadstock, M., Michie, S., & Marteau, T. (2000). Psychological consequences of predictive genetic testing: A systematic review. *European Journal of Human Genetics, 8,* 731–738.

Burke, W., Pinsky, L., & Press, N. (2001). Categorizing genetic tests to identify their ethical, legal, and social implications. *American Journal of Medical Genetics, 106,* 233–240.

Cardeiro, D. (Summer, 2003). Ethnicity and the family history. NCHPEG Newsletter *The Genetic Family History in Practice* found at *http://www.nchpeg.org,* pp. 2–3.

Chemoprevention Working Group. (1999). Prevention of cancer in the next millennium: Report of the Chemoprevention Group to the American Association for Cancer Research. *Cancer Research, 59,* 4743–4758.

Clayton, E. (2003). Ethical, legal, and social implications of genomic medicine. *New England Journal of Medicine, 349,* 562–569.

Collins, F., Green, M., Guttmacher, A., & Guyer, M. (2003). A vision for the future of genomics research. *Nature, 422,* 835–847.

Fitzgerald, M., MacDonald, D., Krainer, M., Hoover, L., O'Neill, E., Unsal, H., et al. (1996). Germline mutations in Jewish and non-Jewish women with early-onset breast cancer. *New England Journal of Medicine, 334,* 143–149.

Grody W. W., Cutting G. R., Klinger K. W., et al. (2002). Laboratory standards and guidelines for population-based cystic fibrosis carrier screening. *Genetics in Medicine, 3,* 149–154.

Hadley, D., Jenkins, J., Dimond, E., DeCarvalho, M., Kirsch, I., & Palmer, C. (2004). Colon cancer screening practices following genetic counseling and testing for hereditary nonpolyposis colorectal cancer (HNPCC). *Journal of Clinical Oncology, 22*(1), 40–44.

Lerman, C., & Shields, A. (2004). Genetic testing for cancer susceptibility: The promise and the pitfalls. *Nature Reviews, 4,* 235–241.

Leung, W. (2000). Results of genetic testing: When confidentiality conflicts with a duty to warn relatives. *British Medical Journal, 321,* 1464–1466.

Lynch, H., & de la Chapelle, A. (2003). Hereditary colorectal cancer. *NEJM, 348,* 919–932.

NCHPEG Working Group of the National Coalition for Health Professional Education in Genetics. (2001). Recommendations of core competencies in genetics essential for all health professionals. *Genetics in Medicine 3*(2), 155–158.

Peterson, S., Watts, B., Koehly, L., Vernon, S., Baile, W., Kohlmann, W., & Gritz, E. (2003). How families communicate about HNPCC genetic testing: Findings from a qualitative study. *American Journal of Medical Genetics, Part C, 119C,* 78–86.

Plantinga, L., Natowicz, M., Kass, N., Hull, S., Gostin, L., & Faden, R. (2003). *American Journal of Medical Genetics, Part C, 119C,* 51–59.

Safer v. Pack, 677A, 2d 1188 (N.J. App.), appeal denied, 683A, 2d 1163 (N.J. 1996).

Tranin, A., Masny, A., & Jenkins, J. (2003). *Genetics in oncology practice. Cancer risk assessment.* PA: Oncology Nursing Society Press.

Yoon, P., Scheuner, M., & Khoury, M. (2003). Research priorities for evaluating family history in the prevention of common chronic diseases. *American Journal of Preventive Medicine, 24*(2), 128–135.

Yoon, P., Scheuner, M., Peterson-Oehlke, K., Gwinn, M., Faucett, A., & Khoury, M. (2002). Can family history be used as a tool for public health and preventive medicine? *Genetics in Medicine, 4*(4), 304–310.

CHAPTER 3
Connecting Genomics to Health Benefits

For the first time in history, humankind can read its genome—its Book of Life. This book is unlike any other, for, in reading it, we will uncover an ever-expanding view of ourselves.

Francis Collins, Director, National Human Genome Research Institute

The utilization of genetic information to improve health benefits of individuals will encompass multiple approaches and requires that health care professionals understand the range of genomic contributions to health and disease manifestation. As basic scientific research identifies the contribution of genetic changes to the progression of disease development, opportunities to intervene with this biologic process will open avenues for individualized care.

Introduction

All along the continuum of care there will be options to improve health outcomes including diagnostics, prevention, medical surveillance, therapeutics, predicting, and monitoring of disease response. But along with those options comes a tremendous challenge as well. This knowledge of "ourselves" integrates more than just the biological influences. The role of behavioral, social, and environmental factors (lifestyle, socioeconomic factors, pollutants, etc.) to modify or influence genetics in the manifestation of disease also needs to be taken into consideration. Research is needed to assess how health strategies that are based on genetic information influence individual behavior, access to services, and costs, and ultimately impact health. Informed nurses will play a key role in translating this new information and connecting genomics to improved health benefits.

Rebecca's Story

I understood cancer as a nurse, but I was about to learn even more as a patient. For over twenty years I had been an oncology nurse. My most recent experience was working with persons who had a hereditary contribution to their family's history of cancer. I understood that hereditary factors do increase someone's risk for cancer, such as with breast–ovarian or colon cancer syndromes. But I had non-Hodgkins lymphoma! Among the many questions asked by my son were whether or not there was a history of cancer in my family and did this increase his risk for cancer. All I could say was that, based on today's knowledge, lymphoma did not appear to be a cancer that had a hereditary contribution to why it occurred. But what I could explain less well was that there are genetic changes that occur in normal cells as they progress from being benign to cancerous. I was about to learn even more about how those genetic changes found in my lymphoma cells could be used in guiding the decisions about my care.

Rebecca

Genetic Approaches to Prevention and Early Detection

The diagnosis of cancer is currently based on the morphological examination of tumor cells along with a description of clinical information. There is research using deoxyribonucleic acid (DNA) microarray technology to provide a molecular classification of cancer into disease categories based on gene expression profiling (Ansell et al., 2003; Staudt, 2002; Tefferi et al., 2002). The initial diagnosis of what kind of cancer a person has can be made based on the cell type found in sites of disease, such as in lymph nodes or the results of a bone marrow biopsy. Sequential acquisition or loss of surface markers occurring during cell differentiation makes it possible to classify the type of malignancy that is present (Kipps, 2002). Classification of cancer utilizing individual gene expression or multiple gene sets is beginning to allow improved accuracy of diagnostic samples.

The diagnosis of my type of cancer rested within the pathologist's hands. Based on screening tests, I knew that I had cancer in my abdomen and throughout my lymph nodes. But the bone marrow test results would tell me more about what was suspected—that I indeed had lymphoma. There are several types of lymphoma—some with worse prognoses than others. So, I was extremely anxious to know the bone marrow test results.

Molecular characterization of B-cell lymphomas (follicular) has advanced with the identification of characteristic chromosomal translocations or rearrangements occurring specifically with certain types of non-Hodgkins lymphoma. These genetic changes consistently alter the regulation of a particular gene (Siebert et al., 2001). Gene expression profiling is becoming useful in determining the common cell of origin of the cancer, assessing the common mechanism of cell transformation, and evaluating uniform clinical behavior. Understanding each of these molecular features provides the opportunity to develop hypotheses as to the possibility of new diagnostic tests as well as molecular targets for treatment design that can be evaluated through clinical trials (Staudt, 2003) (see *http://www.nih.gov*).

Genetic Testing

Genetic testing can be used in various ways. A genetic screening test may be useful in determining the potential risk for development of a disease in one's lifetime. Such tests may be done in a variety of settings and at any point during the lifespan including prenatal, at birth, or throughout childhood to adult years. Some tests screen for the possibility of disease and therefore are predictive of the potential for disease. Other tests indicate that the individual is a carrier of a genetic change that, during reproduction, has the potential to result in disease if his or her partner carries a similar genetic error. Genetic tests may also be used to confirm a diagnosis suspected from clinical findings.

The complexity of interpreting genetic test results can create issues for health care professionals as well as for the patient. Some tests are very sensitive, specific, and are accurate in their usefulness for clinical decision making. Other tests may predict the potential for disease but not take into consideration additional intervening variables that influence disease manifestation in that individual. Some results may rely on multiple layers of test outcomes to identify the implications for the patient (i.e., cystic fibrosis testing, Farrell & Fost, 2002). These genetic tests all have indications for use, limitations, and risks that need to be understood by the health care provider so that the individual patient can weigh such information in making life-altering decisions (Burke, Pinsky, & Press, 2001). The Secretary's Advisory Committee on Genetics, Health, and Society will consider many issues related to genetic testing as part of its focus on the impact of genetic technologies to society (*http://www4.od.nih.gov/oba/sacghs.htm*).

Research Genetic Testing

Not all genetic tests have clinical utility. Extensive laboratory and bioinformatics research identify technological capabilities for testing, but perhaps do not offer any available interventions based on test outcomes. Or perhaps the test identifies only one gene involved in the disease and is not inclusive of the cascade of events that occur within the whole process. Significant research is needed to identify genetic characteristics that are clinically relevant. Persons participating in research that collects genetic information must be provided information that clarifies the benefits, risks, and usefulness of test results to their individual care so that they can truly give informed consent to participate in the study. Many times research results will not be given to the individual because the research genetic test is valuable in hunting for future clinically useful information but has limited clinical meaning until the research is concluded.

Healthy individuals may also be approached to consider participating in research studies that are collecting samples for genetic analysis. Databanks are becoming an important tool to help identify what is genetically different among individuals and what factors contribute to their health or illness. One project initiated in 2003, the HapMap, will chart genetic variation within the human genome. By comparing genetic differences among individuals, the goal is to create a tool to help researchers detect the genetic contributions to many complex diseases *(http://www.genome.gov)*.

The use of markers of biological variation will also be useful in predicting patient response to pharmaceuticals (Evans & McLeod, 2003). Pharmacogenetics and pharmacogenomics are emerging applications of genetic tests that help to personalize clinical care. The implementation of these tools will require significant research to be able to offer more targeted, more effective medicines (Lindpaintner, 2003).

Investigational Genetic Testing

Rebecca understood the importance of contributing to research. She also expressed interest in knowing genetic test results, although results were inconclusive.

Each time I go in for bone marrow biopsies, and periodically throughout the treatment regimen and follow-up, multiple research bloods are drawn. It clearly states in the informed consent that I signed that the results of these research studies will not be given to me. These test samples are being analyzed, along with other patients' samples, to determine the correlation of gene expression measurements in lymphoma with clinical outcome. Although I recognize the preliminary status of the meaning of this information, I am a scientist by training and am curious as to the results. It's really quite fascinating; it would be great if it weren't happening to me!

Clinical Genetic Testing

Clinical genetic testing for the potential of developing cancer is possible for several hereditary forms of cancer. Genes have been identified that, when changed or mutated, increase the individual's potential of developing cancer. This risk is not a diagnosis of cancer, nor do all persons with such mutations "always" get cancer. However, their risk is elevated above the normal population risk. Clinical genetic testing is currently available for several types of cancer, such as breast, ovarian, colon, and melanoma *(http://cancer.gov/cancerinfo/prevention-genetics-causes)*. There are no clinical genetic tests currently available to predict individuals at risk for the development of lymphoma.

Disease and Drug Response

Multiple persons diagnosed with the same cancer respond differently to the same treatments, indicating that perhaps additional factors influence treatment outcomes. Known prognostic factors that negatively impact lymphoma treatment success include age (> 60), LDH > normal, performance status 2 or higher, stage III or IV disease, and extranodal involvement (> one site). Prognostic significance of molecular markers such as protein expression and gene rearrangements are being evaluated as to usefulness in identifying those needing more aggressive treatment. For instance, Bcl-2 protein expression may be a prognostic indicator for disease responsiveness and clinical drug resistance (Gascoyne et al., 1997). A high level of Bcl-2 protein inhibits apoptosis (cell death) and may be useful in designing treatment strategies for lymphoma and predicting treatment outcomes (Cory & Adams, 2002).

Therapeutic Approaches

Having had the experience, as an oncology nurse, of providing information to patients about options for treatment, I knew where to look for summaries of standard and research treatments for lymphoma (PDQ at *http://www.cancer.gov*). I also sought out a second opinion from a "lymphoma expert" to help guide my decision about treatment options. It became clear that standard treatment is insufficient in my case. My physician told me that there is no "cure" for my cancer; it often reoccurs in 3 to 5 years. So I knew I had to do more than get "standard chemotherapy." Clinical trials often go that next step of trying something more than standard treatment. And although the side effects and cancer responsiveness are unknown, I decided that a clinical study *(http://www.clinicaltrials.gov)* was my best shot at living to see my granddaughter grow up. There was a study that included standard chemotherapy up front, but then had the addition of a vaccine or a placebo as the research question. I was fascinated that the genetics technology that I was so passionate about because of its great potential for improved health care in the future was opening a treatment door for me now.

The design of experimental vaccines aimed at enlisting a patient's immune system to attack malignant tumors is offering individualized treatment based on patient-specific cancer cell markers (Bendandi et al., 1999; Neelapu & Kwak,

2002). The development of such vaccines has been difficult because of the inability to distinguish tumor cells from normal cells. Vaccines used to treat cancer take advantage of the identification of unique molecules that exist or are more abundant in cancer cells. This tumor-specific antigen, when identifiable, can be used as a vaccine to induce an individual's immune system to attack that person's cancer cells without harming normal cells (NCI, 2003).

I had to have a peripheral lymph node removed to obtain tissue to provide the starting material for my vaccine. I was concerned that the sample might be insufficient to allow vaccine production. I also would not know if I received a placebo or a real vaccine via the injections because I was in a randomized study. I made the decision to take that chance, but selfishly hoped I was the patient who got the "real thing."

Production of the individualized vaccine is time-consuming and resource intensive. The laboratory effort can take up to six months to complete vaccine development. There are many methods used to develop vaccines including whole cell tumor vaccines, dendritic cell vaccines, and, in this example, idiotype vaccines (NCI, 2003). If all goes as planned, the vaccine is administered by rotating subcutaneous injection sites to upper and lower extremities. Little or no toxicity is expected from the vaccine, but skin reactions are monitored. Studies indicate that responses to vaccines can be improved by use of GM-CSF, a protein that stimulates the proliferation of antigen-presenting cells when administered concurrently with vaccine. The addition of GM-CSF can cause chills, fever, myalgias, skin reactions, and other side effects.

Knowing the molecular biology of lymphoma has provided another option for treatment consideration. Based on a better understanding of the pathway of lymphoid malignancy, the cancer cell surface marker CD20 is now known to be expressed early in B-cell differentiation. This differentiation antigen was recognized as a potential target for immune therapy resulting in the study of Rituximab to assess efficacy in treating CD20-expressing B-cell malignancies (Kipps, 2002). Research results of Rituximab, a monoclonal antibody that attaches itself to specific proteins on cancer cells (the CD20) and then enlists the body's immune system to kill the malignant cells, have been promising (Hainsworth et al., 2002).

Health Modifiers

I've worked my whole career in a clinical oncology setting. Could something I had been exposed to, such as a virus or chemotherapy drugs, have put me at risk for my cancer? Many of my oncology nurse colleagues were also being diagnosed with cancer, and I wondered about the occupational exposures we had all experienced. I read about suspected environmental contributions to lymphoma development, such as pesticides, water contaminants, and hormones. I'd had a hysterectomy for fibroids; perhaps the hormone replacement had modified my body's ability to respond appropriately to fight off the cancer cells. Most likely, according to the literature, my body had responded to an infectious agent, and the cells remained active rather than turning off as they normally would. Whatever the reason(s) for the cancer, it wasn't something I could change now. I had to move on from would've, could've, should've, to dealing with the now.

The National Institute of Environmental Health Sciences has major research activities focused on improving the understanding of how individuals differ in their susceptibility to environmental agents and how these susceptibilities change over time *(http://www.niehs.nih.gov/envgenom/home.htm)*. Environmentally responsive genes, such as those involved with DNA repair, cell cycle control, cell signaling, cell division, homeostasis, and metabolism, are being researched as having a potential role in increased susceptibility to environmental exposures.

Conveying Genetic Risk Information

I, the patient, now became even more aware of the importance of knowledgeable and caring health care providers. I wanted to understand what to expect, when the scheduled treatment would occur, and who to call in an emergency. Once cancer was diagnosed, most of the "control" over my life was put in others' hands. I sought out information to help me gain back some control in my life through participation in decisions about my care. I would often ask each of those administering the vaccine portion of my study about the rationale for this type of treatment, expected side effects, and long-term risks. It was obvious which nurses and physicians understood genetics and upon whom I could rely for translating the complexity of this treatment to me.

Conveying genetic information is not easy.

Monitoring for Residual Disease and Medical Surveillance

Assays of tissue samples can be done periodically to determine response to treatment, as well as to monitor for disease relapse. Immunologic studies that include proliferative response to the vaccine and presence of rearranged Bcl-2 are assessed frequently to determine evidence of response.

There is also research assessing for a genetic signature for the potential for disease metastasis. It has been determined that the metastatic gene expression pattern is already present in some primary tumors. It may be that these types of cancer have a greater potential to spread and this assessment allows an option for identifying those individuals needing more initial aggressive treatment or enhanced medical surveillance.

Each time I go in for follow-up medical surveillance tests, I get anxious. Life as I know it can change immediately with test results that indicate the need for more treatment. But at least I know that, because of the new technology, my body may have a better chance of fighting off the cancer when caught early.

Rebecca

Health Care Resources and Systems

It is tremendously important that managers of health care systems begin to contemplate the resources needed to provide genomic-based care. Education of multidisciplinary teams involved in providing care to individuals, for example, who have cancer need to include genomic concepts. As one researcher summarized, the ability to take tissue samples, identify gene expression patterns, develop hypotheses about potential interventions, and then try them out is occurring a lot faster than he ever expected. The knowledge gained from such research is moving into the clinical setting quickly with the potential to overwhelm available resources (Bloom, 2003). There is the concern that the current health care system will not be able to effectively transform the biomedical research advances into meaningful advances in health care. Synthesis of information requires awareness, time, interest, access, and incentives. The current health care system is not structured to support the type of restructuring needed to take the focus of care from symptom- and disease-based to predictive and individually focused (Guttmacher, Jenkins, & Uhlmann, 2001). The new use of genetics in health care requires that interdisciplinary efforts to design models of genetic services delivery begin now.

Summary

Utilization of genetic information in practice has the potential to improve health outcomes across the continuum of care. These benefits can occur only if health care providers invest time and energy to understand the scientific basis of health and illness and utilize that knowledge to transform clinical care. Utilization of such knowledge will enhance availability of options for individuals, including diagnosis, prevention, and treatment. However, such individualized or personalized care brings with it challenges for professionals, such as communication and interpretation of complex genetic test results. Health care systems and resources will need to be educated and perhaps even restructured to provide care that integrates genomic discoveries.

Chapter Activities

1. There are many issues for society that have been identified with genetic testing by the Secretary's Advisory Committee on Genetics, Health, and Society. Go to *http://www4.od.nih.gov/oba/sacghs.htm* and determine if nurses have offered their viewpoints on any of these issues.
2. Identify a patient that you have seen recently in your practice. Go to *http://www.clinicaltrials.gov* and determine if there are any clinical studies that he or she would be eligible to participate in. What would the risks and benefits be to the patient if he or she would participate in the study?

References

Ansell, S., Ackerman, M., Black, J., Roberts, L., & Tefferi, A. (2003). Primer on medical genomics. Part VI: Genomics and molecular genetics in clinical practice. *Mayo Clinical Proceedings, 78,* 307–317.

Bendandi, M., Gocke, C., Kobrin, C., Benko, F., Sternas, L., Pennington, R., Watson, T., Reynolds, C., Gause, B., Duffey, P., Jaffe, E., Creekmore, S., Longo, D., & Kwak, L. (1999). Complete molecular remissions induced by patient-specific vaccination plus granulocyte-monocyte colony-stimulating factor against lymphoma. *Nature Medicine, 5*(10), 1171–1177.

Bloom, F. (2003). Science as a way of life: Perplexities of a physician-scientist. *Science, 300,* 1680–1685.

Burke, W., Pinsky, L., & Press, N. (2001). Categorizing genetic tests to identify their ethical, legal, and social implications. *American Journal of Medical Genetics, 106,* 233–240.

Cory, S., & Adams, J. (2002). The Bcl2 family: Regulators of the cellular life-or-death switch. *Nature Reviews Cancer, 2,* 647–656.

Evans, W., & McLeod, H. (2003). Pharmacogenomics-drug disposition, drug targets, and side effects. *New England Journal of Medicine, 348*(6), 538–549.

Farrell, P., & Fost, N. (2002). Prenatal screening for cystic fibrosis: Where are we now? *Journal of Pediatrics, 141*(6), 758–763.

Gascoyne, R., Adomat, S., Krajewski, S., Krajewska, M., Horsman, D., Tolcher, A., O'Reilly, S., Hoskins, P., Coldman, A., Reed, J., & Connors, J. (1997). Prognostic significance of

Bcl-2 protein expression in Bcl-2 gene rearrangement in diffuse aggressive non-Hodgkin's lymphoma. *Blood, 90*(1), 244–251.

Guttmacher, A., Jenkins, J., & Uhlmann, W. (2001). Genomic medicine: Who will practice it? A call to open arms. *American Journal of Medical Genetics, 106,* 216–222.

Hainsworth, J., Litchy, S., Burris, H., Scullin, D., Corso, S., Yardley, D., Morrissey, L., & Greco, A. (2002). Rituximab as first-line and maintenance therapy for patients with indolent non-Hodgkin's lymphoma. *Journal of Clinical Oncology, 20*(20), 4261–4267.

Kipps, T. (2002). Advances in classification and therapy of indolent B-cell malignancies. *Seminars in Oncology, 29*(1 Suppl 2), 98–104.

Lindpaintner, K. (2003). Pharmacogenetics and pharmacogenomics in drug discovery and development: An overview. *Clin Chem Lab Med, 41*(4), 398–410.

National Cancer Institute. (2003). Cancer vaccine fact sheet. Found at *http://www.cancer.gov/newscenter/pressreleases/cancervaccines.*

Neelapu, S. S., & Kwak, L. W. (2002). Vaccine approaches to non-Hodgkin's lymphoma therapy, malignant lymphomas, *American Cancer Society Atlas of Clinical Oncology,* 316–329.

Siebert, R., Rosenwald, A., Staudt, L., & Morris, S. (2001). Molecular features of B-cell lymphoma. *Current Opinion in Oncology, 13,* 316–324.

Staudt, L. (2002). Gene expression profiling of lymphoid malignancies. *Annual Review of Medicine, 53,* 303–318.

Staudt, L. (2003). Molecular diagnosis of the hematologic cancers. *New England Journal of Medicine. 348*(18), 1777–1785.

Tefferi, A., Wieben, E., Dewald, G., Whiteman, D., Bernard, M., & Spelsberg, T. (2002). Primer on medical genomics. Part II: Background principles and methods in molecular genetics. *Mayo Clinical Proceedings, 77,* 785–808.

CHAPTER 4

Connecting Genomics to Health Benefits: Genetic Testing

Genetic factors play an important role in maintaining health and preventing illness. Nurses are at the interface between technology, clinical application of new genetic testing and treatments, and the patients and families who make use of and are affected by new genetic approaches to health and illness. Knowledge of the difference between genetic screening, clinical diagnosis of disease, and identification of genetic predisposition to disease is integral to nursing practice. Awareness of the influence of ethnoculture influences and economics in the prevalence and diagnosis of genetic disease, as well as the influence of ethnicity, culture, and economics in the client's ability to use genetic information and services, is a necessity for nurses. Knowing the indications for genetic testing and/or gene-based interventions allows nurses to identify clients for whom genetic testing and services are appropriate and available. Having this knowledge and skill enables the nurse to explain basic concepts of probability and disease susceptibility and the influence of genetic factors in maintenance of health and development of disease, and to seek assistance from and make referrals to appropriate genetics experts and peer support resources. Nurses in all practice settings will participate in coordination of patient care and collaboration with an interdisciplinary team of health professionals to help clients and families receive the genetic information and support they need (NCHPEG, 2001).

Introduction

Genetic testing, now becoming a common tool for diagnosis and disease management in today's health care, did not come into clinical use until the 1960s with the work of Dr. Robert Guthrie who developed a newborn screening test for phenylketonuria (PKU). Dr. Guthrie introduced an innovative system for collection and transportation of blood samples from newborns on filter paper. This cost-effective newborn screening for PKU made possible wide-scale genetic screening allowing for early dietary intervention and prevention of mental retardation (Jorde et al., 2003; *http://www.geneclinics.org*).

During the same decade, the analysis of chromosomes (cytogenetics) became established as a method for diagnosing genetic conditions such as Down syndrome, Turner syndrome, and Klinefelter syndrome. The procedure, amniocentesis, was introduced for prenatal diagnosis of chromosome abnormalities in 1966 and was used primarily in women over 35 years. Prenatal screening for neural tube defects—the AFP screening test—was introduced in the 1970s, allowing for the availability of widespread screening of pregnant women *(http://www.dnapolicy.org)*.

In the following two decades, technological advances, combined with human genome discoveries, led to an explosion of genetic tests for carrier, prenatal, diagnostic, and susceptibility purposes. Now, for example, states routinely collect blood spots from newborns for up to 30 metabolic and genetic diseases. Prenatal screening for neural tube defects and chromosomal abnormalities, such as Down syndrome, is routinely offered to all pregnant women. The American College of Obstetrics and Gynecology (2001) and the American Society of Medical Genetics (ACMG, 2001) currently recommends that all couples be offered cystic fibrosis (CF) carrier screening prior to or during pregnancy so that they will have the option of prenatal diagnosis if both parents are found to be carriers. Other medical societies, such as the American College of Medical Genetics (ACMG, 1998), have developed guidelines for when to offer susceptibility testing for hereditary breast–ovarian cancer.

All of these advances create new opportunities for individuals and families in terms of prevention, early intervention, and management strategies. Genetic testing also brings with it many psychosocial issues for individuals and families, such as how to cope with a new genetic diagnosis, how or whether to share genetic test results with family, and how to make sure that genetic testing information remains private and confidential. Cultural and economic issues are also involved with genetic testing in terms of uptake of genetic testing and availability of and access to genetic testing services.

Nurses are at the interface between the individual, the medical system, and genetic testing opportunities. As the role of genetic testing increases in all areas of clinical practice, nurses will need to have a good understanding of what genetic tests are, the indications for their use, and the professionals and services available if they are to appropriately offer genetic testing and support families through the testing process.

In this chapter, we explore genetic testing and associated psychosocial issues related to three types of genetic testing in common use today: newborn screening, prenatal screening and diagnosis, and carrier testing. Complementary information on genetic testing for diagnostic purposes is further described and discussed in Chapter 5. Chapter 8 reviews and describes presymptomatic testing for Huntington Disease, while Chapter 12 describes predictive testing and issues related to this growing application of genetic testing and gene-based therapies.

We begin this chapter with a story of the initial use of genetic testing—newborn screening. Through Betsy's story, we learn about the impact of genetic testing on Betsy and her family, the resources available to them, and the ongoing concerns for her daughters who have been diagnosed with PKU. Laura's story, in contrast, presents issues with genetic testing that have more recently been made available: preconception and prenatal carrier testing for cystic fibrosis. Each story offers a detailed snapshot of what genetic tests, old and new, are and how they impact individual and family life in terms of health benefits, risks, and limitations.

Betsy's Story: Our Life after Newborn Screening

I have three children and two have PKU. My older daughter is 9, and my younger daughter is 3 years old. Both of my daughters have PKU. My son does not have it. I don't remember being given any information about newborn screening for PKU prior to my older daughter's birth. After she was born, at 48 hours they gave me a sheet about PKU. It said that it happened in 1 in 15,000 births and that if we had relatives with it we should take the test. The information didn't seem to apply to me so I put it away and went ahead and gave permission for the testing.

The pediatrician who saw my daughter in the hospital was not my daughter's regular pediatrician. He was just covering. After two weeks, on February 11, that pediatrician's office called my home. As it turned out, I was on my way to my daughter's pediatrician for the 2-week checkup and missed the phone call. The other pediatrician's office left several messages on our machine over a period of two hours, each becoming more frantic than the one before.

When I got home, I called and the doctor said to me "her PKU came back at 12 and we don't know what to tell you to do." I knew a little about proteins and diet, and I had a book upstairs in the attic. While my husband was looking for the book, I called my regular pediatrician and told her that my daughter had tested positive.

Dr. Z., my pediatrician, called a geneticist she knew and referred us to a large metropolitan hospital in a city near us where there is a metabolic clinic and a geneticist, Dr. M. We went right to the clinic and he did some blood work. The results came back in 2 days, and we met with Dr. M. and his team on February 14. I will never forget that date. Dr. M., the dietitian, and the genetic counselor met with us all day. Well, actually, it was the dietitian and the genetic counselor, mostly. They told me that I could not nurse my daughter any more. They found me a pump and I was told to freeze my milk until my daughter could take a little. At 6 weeks, I could nurse her briefly.

Dr. M. and his team saw us weekly until my daughter was 9 months old. We went to the clinic for blood draws. The blood was drawn in a research room, and I know the room by heart. The lab at the hospital does the studies. Then, my daughter was seen biweekly until she was 2 years old, then monthly through ages 4 and 5. Now that she is 9, we go 3 or 4 times a year to meet with Dr. M.

Mary is the dietitian that I have been working with. I can page her any time, call her, or e-mail her. She has been wonderful. We basically stumbled along until our daughter was 2 years old. When she was 6 months old, we attended a picnic for PKU families at the children's hospital center. This is another of the metabolic centers in our state. At 6 months my daughter was just starting food and I was worrying about what she was getting, with the spitting out of food and all. I talked with the other parents, trying to get some advice, and they just said "you'll figure it out." This was not helpful for me. At that time there were no online resources, so we just fumbled our way through.

Our dietitian became pregnant and when she had her baby and started feeding her baby, she began to understand, in a practical way, where we were coming from and our concerns. She was more flexible than Dr. M. in trying new things. Dr. M. is in his late seventies and he has some fixed ideas. Our state has a PKU camp but it is for 10-year-olds and up, and the focus is on maternal PKU and informing the girls that they must have planned pregnancies. My daughter was only 7 years old so she couldn't be involved. I started searching around online and found a camp on the east coast—a family camp. A family camp for PKU is essential for what you need. We went to the camp. They have counselors for the kids and do camp stuff. The parents met with a PKU chef. She gave us recipes for regular foods. She was a godsend. For the first time, I felt like I had some choices.

This past summer the cost of going to the family camp became too expensive for us, so we went to another camp closer by. It was nice to see our daughter with kids her age. All of the food at this camp was prepared for kids with PKU, so that was great. There is nowhere else they can go and eat anything they want.

I receive PKU formula from our state program. In the beginning, we were getting a 6-month supply. Then we had to go through the pharmacy. We could only pick up one can at a time, so that meant we had to go to the pharmacy every 3 days. One day my husband went to the pharmacy to pick up the formula. It hadn't been quite 3 days and the pharmacist would not give it to him. We had to return 6 hours later! After that, UPS delivered it, and now we get a supply once a month. But it is delivered by an express service and you have to be there to pick it up. If you miss the guy, they ship it back, and there you are without formula.

Our regular insurance does not cover formula. The billing process works like this. The company bills our insurance, and our insurance denies coverage. The bill then goes to our state program, and whatever they don't pay goes to Medicaid. PKU food isn't covered either, but it is getting better. The state program now covers some phenyl-free foods, such as the Flex-10 bar. This is like a power bar for people with PKU. There are also other items available, such as juice and pills. My daughters have one Flex bar each day. But it took 3 months to get the cost covered.

Not being able to get the low-protein food at a reasonable cost is a great challenge for us. The companies make frozen cheese for people with PKU that I could get, but it costs $100 and I can't afford it. I can't put us all on low-protein food so I have to split my food budget. If more monies were available for low-protein food or if our insurance would cover it, then this would help me to keep the girls' phenylalanine levels down. Having to go through all of the different services to get the food is a hassle too. One department covers some things and WIC covers others. I can get 6 cans of formula paid for by WIC and 3 cans by the other department. I can't imagine it is easy to bill. But they have to split up the financial resources. So, I would say a big challenge is getting the food paid for.

The other challenge has been not having a support group for PKU. I have tried to get PKU families together. When I went to the formula convention, I gave my name to all of the dietitians and asked them to talk with their patients about a group. I haven't heard a thing. Another thing that would be good for families is if we had an emergency food shelter like they do in some states. In one state, for example, if a family has PKU formula or food that they don't want, they can donate this to the shelter. Then, if another family has an emergency, they can call the shelter for food and help. Our state needs something like this. No parents have called me to get this going. The parents are the only ones who will make insurance companies stand up and take notice. But I can't do it alone. I need other families with me.

When I was pregnant with my second daughter, people asked me if I was going to have an amniocentesis. I said no because you can't fix PKU, and I know how to handle it. I always wanted four kids, but after my second child was born with PKU we decided not to have any more because we couldn't afford it. My obstetrician never suggested genetic counseling. We did have genetic counseling after my first daughter was born. I know that PKU is a recessive condition and that I have a 1 in 4 chance to have a child with PKU. That was all that entered my thinking. Now, I am glad that I have two children with PKU because the recipes I make are made for two and my oldest daughter always had to eat the same thing until it was gone. Now we can split the recipes up.

Recently, a national newsletter printed a letter from a mom who had just found out that her baby had PKU. She was crying and upset. There was a response from a teen with PKU saying to the mom that it wasn't so bad. I say, "you are not a parent" to this teen. You don't know how it is when you have a perfect baby and someone takes that away when they tell you that your baby has a medical condition. That is a big, challenging, and scary thing. The exchange of letters intrigued me. The teen doesn't have a clue about being a parent. I do.

Betsy

Laura's Story: Finding Out about Cystic Fibrosis in My Family

My father suffered from atypical CF. He had suffered from chronic lung disease that, at that time, was described as "CF like." The disease appeared after he was married; he was 24. He died at the age of 54 before genetic testing was available. We found out that the lung disease was really CF only after he passed away, and we (his children) had genetic testing performed.

When I was pregnant 8 years ago, I underwent genetic testing for CF. I was found to be a carrier of the 10kb+3849 C→T mutation. I had suggested to my brothers and sisters that they be tested as well. To our surprise, two of them were found to be carriers of a different CF gene mutation—the W1282X mutation. My mother is not a carrier, so it became obvious that my father had two different CF mutations, meaning that he was a compound heterozygote. He had one mutation that gives the phenotype of the classic CF disease and a second one that gives a phenotype of a mild disease and leaves men fertile.

One of my sisters had received results of the genetic tests by surface mail, which notified her that she is a carrier of the W1282X mutation. She was very confused and did not know what to do next. Luckily, I could help her. These days, we have genetic services that usually telephone the patient and recommend that his or her partner be tested.

Growing up with a sick and limited father was not an easy experience. It affected many aspects of my life. The worry for his health and the restriction of his lifestyle caused us to change our habits, our trips, and things like that. As we did not know that he actually suffered from CF, we continued to hope for improvement and not always did we understand (or want to believe) that each episode and deterioration is irreversible. I understand now that it was impossible to diagnose him at that time. There was no genetic testing available. But it would have helped us to cope, knowing the exact diagnosis, as we went through the long debate about whether it was wise to go through with a lung transplant. When we finally decided for the transplantation, it was too late.

When I was about to marry, my sister-in-law asked a question that seemed irrelevant at the time. She asked, "is your father's condition genetic?" This had a great impact on me since we never thought it to be genetic. That is when we understood that genetic information can be very helpful on the one hand, and on the other hand lead to discrimination, because that was the real essence of her question.

Looking back on my experience, I would have liked to see more personal attention given to individuals undergoing genetic testing. I would like to see more education of the public about genetic conditions, testing, and the meaning of test results. For this, I think there should be more nurses involved in genetics. This would help implement the knowledge, explanations, and personal follow-up that the individuals and families so desperately need.

Laura

GENETIC TESTING: WHAT IT IS AND HOW IT IS USED CLINICALLY

Genetic Testing Defined

Genetic tests involve laboratory analyses of chromosomes, genes, or gene products (enzymes or proteins) to detect a genetic alteration that can cause or is likely to cause a specific genetic disorder or condition (SACGT, 2000). Genetic testing can be DNA- or RNA-based, chromosomal, or biochemical *(http://ghr.nlm.nih.gov)*. Genetic tests are currently used to predict risk of disease, to screen newborns for genetic conditions, to screen pregnant women for risk of genetic conditions, for prenatal or clinical diagnoses or prognoses, and to direct clinical care. Today there are more than 900 genetic tests performed in more than 500 laboratories

TABLE 4-1 Indications for Genetic Testing

Carrier Testing Performed to determine whether an individual (or couple) carries a recessive allele for an inherited disorder, which can then be passed on to offspring (i.e., cystic fibrosis, Tay-Sachs disease, sickle cell disease).

Preimplantation Diagnosis Performed following in vitro fertilization to diagnose a genetic disorder in an embryo prior to implantation (i.e., cystic fibrosis, Duchenne muscular dystrophy, fragile X syndrome).

Prenatal Diagnosis Performed to diagnose a genetic disorder in a developing fetus (i.e., chorionic villus sampling and amniocentesis to obtain fetal cells for analysis of conditions such as Down syndrome, cystic fibrosis, Duchenne muscular dystrophy).

Newborn Screening Performed in newborns, typically as part of state public health programs, to identify certain genetic conditions for which early diagnosis and treatment are available (i.e., phenylketonuria (PKU), galactosemia, congenital hypothyroidism).

Diagnostic Testing Performed to diagnose or confirm the diagnosis of a genetic condition in an individual. Diagnostic testing is often used to help predict the long-term effects and course of the condition and to determine treatment options. (i.e., fragile X syndrome, velocardiofacial syndrome (22q11 deletion syndrome), Turner syndrome).

Predictive Testing Used to determine the probability of a healthy individual, without a known family history, developing a disease (i.e., breast–ovarian cancer, colon cancer, cardiovascular disease). These tests are generally reserved for individuals who have a family history of the condition, but in the future, may be offered to individuals without a family history.

Presymptomatic Testing Performed to help determine whether or not individuals who have a family history, but who are not currently symptomatic, in fact have a specific gene mutation. (i.e., Huntington Disease, Myotonic dystrophy, some cases of Alzheimer's disease).

(*http://www.genetests.org, http://www.dnapolicy.org,* Jorde et al., 2003).

Table 4-2 Common Genetic Tests Individuals and Families Undergo

Prenatal	Clinical Examples
DNA-based	Prenatal diagnosis—cystic fibrosis, fragile X syndrome
Chromosomal	Prenatal diagnosis—Down syndrome, Trisomy 18
Biochemical	Prenatal screening—Multiple-marker screening for neural tube defects and Down syndrome
Carrier testing	Prenatal screening—Determines if an individual or couple carries a recessive allele for an inherited condition which can then be passed on to children: cystic fibrosis, sickle cell
Newborn Screening	**Clinical Examples**
Widest application of genetic testing in the United States. Available for an increasing number of genetic disorders. Each state has a mandatory number of newborn screening tests.	PKU, Galactosemia, Sickle Cell Anemia, Congenital Hypothyroidism, Homocystinuria
Diagnostic Testing	**Clinical Examples**
Identifies or confirms a diagnosis of a genetic condition by detecting the presence or absence of a particular genetic alteration. May be DNA-based, chromosomal, or biochemical.	Down syndrome (chromosomal analysis), fragile X syndrome (DNA-based), Tay-Sachs disease (biochemical)
Predictive/Susceptibility Testing	**Clinical Examples**
Determines the probability of a healthy individual developing a disease with or without a family history. Usually DNA-based.	Hereditary breast–ovarian cancer, hereditary colorectal cancer (HNPCC, FAP)
Presymptomatic Testing	**Clinical Examples**
Helps determine whether or not individuals who have a family history, but who are not currently symptomatic, have a specific mutation.	Huntington Disease, Myotonic dystrophy

(*http://www.GeneClinics.org;* Jorde et al., 2003). The Secretary's Advisory Committee on Genetic Testing (SACGT, 2000), established in 2000, developed a list of indications for genetic testing (**Table 4-1**). **Table 4-2** lists common uses of genetic tests, while **Table 4-3** describes current terminology used when evaluating the accuracy and use of genetic tests.

TABLE 4-3 Standard Concepts and Accuracy and Effectiveness of Genetic Laboratory Tests

Analytical Validity

How well a test measures the property or characteristic it is intended to measure. In the case of genetic tests the property would be DNA, proteins, or metabolites.

- *Analytical Sensitivity:* an analytically valid test that is positive when the relevant gene mutation is present.
- *Analytical Specificity:* an analytically valid test that is negative when the relevant gene mutation is absent.

Clinical Validity

A measurement of the accuracy with which a test identifies or predicts a clinical condition.

- *Clinical Sensitivity:* a clinically valid test that is positive if the individual being tested has the disorder or predisposition.
- *Clinical Specificity:* a clinical valid test that is negative if the individual does not have the disorder or predisposition.
- *Positive Predictive Value:* a clinically valid test that is positive if the individual being tested has or will get the disorder or condition.
- *Negative Predictive Value:* a clinically valid test that is negative if the individual being tested does not have or will not get the disorder or condition.

Determining the clinical validity of a test is challenging when different mutations within the same gene cause the same disorder and different mutations can result in different degrees of severity of the condition. Furthermore, some gene mutations may or may not lead to disease depending on how penetrant or completely expressed they are.

Clinical Utility

The degree to which benefits are provided by positive and negative test results. A genetic test has utility when the results, positive or negative, provide information that is of value to the person who is tested. Even when no interventions are available to treat or prevent the disorder or condition, there may be benefits associated with knowledge of the result. Social, psychological, and economic harms may also result from such knowledge, especially when there are no privacy and discrimination protections. Determining the clinical utility of a genetic test involves obtaining information about the risks and benefits of both positive and negative results.

(Secretary's Advisory Committee on Genetic Testing, 2000)

Types of Genetic Testing

Biochemical tests analyze the presence or absence of key proteins, which indicate altered or malfunctioning genes. DNA-based tests (gene tests), on the other hand, are more sophisticated techniques to test for genetic traits, predispositions, and disorders. They involve direct examination of the DNA molecule itself. A DNA sample can be obtained from any tissue, for example, a person's blood, body fluid, or tissue, and examined for an alteration that indicates a disease, trait, or condition. The alteration can be relatively large, such as an entire chromosome or piece of a chromosome missing or added. In some cases the alteration is very small, with a missing or altered chemical base within a DNA molecule. Genes can be amplified (too many copies), over-expressed (too active), inactivated, or lost. Sometimes the alteration involves pieces of chromosomes that have become transposed (translocation) or rearranged (inversion).

A variety of techniques are used to examine a person's DNA. Sequence analysis or gene sequencing, a process by which the nucleotide sequence is determined for a segment of DNA, is frequently used. For some types of gene tests, short pieces of DNA called probes, whose sequences are complementary to the altered sequence, are used. These probes seek their complement among the 3 billion base pairs of an individual's genome. If the altered sequence is present in the patient's genome, the probe will bind to it and highlight the mutation. Another type of DNA testing involves comparing the sequence of DNA bases in a person's gene to a normal version of the gene to determine whether a mutation is present. Gene mutation scanning, or mutation screening, involves a two-step process by which a segment of DNA is screened via one of a variety of scanning methods to identify variant gene regions. The variant gene regions are further analyzed (by sequence analysis or mutation analysis to identify the sequence alteration (**Figure 4-1**—from Gene Reviews) *(http://www.geneclinics.org)*.

The cost of genetic testing ranges from several hundred to thousands of dollars depending on the size of the genes, techniques used, and number of mutations tested for. Gene sequencing, for example, can be used to examine an entire gene for alterations and can cost thousands of dollars, while gene testing for a single mutation, such as targeted testing for a CF or hereditary breast cancer gene mutation costs several hundred dollars.

For genetic testing to provide meaningful results, multiple testing methodologies may be required. In addition, other family members may need to be tested. In many testing situations, a genetic consultation is recommended prior to testing (see Chapter 7 for more information). It is important for individuals and families to understand that these services will entail additional costs.

A. PCR amplification of the DNA segment(s) of interest to create enough DNA for analysis

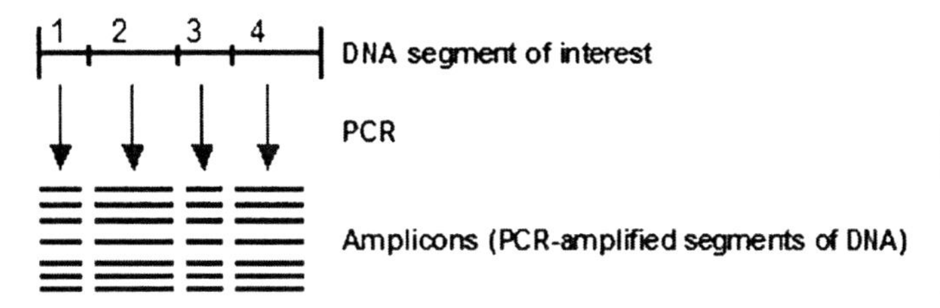

B. Visualization and comparison. Patient and control amplicons are compared using one of several different scanning methods (e.g., SSCP, CSGE, DGGE, DHPLC). In the SSCP example below, gel electrophoresis separates amplicons by mobility.

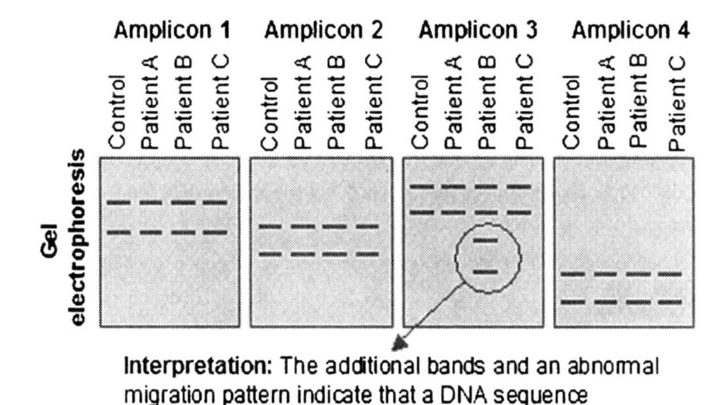

Step 2: In mutation scanning, variant DNA segments (e.g., segments with altered mobility in the SSCP example) may be subjected to further testing, such as sequence analysis, to identify the sequence alteration (mutation).

FIGURE 4-1 Mutation Scanning from http://geneclinics.org

Clinical Genetic Tests

Genetic tests performed for clinical purposes are those tests in which the specimens are analyzed and results reported to a clinical provider or to the patient for the purpose of diagnosis, prevention, or treatment of an individual. In the United States, laboratories performing clinical tests must be approved by Clinical Laboratory Improvement Amendments (CLIA). There is a charge for clinical tests, with the cost varying by complexity. Results are always reported in writing, and the time between analysis and reporting of results is generally based on the complexity of the testing. For example, gene mutation analysis for a single gene mutation usually takes 10 days to complete, while gene sequencing may require as long as 2 months for completion *(http:/www.geneclinics.org)*.

Research Genetic Tests

Research testing, in contrast to clinical genetic tests, is comprised of those tests in which specimens are analyzed for the purpose of gaining a better understanding of the genetic components of a specific condition, or for developing a clinical genetic test *(http://www.geneclinics.org)*. Research laboratories are not subject to CLIA regulations. There is usually no cost to the individual for research genetic testing; however, the test results are generally not provided to the individual or the health care provider. In some research studies, the research laboratory will, at the request of the individual, share test results with a clinical laboratory so that the individual's test results can be confirmed, a formal report issued, and prevention and intervention options considered.

Oversight of Genetic Testing

The Secretary's Advisory Committee on Genetics, Health and Society (formerly the Secretary's Advisory Committee on Genetic Testing [SACGT]) was formed to "explore, analyze and deliberate on the broad range of human health and societal issues raised by the development and use, as well as potential misuse, of genetic technologies and make recommendations to the Secretary of Health and Human Services, and other entities as appropriate." (September 23, 2002, p. 1). The scope of the Committee's work includes assessing approaches to clinical genetic testing, examining current patent policy and licensing practices and their impact on genetic testing, and serving as a public forum for discussion of emerging scientific, ethical, legal, and social issues raised by genetic testing and technologies. The Committee consists of a core of 13 members and includes a nurse. Meetings are held twice a year and are open to the public *(http://www.sacghs.org)*.

Information Provided by Genetic Testing

Genetic tests provide genetic information about an individual, can confirm a diagnosis if symptoms are present, and can direct a physician to appropriate management and interventions. Testing for the fragile X-gene mutation is an example of a diagnostic genetic test (see Chapter 6). Diagnosis of fragile X syndrome in a person helps families understand the underlying cause of their child's intellectual and physical differences, and allows them to tailor educational programming to the child's needs. Genetic tests can determine whether a person is a carrier of a particular mutation for a disease. A person who is a carrier does not have the condition but can pass on the altered gene to his or her children. Carrier testing of couples for a cystic fibrosis gene is an example of this type of testing. Prenatal testing helps expectant parents to learn whether their unborn child will have a genetic condition such as a neural tube defect or Down syndrome. Newborn screening involves genetic testing to identify infants at increased risk for metabolic and other genetic conditions for which further diagnostic testing and interventions can be done *(http://genes-r-us.uthscsa.edu/index.htm)*.

Genetic tests for adult-onset conditions such as Alzheimer's disease and hereditary cancers are receiving increased attention and are the subject of much discussion and debate. This type of testing is targeted to healthy and presymptomatic individuals who are identified as being at high risk due to a significant family history of the condition. The testing, called susceptibility or predictive testing, gives only a probability of developing the condition. One of the major limitations of susceptibility testing is the difficulty in interpreting a positive result, since some individuals who carry a disease-associated mutation never develop the disease. This is believed to be due to the disease mutations working together with other, as yet unidentified, mutations or with environmental influences to cause disease (Jorde et al., 2003). The benefits, limitations, and nursing considerations regarding susceptibility testing were covered in Chapter 2.

Other uses of genetic testing include parentage testing (paternity or maternity testing) and DNA banking for genetic testing. Paternity testing involves obtaining DNA sequences from a particular child and alleged parent for the purpose of comparison to estimate the probability that the child and parent are related. The usual indication for parentage testing is social; however, establishing parentage is sometimes necessary to be able to interpret family DNA studies correctly. Parentage testing can be expensive and is not usually covered by medical insurance *(http://www.geneclinics.org)*.

DNA banking involves obtaining DNA from cells and freezing or refrigerating them for future testing. DNA material can be stored indefinitely. DNA banking is offered to individuals with a genetic condition for which no genetic testing is yet available, to individuals who do not currently wish to pursue genetic testing but may choose to do so in the future, and to terminally-ill individuals with a known or suspected genetic condition. The storage cost of DNA banking is not significantly high; however, it is not usually covered by insurance. Individuals and families participating in DNA banking need to learn about the laboratory's DNA banking policies and procedures, and to determine at the time of DNA banking who will have access to the sample for testing purposes after the donor's death. The GeneClinics

Box 4-1 Concerns Raised by Individuals and the Public about Discrimination and Stigmatization Related to Genetic Testing

Genetic information gained from genetic testing can be identified at any point during the life span. It has the power to alter major life decisions such as whether to have children, and employment and lifestyle choices. Genetic information also creates the potential for misuse. It raises important questions about the meaning of "normal" and the distinction between diseases and traits. Discrimination occurs when people are treated differently based on genetic information resulting from test results by family, society, employers, or insurers. For example, when a pharmacogenetic test reveals that a person is a "nonresponder," is this a disease state? Does this information mean that a person will be denied certain medications or treatments in the future based on this test result? Genetic testing raises important questions on identity, difference, and social tolerance. Another important concern facing our society is to what extent, if any, genetic traits, predispositions, or conditions should provide a basis for determining access to social goods, such as employment or insurance. Several laws at the federal and state levels have been put into place to protect people against genetic discrimination. Genetic testing is a rapidly growing field, however, and these laws may not cover every situation.

Individual Concerns: Equity and Discrimination

- Genetic testing and screening may promote or increase discrimination in employment and insurance.
- Genetic technology will only be available to the affluent.
- Private companies may use a person's genetic information for lucrative purposes.

Societal Concerns: Potential for Stigmatization of Groups

- Ethnic minorities
- Employees
- Individuals with certain conditions or susceptibilities
- Children

(Grady, 1998; Scanlon & Fibison, 1995)

Reliable Information Available about Genetic Discrimination

National Human Genome Research Institute—
http://www.genome.gov/10002077

Genetic Alliance—
http://www.geneticalliance.org/geneticissues/discrimresources.html

U.S. Department of Energy Office of Science—
http://www.ornl.gov/TechResources/Human_Genome/elsi/legislat.html

Web site *(http://www.geneclinics.org)* provides a listing of laboratories currently offering DNA banking for clinical and research purposes. Most of the laboratories have an informed consent form "Consent for Banking of DNA Sample" that outlines the terms of DNA banking including use of the DNA sample for other research.

Genetic testing has many benefits in terms of indicating who may benefit from prevention and early intervention and treatment to improve health. However, there remain many uncertainties surrounding test interpretation, current lack of available treatments for certain diseases, the potential for test results to provoke anxiety, and the concerns regarding discrimination and social stigmatization. In Laura's story, for example, she recounts how her sister-in-law inquired about whether her father's condition was genetic, and Laura felt that the "real essence of her question" was related to concern that her father's condition would be inherited in the family, creating the possibility of discrimination. **Box 4-1** lists concerns raised by individuals and society about discrimination and stigmatization related to genetic testing.

A central role for all nurses is learning how to assist individuals in making decisions about genetic testing with provision of sufficient information (Grady & Collins, 2003). Nurses need to be able to offer reliable information about genetic tests to individuals and families and to provide patient and family education to help them understand the complex issues related to genetic testing. Nurses also participate in ensuring informed health choices and consent, advocate for privacy and confidentiality of genetic test results and information, and practice nondiscrimination (ISONG Position Statement, Informed Decision-Making and Consent, 1999). **Table 4-4** presents the topics discussed during the informed decision-making process.

TABLE 4-4 Topics Discussed During Informed Decision-Making and Consent Process for Genetic Testing

- Purpose of the test
- Reason for offering the test
- Type and nature of genetic condition being tested
- Accuracy of genetic test
- Risks associated with genetic testing, including unexpected results
- Acknowledgment of the right to refuse testing
- Other available testing options
- Available treatment and intervention options
- Further decision making that may be needed upon receipt of results
- Consent to use patient's DNA or sample for further research purposes
- Availability of additional counseling and support services

(International Society of Nurses in Genetics, Inc., 1999; Scanlon & Fibison, 1995)

CLINICAL USES OF GENETIC TESTING: NEWBORN SCREENING, CARRIER TESTING, PRENATAL SCREENING AND DIAGNOSIS

Newborn Screening

Newborn screening is the earliest and most widespread use of genetic testing. Since Dr. Guthrie's development of the first method to screen newborns for PKU in the early 1960s, newborn screening for PKU is now performed in all 50 states (*http://genes-r-us.uthscsa.edu*). Newborn screening programs consist of the following components: (1) initial screening, (2) follow-up testing of infants who have screened positive, (3) diagnosis and confirmation of true positive test results, (4) short- and long-term management and care, (5) parent and family education, and (6) program evaluation (Newborn Screening Task Force, 2000).

The primary focus of newborn screening is to identify infants who are likely to have a particular genetic condition for which early diagnosis and treatment is beneficial, and to ensure that every newborn receives appropriate services in a timely manner (Lloyd-Puryear & Forsman, 2002). To achieve this goal, newborn screening programs have a systematic and coordinated approach to screening, follow-up, diagnosis, treatment and management, parental education, and program evaluation. With PKU, for example, early dietary intervention, as described in Betsy's story, prevents the harmful buildup of phenylalanine, which if left untreated, leads to mental retardation (see **Box 4-2**). Or, children with sickle cell diseases (SCDs) can receive antibiotic treatment to reduce the risk of bacterial infections (Jorde et al., 2003).

Newborn screening is not diagnostic. Rather, after a screening test is positive, tests that are more definitive are performed to verify that the condition is present. The screening focuses on identification of genetic conditions prior to the onset of symptoms so that treatment can be initiated to prevent disability. A blood sample (heel stick) for the newborn screening program is generally obtained on each infant prior to hospital discharge to detect genetic conditions, such as PKU, in the preclinical period. Once a diagnosis is made, the infant and family are referred to a metabolic treatment center for education, counseling, and management (Committee on Genetics, 1996; GAO, 2003).

Newborn screening programs for metabolic conditions are the best established of the screening programs. Each state public health department oversees and administers these programs. The specific conditions to be included in each newborn screening panel are determined by the state (**Box 4-3**). The March of Dimes currently recommends that all babies born in the United States be screened for ten different conditions (**Table 4-5**). All states currently screen for PKU and most states now screen for SCD. Early diagnosis of SCD and treatment with prophylactic

Box 4-2 Phenylketonuria (PKU): What It Is and How It Is Inherited in Families

General Information

Phenylalanine hydroxylase deficiency results in an intolerance to the dietary intake of the essential amino acid phenylalanine, producing a spectrum of conditions including phenylketonuria (PKU), non-PKU hyperphenylalaninemia (non-PKU HPA), and variant PKU. Classic PKU is caused by a complete or near-complete deficiency of phenylalanine hydroxylase activity. In the absence of dietary restriction of phenylalanine, most children with PKU develop profound and irreversible mental retardation. Non-PKU HPA has a much lower risk of intellectual impairment in the absence of treatment, while variant PKU is intermediate between PKU and non-PKU HPA.

Newborn screening for PKU and variants has been routine throughout the United States since the mid-1960s and in most other developed countries since the 1970s. The test became routine because of the excellent prognosis for individuals with PAH deficiency who are treated early, and the high risk for severe and irreversible brain damage for individuals who are not treated. In most countries, a parental right of refusal for this test exists, i.e., for religious reasons. This right is rarely exercised by parents.

Clinical Symptoms

Hyperphenylalaninemia affects both males and females. There are three types of hyperphenylalaninemia. PKU is the most severe of the three types, and in an untreated state is associated with plasma phe concentrations of about 1000 μ mol/l and a dietary phe tolerance below 500 mg/day. PKU is associated with a high risk of severely impaired intellectual functioning when not treated. Non-PKU HPA, on the other hand, is associated with plasma phe concentrations consistently above normal (above 120 μ mol/l) but below 1000 μ mol/l when the individual is on a normal diet. Individuals with non-PKU HPA have a much lower risk of impaired intellectual development in the absence of treatment. Variant PKU includes those individuals who do not fit the description for either PKU or non-PKU HPA.

Pathology

Phenylalanine hydroxylase (PAH) deficiency is considered a multifactorial condition in that both environment (dietary intake of phe) and genotype (mutation of the PAH gene) are necessary to cause disease. Variability of the metabolic phenotypes in PAH deficiency is caused primarily by different mutations within the PAH gene. An individual's genotype does predict his or her biochemical phenotype, but does not predict the individual's clinical phenotype (occurrence of mental retardation).

Box 4-2 continued

PAH deficiency is therefore considered a "complex" condition at the cognitive and metabolic levels. It is becoming difficult to assess the various clinical phenotypes given that most individuals with PAH deficiency in developed countries are treated successfully. The PAH gene is located on chromosome 12q23.2.

Prognosis

Before the 1950s, there was no treatment for PKU. In the late 1950s it was shown that individuals with PKU responded to dietary restriction of the essential nutrient, phenylalanine. By the 1960s, a microbial inhibition assay was used for mass screening of newborns, providing early diagnosis and access to successful treatment. Hyperphenylalaninemia is now treatable, and affected individuals can lead normal lives. Research to improve current treatment with restrictive phenylalanine diet supplemented by medical formula is ongoing.

While it was once thought that dietary restriction could be stopped in childhood, it is now known that individuals with hyperphenylalaninemia must remain on the restricted diet for life. This is especially important for women who have the condition. If a woman has high plasma phe concentrations during pregnancy, the developing baby is at significant risk for congenital heart defect, intrauterine and postnatal growth retardation, microcephaly, and mental retardation. Because of this risk, women with hyperphenylalaninemia are counseled about using reliable methods of contraception to prevent unplanned pregnancies. Support of women with PKU starting before conception and continuing after delivery is essential for optimal outcome.

Inheritance

Inheritance in families is autosomal recessive, affecting males and females equally. Unaffected parents of a child with PKU or variant are obligate carriers. Each parent carries at least one disease-causing PAH allele. Such carriers are asymptomatic and never develop hyperphenylalaninemia. Carrier parents have with each pregnancy:

- a 25% chance to have a child with PKU or variant;
- a 50% chance to have a child who is a carrier like themselves, and
- a 25% chance to have a child who inherits the normal alleles and who is neither affected nor a carrier.

Prevalence

One in 10,000 individuals of Caucasian or East Asian ancestry is affected. This is in comparison to 1 in 4,500 individuals of Irish ancestry and 1 in 200,000 individuals of Finnish or Ashkenazi Jewish ancestry.

Box 4-2 continued

Diagnosis

Infants with PKU or variant show no physical signs or symptoms of hyperphenylalaninemia. The main route for phenylalanine metabolism is hydroxylation of phenylalanine to tyrosine by the enzyme phenylalanine hydroxylase (PAH). The diagnosis of primary phenylalanine hydroxylase deficiency is based on the detection of an elevated plasma phenylalanine (phe) concentration and evidence of normal BH4 cofactor metabolism. PKU and variants are most commonly diagnosed upon routine screening of newborns and can be detected in virtually 100% of cases using the Guthrie card blood spot obtained from a heel prick. Following a positive newborn screening result, biochemical testing is performed for diagnosis. Individuals with PKU show phe concentrations that are persistently above 120 μ mol/l in the untreated state.

Relevant Testing

Newborn screening and diagnosis of PKU and variants rests on biochemical testing. Once the diagnosis is established by biochemical testing, direct DNA testing of the PAH mutation can be used to identify the disease-causing alleles and for genetic counseling purposes. DNA testing is used for confirmatory testing, prognosis, carrier testing, and prenatal diagnosis. Some laboratories offer testing of a panel of 4 to 15 of the most common PAH mutations, with a detection rate of about 30–50%. Mutation analysis is available on a clinical basis. Mutation scanning is also available on a clinical basis and detects virtually all point mutations in the PAH gene. A few laboratories worldwide are performing direct gene sequencing on a routine clinical basis for mutation identification. Entire gene sequencing is able to identify gene mutations 99% of the time.

Published Statements

U.S. Preventive Services Task Force (1996) Screening for Phenylketonuria

Canadian Task Force on Preventive Health Care (1994) (Reaffirmed 1998) Screening for Phenylketonuria

National Institutes of Health Consensus Development Statement Panel (2000) Consensus Statement: Phenylketonuria: Screening and Management

References: Jorde et al., 2000; *http://www.geneclinics.org; http://genes-r-us.uthscsa.edu*

Box 4-2 continued

Support Resources

Children's PKU Network
3790 Via De La Valle, Suite 120
Del Mar, CA 92014
Phone: 800-377-6677 or 858-509-0767
FAX: 858-509-0768
E-mail: pkunetwork@aol.com
www.pkunetwork.org

March of Dimes Birth Defects Foundation
1275 Mamorneck Avenue
White Plains, NY 10605
Phone: 1-888-663-4637
www.marchofdimes.org

PAH/PKU Resource Booklet for Families
www.mcgill.ca/pahdb/handout/handout.htm

Box 4-3 Newborn Screening and Public Health

Newborn screening is considered a state and public activity. Each state is responsible for designing and implementing its own newborn screening program, including making decisions about which disorders to include in its newborn screening panel and program. The number of genetic and metabolic conditions included in state newborn screening programs varies from 4 to 36. Most states screen for 8 or fewer conditions. The state health departments or boards of health, with input from advisory committees, generally have the authority to decide which disorders to include in the program. Screening for some disorders may be mandated by state law. The decision to include a specific disorder(s) is based on the following similar criteria in each state: (1) how often the condition exists in the general population, (2) whether an effective screening test exists, and (3) whether the condition is treatable. Other considerations include the cost of screening for additional conditions, the cost of performing more tests and acquiring and implementing new technology, and follow-up and management of abnormal results (GAO report on Newborn Screening, 2003).

Box 4-3 continued

At present, there are no federal guidelines regarding the set of disorders that should be included in state newborn screening programs. The exception is the federal regulation that all newborns be screened for PKU, congenital hypothyroidism, and sickle cell disease.

Variation in patient and provider education and notification of test results also exists, in addition to variation among states regarding the number and type of newborn screening tests. Although all states provide education on their newborn screening programs for parents and providers, for example, fewer than one fourth of the states inform parents of their option to obtain testing for additional genetic and metabolic conditions not included in the state's program. Furthermore, all state programs notify a health care provider, such as a physician or hospital, about an abnormal newborn screening result; however, fewer than half routinely notify parents directly of the abnormal results. Follow-up on all abnormal results is a state-sponsored activity and includes obtaining additional laboratory information, referral of the infant and family for treatment, and confirming that treatment has been initiated.

Parental consent is not usually required for newborn screening before screening occurs. The option for parental dissent and exemption from screening is allowed in 33 states, usually for religious reasons, and in 13 states, for any reason. Newborn screening results in more than half of the states by state statute and regulation is considered confidential. However, the confidentiality provisions are frequently subject to exceptions, and these vary among the states. Exceptions include disclosure of information for research purposes and information for law enforcement to establish paternity. Only 17 out of the 50 states have a genetic privacy statute in place that provides specific penalties for violating genetic privacy laws (GAO, 2003).

To examine the issues of variability, the GAO was commissioned to research and write a report and summary of state newborn screening programs' current practices (GAO, March 2003). In addition, to address the identified variations and to create more uniformity among states, the Health Resources and Services Administration is funding a project to assist in the development of a recommended set of conditions for which all states should screen. The recommendations will include the criteria for selection of these conditions (GAO, 2003; National Newborn Screening and Genetics Resource Center).

penicillin decrease risk of death from bacterial infection and sepsis. Families with newborns diagnosed with SCD can now participate in SCD management programs. Parent education is a major component of these programs and has been shown to prevent sepsis in children who are maintained on penicillin prophylaxis (Day, Brunson, & Wang, 1992).

TABLE 4-5 March of Dimes Recommendations for Newborn Screening

Phenylketonuria (PKU)
Congenital Hypothyroidism
Congenital Adrenal Hyperplasia (CAH)
Biotinidase Deficiency
Maple Syrup Urine Disease (MSUD)
Galactosemia
Homocystinuria
Sickle Cell Anemia
Medium Chain Acyl-CoA Dehydrogenase Deficiency
Hearing Loss

(March of Dimes, 2002)

Family Education and Support

Education of prospective and new parents and providers about newborn screening is an important component of newborn screening. Nurses share in the responsibility of such education. Education of parents about and receiving permission for newborn screening is a complex process. Betsy's story clearly illustrates some of the gaps that exist. Ideally, parent education and informed consent includes shared decision making, allowing parents the opportunity to learn of the risks, benefits, and limitations of newborn screening. However, most state newborn screening laws only make accommodations for parents who decline newborn screening. The rationale behind such laws is that the screening and potential detection of a serious illness are in the best interests of the child, and parents' disagreement with this premise should not get in the way of the screening process (Lloyd-Puryear & Forsman, 2002).

In 2000, the Newborn Screening Task Force recommended that parental education receive greater attention. This approach would help to improve parents' understanding of the screening process, decrease anxiety with a positive result, and increase adherence to further testing and follow-up. The task force emphasized that greater attention to parent education and shared decision making was necessary because of the addition of new DNA-based tests and screening for other conditions for which treatment may not be available or for which the efficacy of the treatment is not yet known (Newborn Screening Task Force, 2000).

Another important recommendation of the Newborn Screening Task Force was that it is essential for screening programs to have a way to provide medical foods and formulas and that each state assume responsibility to ensure adequate funding is available for all parts of the program. As Betsy points out in her story, this is an ongoing and costly issue for her and her family. In 2000, the National Institutes of Health convened a consensus conference on PKU. Its recommendation was that ultimately the provisions for treatment and management—medical foods and formula—should become part of any medical benefits package (NIH Consensus Conference, 2000).

Expanded Newborn Screening

Cystic fibrosis newborn screening In some states, newborn screening programs have expanded, adding tests for more metabolic conditions including CF. Newborn screening for CF is available through dried-blood analysis for immunoreactive trypsinogen and DNA analysis. A two-tiered screening procedure has been implemented to reduce the incidence of false positive results and required diagnostic work-ups. However, in addition to identifying infants with CF, the results also indicate CF carrier status. The only definitive diagnostic test for CF is the quantitative pilocarpine iontophoresis sweat test, which cannot be performed until 3 to 4 weeks of age when adequate sweat collection is possible. Some children screened show sweat chloride results within the normal range with one CFTR mutation detected indicating absence of CF, but confirming carrier status. Studies of the impact of CF newborn screening have shown that parents may not fully understand the test results and their genetic risk implications, and may experience high levels of anxiety. Parents are offered genetic counseling to clarify the meaning of these results for themselves and their baby (Parad & Comeau, 2003).

In the United States, CF neonatal screening is being evaluated in pilot projects in several states, including Colorado, Wisconsin, and Massachusetts, as a part of research protocols investigating the possible improvement of nutritional and respiratory outcomes with early identification and initiation of treatment before symptoms appear. At this time, the long-term benefits of early intervention for CF have not been completely clarified. Other concerns also remain about widespread implementation of CF newborn screening. For example, concern regarding the possible impact of carrier status on insurance coverage as well as subsequent reproductive decision making has been raised. Identification of newborns with CF means that parents will learn that they are obligate carriers of CF gene mutations. This allows parents the opportunity to use this knowledge for reproductive planning for future pregnancies. On the other hand, early screening has the potential to increase parental stress and anxiety, particularly with the impact of false positive newborn screening results. For these reasons, many states have not yet mandated CF newborn screening (Baroni, Anderson, & Mischler, 1997; Farrel, 2000; Parad & Comeau, 2003).

Tandem mass spectrometry A number of states are implementing a new technology—tandem mass spectrometry (MS-MS) to expand their newborn screening programs.

MS-MS has the capability to screen for many conditions. For example, in addition to PKU, MS-MS can identify at least 10 other amino acid conditions and 20 more organic acid and fatty acid oxidation disorders. Only one or two small blood spots are required for MS-MS, and the false positive rate is reduced with MS-MS methodology, leading to fewer repeat specimens and reduced parental anxiety that false positive results can elicit in families (Levy, 1998).

Although MS-MS makes it possible to screen for many additional conditions, there has been some controversy regarding its incorporation into newborn screening because its use is changing the principles under which many public health screening programs operate (Newborn Screening Task Force, 2000). MS-MS has the capability to detect relatively rare conditions that are not prevalent enough in the population to be considered public health screening. Furthermore, for some of the conditions, the significance to the health of the newborn is not known, and there is no available treatment. For other conditions, there is currently no available treatment that can be initiated early enough to prevent neurologic damage. These issues have prevented some states from expanding their newborn screening to include MS-MS (Lloyd-Puryear & Forsman, 2002).

The use of DNA-based screening to detect specific mutations such as those for adult-onset conditions like Alzheimer's and Huntington Disease as part of newborn screening is being considered. Benefits noted include early identification even if there is no treatment, the avoidance of unnecessary medical procedures, and inaccurate diagnoses, for example, confusing symptoms of Huntington Disease with depression. There remains limited knowledge about such screening, however. As an example, several conditions for which screening is available have a predominant gene mutation, but multiple genotypes lead to the same phenotype. If all newborns were tested for the predominant gene mutation alone, the screening would miss a substantial number of other mutations and leave many at-risk newborns inaccurately undetected. At present, the exception to this problem is DNA-based analysis for sickle cell disease (Lloyd-Puryear & Forsman, 2002).

Newborn hearing screening It is known that the effects of gene mutations cause at least 50% of all profound hearing loss. Human genome discoveries have identified specific genes causing hearing loss such as the Connexin 26 gene, making it possible to identify genetic causes of hearing loss in the newborn period in some families (American College of Medical Genetics [ACMG], 2000). Newborn screening for hearing loss is now mandated in many states. Infants who fail their newborn hearing screen are referred for further evaluation of hearing loss, and when confirmed, parents are offered genetic counseling to discuss genetic causes, available gene testing for hearing loss, and future reproductive implications.

It is important for nurses to recognize that not all parents will be interested in genetic counseling and testing for reproductive purposes. For some families who are members of the deaf community, for example, hearing status is a part of their identity, and genetic testing and counseling regarding deafness is not acceptable (Arnos, Israel, & Cunningham, 1991). Family values, culture, and expectations need to be assessed when providing information about the availability of genetic counseling and testing in newborns with hearing loss (Williams & Lea, 2003).

Appropriate and timely diagnoses continue to be challenges in many states where newborn hearing screening occurs. Barriers to receiving early intervention include lack of well-established linkages between the hospital-based screening programs and early intervention programs, and the data management and tracking systems for following newborns through the screening and diagnostic process. State programs are assuming greater responsibility for tracking and follow-up of infants at risk for hearing loss, leading toward more firmly established systems and linkages for timely follow-up and diagnosis (Lloyd-Puryear & Forsman, 2002).

Nurses who care for infants and families need to keep up-to-date as newborn screening continues to evolve and change. Important roles for nursing will continue to be educating families about the benefits and limitations of newborn screening and participating in the ongoing dialogue to determine newborn screening policy (Lloyd-Puryear & Forsman, 2002).

Carrier Testing

Carrier testing is performed to identify individuals who have a gene mutation for a specific genetic condition that is inherited in an autosomal recessive or X-linked recessive manner. Individuals who are carriers usually do not have symptoms related to the gene mutation, but are at increased risk to have children with the particular genetic condition. Carrier testing is offered preconceptionally and prenatally to individuals who have family members with a particular genetic condition such as CF. As Laura's story illustrates, carrier testing for CF is also offered when a family member is identified as a carrier. Individuals of ethnic or racial groups known to have a higher carrier rate for a particular condition are also offered carrier testing. **Table 4-6** provides a listing of the various carrier tests offered to individuals and couples from particular ethnic and racial groups.

Prior to carrier testing of family members, molecular testing of an affected family member may be required to identify the particular mutation in the family. In Laura's family, the specific CF mutations in her family were only identified through carrier testing, as her father had never had genetic testing for CF. In other situations, DNA testing may not be the primary way of determining carrier status. This is the case with Tay-Sachs disease where the initial testing involves measurement of the enzyme Hexosaminidase-A.

The ability to identify carrier status allows for reproductive choices. These include prenatal diagnosis when both parents are found to be carriers of the same recessive gene mutation or when a female is identified to be a carrier of an X-linked

TABLE 4-6 Carrier Testing for Various Populations

- Cystic Fibrosis: all couples, but especially Northern European Caucasian and Ashkenazi Jewish
  - Test: DNA mutation analysis
- Tay-Sachs Disease: Ashkenazi Jewish
  - Test: Serum or Leukocyte Hexosaminidase A and DNA mutation analysis
- Canavan Disease: Ashkenazi Jewish
  - Test: DNA mutation analysis
- Sickle Cell Anemia: African-American, Puerto Rican, Mediterranean, Middle Eastern
  - Test: Hemoglobin electrophoresis
- Alpha-thalassemia: Southeast Asian, African-American
  - Test: Complete Blood Count and Hemoglobin electrophoresis

(Nussbaum, McInnes, & Willard, 2001)

recessive condition such as hemophilia. Couples need to be provided with accurate information about the risk of carrying a gene for a particular condition and the sensitivity of the carrier testing. Genetic education and counseling should be offered prior to carrier testing so that couples can consider the various personal and social concerns that may result from pursuit of carrier testing. As Laura points out in her story, couples need to be given accurate information about the testing, the condition being tested for, and the meaning of both positive and negative test results.

Studies of decision making about whether to have or not to have carrier testing and the psychological effects of carrier testing have been conducted during the past decade. Studies have shown that the uptake of carrier testing tends to be high, especially when made accessible and recommended by a physician. The diseases for which carrier testing is offered, their potential severity, and the ability to treat the conditions have been found to be important considerations in the decision-making process (Eng et al., 1997; Kronn, Jansen, & Ostrer, 1998; Lerman et al., 2002). In studies of psychological effects, some participants testing positive reported feelings of stigmatization and hopelessness regarding the health of their children (Williams & Schutte, 1997). However, there has been no evidence for long-term, negative effects of testing in carriers. Nurses can facilitate future studies that can further explore psychological effects and the use of new methods to increase access to testing, such as videotapes, Web sites, and home testing (Lerman et al., 2002).

Cystic fibrosis is a common condition for which carrier testing is now offered to all couples who are planning pregnancy or seeking prenatal care. The National Institutes of Health (1997), the American College of Obstetrics and Gynecology, and the American College of Medical Genetics (ACOG & ACMG, 2001) have developed guidelines that recommend CF carrier testing be offered in the following situations: (1) individuals with a family history of CF, (2) reproductive partners of a person who has CF, (3) couples in whom one or both partners are Caucasian, and (4) couples with a high risk of having a child with CF, including individuals who are Caucasian or of Ashkenazi Jewish ancestry. CF carrier testing is also to be made available to couples in other racial and ethnic groups who are at lower risk and for whom the CF carrier testing is less sensitive (ACOG & ACMG, 2001).

Prenatal Screening and Diagnosis

Prenatal screening involves many of the same principles as newborn screening. After a prenatal screening test is positive, more definitive testing can be done to confirm that the condition is present. In contrast to newborn screening where identification of diseases prior to the onset of symptoms and treatment is available, prenatal screening may lead to diagnosis of conditions for which there may be no treatment available before birth, and the only options may be to continue or to end the pregnancy (Grant, 2000; Williams & Lea, 2003).

In the United States, approximately 70% of women choose to have multiple-marker maternal serum or ultrasound screening for common birth defects during their pregnancy (Filly, 2000; Palomaki et al., 1997). The multiple-marker maternal serum screen is offered to all pregnant women for the purpose of identifying women who are at risk for having a baby born with a neural tube defect such as spina bifida, and Down syndrome (trisomy 21). This type of genetic testing involves biochemical measurement of serum markers including alpha-fetoprotein, a protein made by the fetal liver and human chorionic gonadotropin (hCG), which is a substance produced by the placenta. Estriol (UE3) and dimeric inhibin (DiA) are other markers produced by the placenta that can be measured in maternal serum. These two additional markers are often included in the screening for Down syndrome and other chromosomal abnormalities (Canick et al., 1998; Wald et al., 1998).

It is of critical importance to perform the multiple-marker screening test at the appropriate time during a pregnancy. Levels of these biochemicals are measured in a pregnant woman's blood at 16 to 18 weeks of pregnancy. Analysis incorporates other factors that may influence the test result such as maternal age, race, presence of maternal insulin-dependent diabetes, multiple gestation, and family history of neural tube defect or Down syndrome (Palomaki, 1986).

When the maternal multiple-marker screening test shows a pattern indicating increased risk, further diagnostic testing such as high-resolution ultrasound evaluation and amniocentesis are offered. Multiple-marker screening can detect 60% of pregnancies with Down syndrome and 80% of pregnancies with an open neural tube defect (American College of Obstetrics and Gynecology, 1991; Canick & Kellner, 1999), and most cases of trisomy 18. Women who receive positive multiple-marker screening results are generally referred to a prenatal diagnostic center for counseling and diagnostic tests. Women who choose to undergo further diagnostic testing need continued contact and support from their primary pregnancy provider due to increased stress and anxiety (Lewis, 2002).

Multiple-marker screening, like newborn screening, gives false positive results. It is important to reassure women that most who receive positive results will go on to have normal pregnancy outcomes (Grant, 2000). It can be explained that further evaluation is needed to clarify the significance of a positive maternal serum screening test. When further evaluation does reveal a fetal abnormality, women and their families are offered genetic counseling to consider prenatal or early infant therapy, preparation for the birth of a child with special needs, or ending their pregnancy (ACOG, 1991).

Unlike newborn screening, which is mandatory in all states, maternal multiple-marker screening is voluntary. Translating the complexities of maternal multiple-marker screening into understandable terms so that women can make a meaningful decision whether or not to have the test is an important nursing role. The discussion involves sensitivity to a woman's cultural values and health beliefs. Appropriate education to help pregnant women and their families make informed choices and understand the outcomes and additional testing and steps that may be involved is an important role. The discussion of prenatal screening with couples who have had a previous child with a genetic condition such as Down syndrome, neural tube defect, CF, or in Betsy's case, PKU, is especially sensitive. The decision to have a prenatal screening test may be complicated by having lived with a child who has the condition, and contemplation of the decision to pursue pregnancy termination. Nurses can offer all choices to women including continuing the pregnancy and preparation for a child with a birth defect or genetic condition (Williams & Lea, 2003). As Betsy points out in her story, there was no way to fix PKU prior to birth, and she already knew how to manage PKU with her first daughter. This experience influenced her decision not to pursue further diagnostic testing.

Prenatal diagnostic tests are also offered when a woman and her family have an increased risk for having a child with a genetic condition. Examples of this indication include maternal age of 35 years or older and a family history of a genetic condition or birth defect. Routine prenatal diagnostic procedures include amniocentesis and chorionic villus sampling. More specialized procedures may be needed and these include periumbilical blood sampling, placental biopsy, and fetoscopy with skin biopsy. All prenatal diagnostic test procedures have an associated risk to the fetus and the pregnancy. For this reason, informed consent is required, usually in conjunction with genetic counseling (*http://www.geneclinics.org;* Lewis, 2002).

Women and couples choose to have prenatal diagnosis for a number of reasons. One major reason is to provide reassurance and reduce anxiety for couples who have an increased risk for a particular condition in their baby. A negative prenatal diagnostic test result for the condition in question will spare a couple and family months of needless worry. For some women and their partners, a pivotal consideration is whether there are interventions available for the fetus and/or newborn if the test result is positive. The availability of prenatal diagnosis for some couples at risk for having a child with a particular genetic condition allows them to consider pregnancy, where in the past they might have remained childless. Another important reason is to provide the woman and her partner with information that they can use to make decisions about continuing a pregnancy and future family planning. For conditions that require special interventions at birth, couples who choose to continue a pregnancy can make appropriate plans for the delivery including the specific hospital, available specialists, and the preparation of their home for management following birth. For some conditions, prenatal diagnosis may permit prenatal treatment prior to delivery, thereby preventing irreversible complications before birth (Giarelli, Lea, Jones, & Lewis, in press).

Choosing pregnancy termination because of the presence of a genetic condition or abnormality may or may not be a viable alternative for a woman and her partner. This is a highly personal decision that rests on personal values, and religious and cultural beliefs. The decision to end a pregnancy is often made within the context of an extended family or community that has specific values and beliefs. The two sets of beliefs—those of the couple and those of the community—may or may not be in alignment. A couple may wish to end a pregnancy but feel that their family and community would oppose this choice and therefore elect to continue the pregnancy. In the context of this situation, if they choose to end the pregnancy, they may decide to tell their friends and family that they have experienced a spontaneous pregnancy loss. On the other hand, a couple may fear social stigmatization and discrimination if they elect not to end a pregnancy when found to have a genetic condition, and when their family and most of their friends would choose to do so (Giarelli, Lea, Jones, & Lewis, in press).

New Reproductive Technologies

As of this writing, more than 1 million children have been born as a result of assisted reproductive technologies (ART). ART technologies include in vitro fertilization, intracytoplasmic sperm injection, embryo cryopreservation, and babies born after preimplantation diagnosis (Genetics and Public Policy Center, 2003). **Table 4-7** provides a description of each of these technologies. Preimplantation genetic diagnosis (PGD), as an example, is performed on early embryos resulting from in vitro fertilization with the goal of decreasing the chance that a specific genetic condition will occur in the fetus. PGD is offered to couples with a high

TABLE 4-7 Assisted Reproductive Technologies (ART)

ART: a set of medical and laboratory procedures used to help couples who are having difficulty conceiving and couples who are at risk for having a child with a genetic condition such as cystic fibrosis.

Common ART Procedures

In vitro Fertilization (IVF) and Embryo Transfer (ET): most common ART procedures. IVF-ET involves stimulating a woman's ovaries to produce mature eggs; retrieving the eggs; fertilizing the egg with sperm; and implanting the fertilized, developing embryos into the woman's uterus. Fertilization of the egg is done in vitro, rather than inside of the woman's body.

Intracytoplasmic Sperm Injection (ICSI): an in vitro process of using a microscopic glass needle to inject a single sperm directly into a mature egg to achieve fertilization. IVF procedures are used for egg retrieval and embryo transfer.

Preimplantation Genetic Diagnosis (PGD): used by both fertile and infertile couples at risk for passing on a genetic condition to their children. Two types of PGD are available. One involves testing one or two cells from a developing embryo at 2 to 4 days after fertilization but prior to implantation for single gene mutations, sex-linked genetic conditions, and chromosome abnormalities. The other method involves testing a polar body after biopsy. The polar body is a product formed during egg maturation that contains a single set of maternal chromosomes. During polar body biopsy analysis, the maternal chromosomes are tested from the egg.

Embryo Cryopreservation: the process of freezing and storing human embryos following an IVF cycle. Embryo cyropreservation permits couples to save their embryos for future pregnancy attempts, and avoid the costs and invasive procedures associated with subsequent IVF. Embryo cryopreservation may be used for donation to other couples attempting to have children and for the purpose of scientific research.

(American Society of Reproductive Medicine at *http://www.asrm.org;* Jones & Fallon, 2002)

chance of having a child with a serious genetic condition such as Duchenne muscular dystrophy, CF, and Tay-Sachs Disease. PGD provides couples with an alternative to prenatal diagnosis and ending a pregnancy in which a genetic condition is present.

Interested individuals and couples should know that PGD is currently offered at a few specialty centers and is available for a limited number of conditions. In some cases, PGD is not possible due to difficulties with obtaining eggs or early embryos and problems with DNA analysis procedures. Traditional prenatal diagnostic methods such as amniocentesis may be recommended for pregnancy monitoring due to possible errors in PGD. The cost of PGD is significantly high and is usually not covered by insurance *(http://www.geneclinics.org)*.

Nurses need to have accurate and current knowledge about PGD so that they can appropriately support and refer couples interested in this reproductive technology. Reproductive technologies such as PGD offer new opportunities to assist individuals in the conception of a child. There are also attendant biopsychosocial issues that need nursing consideration as a part of the informed decision-making and consent process. Jones and Fallon (2002) outline several important issues in reproductive decision making for couples considering PGD or other ART. These include the definition of family, knowledge of genetic information about themselves, privacy of genetic information, and the necessity of having a biologic child. Since reproductive decisions of this nature will irreversibly change people's lives, nurses involved with couples in their decision making can offer support through the use of anticipatory guidance (Jones & Fallon, 2002).

Couples should also be informed that the effects of PGD and other ART procedures on the health and development of the resulting children remains unclear. Some medical studies suggest that children born as results of ART are as healthy as children conceived naturally. Other studies have associated ART with a higher incidence of birth defects, genetic conditions, and possibly cancer. Ongoing discussion and research continues regarding the long-term effects of ART (Genetics and Public Policy, 2003).

Nursing Care and Prenatal Screening and Diagnosis

Nurses have several important roles during prenatal screening and diagnostic processes. These include ensuring informed decision making and consent, supporting the woman and family's access to accurate and understandable information as they consider their testing options and subsequent decisions regarding pregnancy, and decision-making support (Williams & Lea, 2003).

Informed decision making and consent Informed decision making and consent is a communication process whereby the individual receives appropriate and necessary information to make a decision about a significant health care intervention. Support from health care providers, community, and family resources may help ensure a successful outcome to the decision made. The final step in the process is the written acceptance or rejection of the recommended test, evaluation, or procedure. The nurse ensures that the woman and partner have complete and accurate information about the condition being tested for, the risks and benefits of the procedure, and available alternatives and interventions. Nursing assessment includes the ability of the woman and her partner to understand, to process, to evaluate the information, and to make a voluntary decision (ISONG, 1999). There are a number of factors the nurse considers that may interfere with informed decision making and consent (**Table 4-8**).

When each component of the informed decision-making process is met, the woman or couple then has the ability and information to decide to pursue or not to pursue the test, evaluation, intervention, or procedure. The International Society of Nurses in Genetics (ISONG, 1999) has created a Position Statement regarding the role of the nurse in informed decision making and consent that can serve as a guide for nurses involved in the testing process (**Box 4-4**).

Supporting access to reliable and understandable information Although prenatal diagnosis has been available for more than 30 years, health care professionals may not be well-informed regarding the specific conditions identified, and literature they consult may have out-of-date information (Abramsky et al., 2001). When a prenatal diagnosis of a genetic condition or birth defect is made, families are especially in need of reliable information regarding the type of condition, the range and severity of problems, management options, support organizations, and current understanding of the long-term progno-

Table 4-8 Patient Competencies Needed to Make Informed Genetic Health Decisions and Factors That May Influence Informed Decision Making

The patient must be able to

- Understand the information presented
- Consider options
- Evaluate potential risks and benefits of genetic interventions
- Communicate health choices

Factors That May Influence Informed Decision Making

- Hearing or language deficits
- Intellectual disabilities
- Effects of medications
- Cultural and family beliefs and practices
- Which and how many health care providers are offering information about genetic testing
- The setting in which the informed consent is being obtained
- Whether the patient is alone or with family or friends
- The presence of concurrent illness or health problems
- Insurance policy restrictions

(Bove, Fry, & MacDonald, 1997)

Box 4-4 International Society of Nurses in Genetics, Inc. (ISONG) Position Statement: Informed Decision Making and Consent: The Role of Nursing

Background

The indispensable initial step in the preparation for genetic testing is the process of "informed consent." A written statement describing the risk and benefits of genetic testing is presented to the client to read and sign before the evaluation and/or test is performed. It is intended as a safeguard to ensure the client's autonomy and to provide an opportunity to learn and understand information with respect to both the positive and negative consequences of genetic testing. In the more conventional use, the informed consent process focuses on the provider conveying information about the genetic test and the client signifying acceptance of the testing by a signature on the written consent form.

ISONG supports a more active selection process with an emphasis on the informed decision-making authority of the client to choose either to accept or reject genetic testing. Pivotal to accomplishing this process is a dialogue between the client and the providers in a joint endeavor to facilitate informed decision making and consent. This educational and informative dialogue should occur at the level of language and comprehension of the competent client.

Nurses should encourage clients to seek information and identify concerns prior to giving informed consent. The nursing process can be universally utilized to assist clients contemplating any type of genetic testing and to ascertain whether essential elements of informed consent are present in the decision-making process.

Informed Decision Making and Consent for Genetic Testing: The Role of the Nurse

Genetic testing can now be used for diagnosis, management, treatment, or health and reproductive decision making. The benefits of genetic testing are many and range from early detection for treatable disorders to prevention by health planning before the onset of symptoms for those who are at risk for a genetic disorder. Genetic testing should be carried out within the context of voluntariness, informed consent, and confidentiality. Nurses, as the omnipresent health care provider, have a central role in providing information and support to individuals, families, and communities in the multiphase processes of genetic testing. With genetics knowledge, nurses can advocate, educate, counsel, and support patients and families during the informed decision-making and consent process.

Box 4-4 continued

It Is the Position of the International Society of Nurses in Genetics That:

- All professional nurses are responsible for alerting clients to their right for an informed decision-making and consent process prior to genetic testing (1).
- All professional nurses should advocate for client autonomy, privacy, and confidentiality in the informed decision-making and consent process.
- All professional nurses should ensure that the informed decision-making and consent process includes discussion of benefits and risks including the potential psychological and societal injury by stigmatization, discrimination, and emotional stress, in addition to, if any, potential physical harm.
- All professional nurses should be aware of the criteria that delineate research versus clinical uses of genetic tests, and advise clients of the status of a specific test.
- All professional nurses who have an established relationship and are providing ongoing care to a client contemplating genetic testing should augment the informed decision-making and consent process by assisting the client in the context of the client's specific circumstances of family, culture, and community life.
- All professional nurses should integrate into their practice the guidelines for practice (e.g., informed consent, privacy and confidentiality, truth telling and disclosure, and nondiscrimination) identified by the American Nurses Association (2).
- All advanced practice nurses in preparation for providing genetic services should receive appropriate education that includes knowledge of the implications and complexities of genetic testing, ability to interpret results, and knowledge of the ethical, legal, social, and psychological consequences of genetic testing.
- All health care professionals should collaborate to maximize the potential for the client to make an informed decision.

References

1. International Society of Nurses in Genetics (ISONG). (January 27, 2000). ISONG Testimony to Secretary's Advisory Committee on Genetic Testing (SACGT).
2. Scanlon, C., & Fibison, W. (1995). *Managing Genetic Information: Policies for U.S. Nurses.* American Nurses Association: Washington, D.C.

Approved 9/30/00

sis. Genetic counseling with genetics professionals knowledgeable about this information is generally recommended. Genetics professionals can refer families to other professionals—neurologists, plastic surgeons, cardiologists—for further information about the condition, as well as families who are available to meet and talk with the parents who wish to learn of another family's experience. Couples and families can be referred to the Genetic Alliance *(http://www.geneticalliance.org)* for reliable information and support resources. Nurses can make sure that patients receive accurate and up-to-date information and prepare the woman and partner for what they can expect during genetic evaluation and counseling (Williams & Lea, 2003).

Decision-making support Prenatal diagnosis of a birth defect or genetic disorder presents additional decisions for families, one decision being pregnancy termination. Nurses can use the nursing intervention of decision-making support when a woman or couple is considering options regarding continuing or ending a pregnancy. The framework for this type of decision-making support is communication of information in a nondirective manner so that the woman and family feel free to come to their own decision. Components of decision-making support include helping the woman and family obtain information about alternative decisions, determining the positive and negative aspects of each choice, helping the woman and her family explore their personal beliefs and values, respecting the woman's right to have or not to have information, and helping the woman explain her decision to others (Iowa Intervention Project, 2000; Williams & Lea, 2003).

POTENTIAL RISKS OF GENETIC TESTING

Genetic testing for newborn screening, carrier testing, and prenatal screening and diagnosis poses potential medical, psychological, and socioeconomic risks to individuals. For many individuals the emotional impact of positive test results can cause anxiety and fear, pervasive worry, anger, depression, and despair. Betsy describes feeling overwhelmed with the news that her first daughter had PKU. At the end of her story, when speaking about the teenage girl's letter, Betsy expresses her fears saying, "You don't know how it is when you have a perfect baby and someone takes that away when they tell you that your baby has a medical condition. That is a big, challenging, and scary thing." Nurses can use the coping nursing intervention (Iowa Intervention Project, 2000) to help individuals and families assimilate new information. Nurses can further support these families by making a referral to other health professionals such as psychologists, family therapists, and psychiatrists.

Genetic test results can reveal information about an individual and that individual's family. The test results may alter family dynamics in many ways. For Betsy, for example, learning that her daughter had PKU during newborn screening meant that she and her husband were carriers. They were learning all at once about a new diagnosis in their baby and about their own genetic makeup. As another example, when a newborn screening test shows that a baby is a carrier for sickle cell trait, it means that one of the parents is also a carrier. In this situation, newborn screening test results may inadvertently reveal unexpected information about a child's paternity (SACGT, 2000).

In Laura's story regarding carrier testing, her family did not know that her father had CF until after his death. Carrier testing for CF revealed that she had a CF mutation. This knowledge led to her sister having carrier testing for CF, and her sister had a different CF mutation. The information about the two sisters revealed information about their father, who had an unusual type of CF with two different CF mutations, and their mother, who was not a carrier. Furthermore, the information about Laura's CF carrier status raised certain questions for her sister-in-law: "Is the condition in your father genetic?". Laura expresses that her sister-in-law may possibly discriminate against her because of the family history of CF.

Genetic test results may pose risks for individuals concerning insurance or employment discrimination. As Betsy points out, her insurance company does not pay for many of the food items needed by her daughter. Genetic test results may also raise concerns for groups if they lead to discrimination and stigmatization. This is especially relevant for minority groups who have experienced other forms of discrimination. This was unfortunately the experience for many African-Americans in the 1970s with sickle cell anemia screening. Many of the screening programs were based on inadequate and inaccurate knowledge of the genetics of SCD. Many individuals who were carriers were incorrectly diagnosed as having SCD. Carriers were stigmatized and faced discrimination in employment, and health and life insurance (SACGT, 2000).

Genetic discrimination is also of concern for individuals from the Jewish community because certain genes, which have been linked with higher rates of breast cancer, have been found in Ashkanzi Jewish women. In the fall of 2001, a Woman's Task Force on Public Policy was created to lobby on medical and health-related issues that are of concern to the Jewish community. Because of the lobbying, several bills have been introduced to Congress that would prevent employers from obtaining genetic testing information about job applicants. Another bill aims to prohibit insurers and health management organizations from discriminating against individuals based on their genetic traits for physical or mental disease, defects, or disabilities *(http://www.thejewishweek.com)*.

CULTURAL AND ECONOMIC ISSUES AND GENETIC TESTING

Ensuring that genetic testing is accessible and useful for all individuals is an essential nursing role. The Ethical, Legal and Social Issues Program of the National Human Genome Research Institute has as one of its goals to examine how concepts of race, ethnicity, gender, and socioeconomic factors affect the use and interpretation of genetic testing, information, and services. Awareness that genetic testing is available is an important early stage of knowledge acquisition for individuals in their decision to adopt a new health behavior. Understanding differentials in awareness of genetic testing can provide nurses and other health care providers with insight into intervention approaches that are appropriate for those in whom genetic testing may be beneficial, especially those individuals who may be vulnerable on the basis of inequitable access to genetic testing (Honda, 2003).

A number of factors have been shown to correlate with awareness or knowledge of genetic testing (**Table 4-9**). Educational level, for example, has been shown to be a consistent predictor of awareness. Race and ethnicity have been shown to be strong predictors of awareness of genetic testing for cancer in some studies (Honda, 2003). To date, little research has been conducted examining the role of immigration status and awareness of genetic testing. This is a much-needed area of investigation as immigration status has been increasingly documented as a strong predictor of health and a possible confounding aspect of the relationship between race, ethnicity, and health (Heron, Schoeni, & Morales, 2002). Nurses can contribute to research efforts that explore why racial and ethnic disparities persist in awareness of genetic testing, and to identify culture-specific barriers to the significant lack of awareness among recent immigrants and ethnic minorities.

One important area currently under investigation is how genetic research can be communicated to minority communities in meaningful ways. Several culturally relevant projects involve assessment of literacy levels and preferred learning methods, for example, and the development of educational materials for specific ethnic populations. Barriers to the use of genetic testing and services are other areas under

Table 4-9 Identified Factors That Correlate with Greater Awareness or Knowledge of Genetic Testing for Hereditary Breast–Ovarian Cancer

- Positive Family History: having a first-degree relative with breast cancer
- Upper Income–Higher Socioeconomic Status
- Marital Status: being married
- Younger Age
- Educational Level: well-educated
- Race–Ethnicity: Caucasian most likely and African-American least likely to be aware or have knowledge
- High Cancer Worry: especially in Caucasians
- Having Had a Biopsy
- Perceived Risk: when perceived is higher than actual risk for breast cancer
- Access to Testing: drop in interest if woman has to pay

(Armstrong, Weber, Ubel, Guerra, & Schwartz, 2002; Bosompra, Flynn, Ashikaga, Rairikar, Worden, & Solomon, 2000; Durfy et al., 1999; Hutson, 2003; Lipkus et al., 1999; Meischke, Bowen, & Kuniyuki, 2001)

investigation and include pilot testing of culturally sensitive training manuals for traditional and nontraditional health care providers (Aguilar et al., 2001).

It is anticipated that there will be continued growth of the number of immigrants to the United States, with the highest rates of increase in Asian and Hispanic populations (U.S. Bureau of the Census, 1996). This presents a major challenge and responsibility in promoting understanding of and equal access to genetic testing information for all of our society. Cultural influences including race, ethnicity, and immigration status are significant contributors to the disparity in awareness and use of genetic testing (Honda, 2003). Effective strategies to educate individuals about genetic testing will involve all nurses. In some clinical settings, for example, nurse practitioners may have a closer relationship with individuals and families that they care for, and can offer more in terms of helping people before, during, and after the genetic testing process. Nursing assessments regarding individual and family cultural and health beliefs are essential before communicating information about genetic testing and health risks. Nurses can seek out culturally relevant educational materials and methods *(http://www.culturediversity.org/basic.htm)* when conveying genetic testing information. Developing a plan of care that blends cultural beliefs and practices with interventions that are acceptable to the individual and family is central to providing culturally competent genetic health care (Lea et al., 2002).

Summary

Nurses in all areas of clinical practice will become increasingly involved in the genetic testing process. Newborn screening is expanding to include many more disorders that can be treated in infancy. Nurses have a central role in educating women and families about the newborn screening process, including the importance of follow-up of testing and management. Prenatal screening and diagnostic testing are now widely used to identify genetic conditions and birth defects. Nurses can assist women and couples to make informed decisions about whether or not to pursue such testing by providing balanced information regarding the benefits and limitations of prenatal testing. Knowledge of the components of the informed decision-making and consent communication process are essential for all nurses involved in genetic testing. The nursing intervention of decision-making support and the use of nondirective education and counseling help families to receive the necessary information and support during times of crisis (i.e., when a fetal abnormality is diagnosed). Newer prenatal techniques such as preimplantation diagnosis can now be made available to couples who may not otherwise have had children. Cultural and economic issues may prevent many individuals from seeking or understanding genetic testing. Nurses can contribute to ongoing research regarding cultural and economic barriers to genetic testing for minority populations and in developing innovative and culturally relevant ways to help people before, during, and after the genetic testing process.

Chapter Activities

1. Mrs N. is a 35-year-old Hispanic woman who is in the first trimester of her third pregnancy. She recently moved to the area. She had her two other children in Mexico. Both were home births and she had no prenatal care. She has come to the prenatal clinic for the first time. She brings her mother and sister with her. What genetic testing would be appropriate to discuss with Mrs. N.? In your nursing assessment, what would you want to learn about Mrs. N. before talking with her about genetic testing and prenatal care? What resources could you use to explain genetic testing?
2. Mr. and Mrs. B. are considering pregnancy. Mrs. B. is 29 years old and in good health. She and Mr. B. are of Scottish and English ancestry. They are concerned about birth defects and genetic conditions in their children because Mr. B.'s sister has a daughter who has CF. Mrs. B. tells you, "We want to do everything we can to have a healthy baby." What genetic testing would you discuss with the couple? What other support/referrals could you make?

References

Abramsky, L., Hall, S., Levitan, J., & Marteau, T. (2001). What parents are told after prenatal diagonsis of a sex chromosome abnormality: Interview and questionnaire study. *British Journal of Medicine, 322,* 463–466.

Aguilar, M., Visio, P., Kolb, S., Livingstone, J., Aguirre, C., & Kay, C. (2001). Cultural and linguistic considerations in development of genetic educational materials in a predominately Mexican-American population in South Texas. Presented at the National Institutes of Health's A Decade of ELSI Research, Bethesda, MD.

American College of Medical Genetics (ACMG). (2000). Statement of the American College of Medical Genetics on universal newborn hearing screening. *Genetics in Medicine 2*(2), 149–150.

American College of Medical Genetics (1998). Genetic susceptibility to breast and ovarian cancer: Assessment, counseling and testing guidelines. New York State Department of Health and the American College of Medical Genetics Foundation.

American College of Obstetrics and Gynecology (ACOG). (1991). Alpha-fetoprotein. *ACOG Technical Bulletin, 154.*

American College of Obstetrics and Gynecology & American College of Medical Genetics (2001). Preconception and prenatal carrier screening for Cystic Fibrosis: Clinical and laboratory guidelines. Washington, D.C.: American College of Obstetricians and Gynecologists.

Armstrong, K., Weber, B., Ubel, P. A., Guerra, C., & Schwartz, J. S. (2002). Interest in BRCA1/2 testing in a primary care population. *Preventive Medicine, 34*(6), 590–595.

Arnos, K., Israel, J., & Cunningham, M. (1991). Genetic counseling for the deaf: Medical and cultural considerations. *Annals of the New York Academy of Science, 630,* 212–222.

Baroni, M. A., Anderson, Y. E., & Mischler, E. (1997). Cystic fibrosis newborn screening: Impact of early screening results on parenting stress. *Pediatric Nursing, 23*(2), 143–151.

Bosompra, K., Flynn, B. S., Ashikaga, T., Rairikar, C. J., Worden, J. K., & Solomon, L. J. (2000). Likelihood of undergoing genetic testing for cancer risk: A population-based study. *Preventive Medicine, 30*(2), 155–166.

Bove, C., Fry, S., & MacDonald, D. (1997). Presymptomatic and predisposition genetic testing: Ethical and social considerations. *Seminars in Oncology Nursing, 13*(2), 135–140.

Canick, J., Knight, G., Palomaki, G., Haddow, J., Cuckle, H., & Wald, N. (1998). Low second trimester maternal serum unconjugated oestriol in pregnancies with Down syndrome. *British Journal of Obstetrics and Gynecology, 95,* 330–333.

Canick, J., & Kellner, L. (1999). First trimester screening for aneuploidy: Serum biochemical markers. *Seminars in Perinatology, 28*(5), 359–368.

Committee on Genetics. (1996). Newborn screening fact sheets. *Pediatrics, 98*(3), 473–501.

Day, S., Brunson, G., & Wang, W. (1992). A successful education program for parents of infants with newly diagnosed sickle cell disease. *Pediatric Nursing, 7*(10), 52–57.

Durfy, S. J., Bowen, D. J., McTiernan, A., Sporleder, J., & Burke, W. (1999). Attitudes and interest in genetic testing for breast and ovarian cancer susceptibility in diverse groups of women in western Washington. *Cancer Epidemiology, Biomarkers & Prevention, 8,* 369–375.

Eng, C. M., Schechter, C., Robinowitz, J., Fulop, G., Burgert, T., Levy, B., Zinberg, R., & Desnick, R. J. (1997). Prenatal genetic carrier testing using triple disease screening. *Journal of the American Medical Association, 278,* 1268–1272.

Farell, P. (2000). Improving health of patients with cystic fibrosis through newborn screening. *Advances in Pediatrics, 47,* 79–115.

Filly, R. (2000). Obstetrical sonography: The best way to terrify a pregnant woman. *Journal of Ultrasound in Medicine, 19,* 1–5.

GAO Report to Congressional Requesters. (March 2003). Newborn screening: Characteristics of state programs. GAO-03-449.

Genetics and Public Policy Center, 2003. *http://www.dnapolicy.org/genetics/testing.*

Giarelli, E., Lea, D. H., Jones, S., & Lewis, J. A. (in press). Chapter 3: Genetic technology: The frontier of healthcare. In *Conversations on Ethics in Nursing.*

Grady, C. (1998). Chapter 9: Ethics, genetics, and nursing practice. In *Genetics in clinical practice.* D. Lea, J. Jenkins, & C. Francomano (Eds.). Boston: Jones and Bartlett.

Grady, P., & Collins, F. (2003). Genetics and nursing science. *Nursing Research, 52*(2), 69.

Grant, S. (2000). Prenatal genetic screening. *Online Journal of Issues in Nursing, 5*(3). [Online]. Available: *http://nursingworld.org/LJIN/TOPIC13/tpc13_3.htm.*

Heron, M. P., Schoeni, R. F., & Morales, L. S. (2002). Health status of older immigrants in the United States. Labor and population program: Working Paper Series, 2002. Santa Monica, CA: The Rand Corporation.

Honda, K. (2003). Who gets the information about genetic testing for cancer risk? The role of race/ethnicity, immigration status, and primary care clinicians. *Clinical Genetics, 64,* 131–136.

Hutson, S. P. (2003). Attitudes and psychological impact of genetic testing, genetic counseling, and breast cancer risk assessment among women at increased risk. *ONF, 30*(2), 241–246.

International Society of Nurses in Genetics, Inc. (1999). Position Statement: Informed Decision-Making and Consent: The role of nursing. *http://www.isong.org*

Iowa Intervention Project. (2000). *Nursing interventions classification* (3rd ed.). St. Louis: Mosby.

Jones, S. (2004). The confluence of two clinical specialties: Genetics and assisted reproductive technologies. *Medsurg Nursing: official journal of the Academy of Medical–Surgical Nurses, 13*(2), 114–121.

Jones, S. L., & Fallon, L. A. (2002). Reproductive options for individuals at-risk for transmission of a genetic disorder. *Journal of Obstetrics, Gynecology and Neonatal Nursing, 31*, 193–199.

Jorde, L. B., Carey, J. C., Bamshad, M. J., & White, R. L. (2003). *Medical Genetics* (3rd ed.). St. Louis, MO: Mosby.

Kronn, D., Jansen, V., & Ostrer, H. (1998). Carrier screening for cystic fibrosis, Gaucher disease, and Tay-Sachs disease in the Ashkenazi Jewish population: The first 1000 cases at New York University Medical Center, New York, NY. *Archives of Internal Medicine, 158*, 777–781.

Lea, D. H., Calzone, K. C., Masny, A., & Parry Bush, A. (2002). *Genetics and Cancer Care: A Guide for Oncology Nurses.* Pittsburgh, PA: The Oncology Nursing Society.

Lerman, C., Croyle, R. T., Tercyak, K. P., & Hamann, H. (2002). Genetic testing psychological aspects and implications. *Journal of Consulting and Clinical Psychology, 70*(3), 784–797.

Levy, H. L. (1998). Newborn screening by tandem mass spectrometry: A new era. *Clinical Chemistry, 44*(12), 2401–2402.

Lewis, J. L. (2002). Genetics in perinatal nursing: clinical applications and policy considerations. *JOGNN, 31*, 188–192.

Lipkus, I., Iden, D., Terrenoire, J., & Feaganes, J. (1999). Relationships among breast cancer concern, risk perceptions, and interest in genetic testing for breast cancer susceptibility among African American women with and without a family history. *Cancer Epidemiology, Biomarkers, and Prevention, 8*, 533–539.

Lloyd-Puryear, M. A., & Forsman, I. (2002). Newborn screening and genetic testing. *Journal of Obstetrics, Gynecology and Neonatal Nursing, 31*(2), 200–207.

March of Dimes. (2002). Newborn screening tests (fact sheet). White Plains, NY: Author.

Meischke, H., Bowen, D., & Kuniyuki, A. (2001). Awareness of genetic testing for breast cancer risk among women with a family history of breast cancer: Effect of women's information sources on their awareness. *Cancer Detection and Prevention, 25*(4), 319–327.

National Institutes of Health. (1997). Genetic testing for cystic fibrosis. *NIH Consensus Statement, 15*(4), 1–37.

National Institutes of Health. (2000). Phenylketonuria (PKU): Screening and management. *NIH Consensus Statement, 17*(3), 1–33.

National Newborn Screening and Genetics Resource Center. Available at *http://genes-r-us.uthscsa.edu*

NCHPEG Working Group of the National Coalition for Health Professional Education in Genetics. (2001). Recommendations of core competencies in genetics essential for all health professionals. *Genetics in Medicine 3*(2), 155–158.

Newborn Screening Task Force. (2000). Serving the family from birth to the medical home: A Report from the Newborn Screening Task Force. *Pediatrics, 106,* S383–S427.

Nussbaum, R. L., McInnes, R. R., & Willard, H. F. (2001). *Genetics in Medicine 6.* Philadelphia: W.B. Saunders Company.

Palomaki, G. (1986). Collaborative study of Down syndrome screening using maternal serum alpha-fetoprotein and maternal age. *The Lancet, 19*(27), 1460.

Palomaki, G., Knight, G., McCarthy, J., Haddow, J., & Donhowe, J. (1997). Maternal serum screening for Down syndrome in the United States: A 1995 survey. *American Journal of Obstetrics and Gynecology, 176,* 1046–1051.

Parad. R. B., & Comeau, A. M. (2003). Newborn screening for cystic fibrosis. *Pediatric Annals, 32*(8), 528–535.

Scanlon, C., & Fibison, W. (1995). *Managing genetic information: Implications for nursing practice.* Washington, D.C.: American Nurses Association.

Secretary's Advisory Committee on Genetic Testing (SACGT). (December 1, 1999–January 31, 2000). *A public consultation on oversight of genetic tests.* Bethesda, MD: National Institutes of Health. *http://www4.od.nih.gov/oba/sacgt.htm*

Secretary's Advisory Committee on Genetics, Health, and Society. (2002). *Charter.* Washington, DC: The Secretary of Health and Human Services. *http://www4.od.nih.gov/oba/sacghs.htm*

U.S. Bureau of the Census. (1996). *Current Population Reports: Population Projections of the United States by Age, Sex, Race, and Hispanic Origin—1995–2050.* Washington, DC: US Government Printing Office, 1996.

Wald, N., Cuckle, H., Densem, J., Nancholhal, K., Roysston, P., Chard, T., Haddow, J., Knight, G., Palomaki, G., & Canick, J. (1998). Maternal screening for Down syndrome in early pregnancy. *British Medical Journal, 297,* 883–887.

Williams, J. K., & Lea, D. H. (2003). *Genetic issues for perinatal nurses* (2nd ed.). Wieczorek, R. R. (Ed.) White Plains, NY: March of Dimes.

Williams, J. K., & Schutte, D. (1997). Benefits and burdens of genetic carrier information. *Western Journal of Nursing Research, 19*(1), 71–81.

CHAPTER 5

Connecting Genomics to Health Outcomes: Pharmacogenetics and Pharmacogenomics

Human genome discoveries are creating new approaches to diagnosing, treating, and managing human health and disease. Nurses will be increasingly involved in assessing patients, administering, and monitoring the effects of a range of therapeutic approaches to the treatment of disease, including pharmacogenetics or prescription of drugs to match individual genetic profiles. This will include educating individuals about these new applications and explaining the influence of genetic factors in maintenance of health and development of disease. Recognition of the role of behavioral, social, and environmental factors including lifestyle, socioeconomic factors, pollutants, and other environmental influences and how these interact with genetic factors will be necessary in caring for individuals' and families' total health. It is therefore essential that nurses become familiar with the indications for genetic testing in the context of individualizing treatment to increase efficacy and avoiding harmful side effects.

Introduction

New genome knowledge is leading to more focused treatments, allowing for identification of drugs and drug dosages that may be more effective based on a person's genotype. These discoveries offer hope for improved treatment for many common conditions such as cancer, cardiac disease, and psychiatric illness. For nurses and other health care providers, this means an increasing use of genetic testing to identify genetic differences (polymorphisms) that can help in the design of individualized treatment and management plans.

In this chapter, we focus on pharmacogenetics and pharmacogenomics and their applications in clinical practice. Through Mary's story, we learn about the difficulties facing individuals and families living with psychiatric illness, such as depressive disorders, with regard to inadequate treatments and lack of recognition and understanding that this is a chronic and sometimes hereditary condition. As Mary begins coping with her own depression following a traumatic delivery of her second child, she learns of the newly available genetic testing that can help her find the most therapeutic treatment and dosage of medication. Mary's story illustrates the ongoing evolution of our understanding of the causes and treatment of psychiatric illness. **Box 5-1** provides a brief summary of the psychiatric illnesses—the mood disorders. As Dr. Francis Collins, Director of the National Human Genome Research Institute stated in an address in 2001: "I would predict that mental illness will be one of

Box 5-1 Psychiatric Conditions: The Role of Genetics in Mood Disorders

General Information

Major depression has been identified as the leading cause of disability among individuals age 5 and older. It is the second leading source of disease burden, surpassing even cardiovascular diseases, lung cancer, diabetes, and dementia. The significant impact of mood disorders on distress in those affected and in family members, lifetime disability, and suicide underscores the importance of research to discover the causes and to inform prevention and treatment.

Clinical Symptoms

Bipolar disorder is characterized by episodes of mania—a distinct period of abnormality and persistently elevated, expansive, or irritable mood—and depression. The depression of bipolar disorder is similar to major depressive disorder, which consists of depressive mood, significant weight loss, sleep disturbances, psychomotor retardation, fatigue or loss of energy, feelings of worthlessness, and recurrent thoughts of death.

Prevalence

The lifetime prevalence rate of bipolar disorder is estimated to be 0.5 to 1.0% (1 in 100 to 1 in 200). The rates of major depression, on the other hand vary considerably among studies, from 2 to 25%, most likely due to different cultural or environmental influences. In bipolar disorder, the prevalence rates are the same for males and females. For major depression, the rates are double in females as compared with males. The rates of depression for females appear to be elevated considerably during reproductive years. By menopause, the rates in males and females are relatively equal.

Inheritance

Many studies have been conducted over the years demonstrating conclusively that genetic factors are involved in the susceptibility to mood disorders, primarily bipolar disorders. A high magnitude of familial aggregation of bipolar disorder, similar to

Box 5-1 continued

that found for many of the major diseases for which the genetic basis has been identified has been observed. This is in contrast to major depression where only a moderate influence of familial aggregation on nonbipolar mood disorders has been identified. Studies of children with parents who have bipolar disorder, for example, have revealed an increased risk of this disorder.

Genome scans of bipolar disorder have not shown conclusively any specific region in the genome, with the exception of two loci (4p12-13 and 13q31-33). The most striking observation is that no two studies employed the same identical ascertainment procedures and that there was substantial diversity in sampling and methods. Candidate gene studies have shown only weak associations with several loci.

Management and Treatment

The main treatment of bipolar disorder is the use of mood stabilizers. Depending on the pattern and nature of the disorder, antidepressive or antipsychotic medication may be required intermittently. Electroconvulsive treatment (ECT) is used in the face of acute suicidal risk or uncontrollable mania. Hospitalization is needed when it is not possible to manage the individual in outpatient settings. Newer genetic research—pharmacogenetics and pharmacogenomics—is paving the way for development of genetic tests that can help in prescribing the right medication in the proper dose to each individual.

National Institutes of Mental Health Human Genetics Initiative: Mood Disorders

The National Institutes of Mental Health Genetics Initiative *(www.grb.nimh.nih.gov/gi/html)* is an effort begun in the late 1980s to establish a national resource of clinical and diagnostic data and biomaterials (DNA and cell lines) for distribution to the scientific community to further research. Other initiatives include initiation of multisite collaborative genetics research projects, collection and dissemination of clinical data and DNA, and development and dissemination of other research resources for genetic analysis, including SNPs in genes of relevance to nervous system functioning.

References

Merikangas, K. et al. (2002). Workgroup reports: NIMH strategic plan for mood disorders research. Future of genetics of mood disorders research. *Society of Biological Psychiatry, 52,* 457–477.

McInnis, M. G. & DePaulo, J. R. (2000). Chapter 109. Major mood disorders. In Rimoin, D. L., Connor, J. M., Pyeritz, R. E., & Korf, B. R. *Principles and Practices of Medical Genetics* (4th ed.). London, England: Churchill Livingstone (2002).

the major beneficiaries of our genetic understanding because we are on the brink of discovering genes for schizophrenia, bipolar disease, and obsessive-compulsive disorder." (Dr. Francis Collins, 2003).

Mary's Story: Sometimes the End Is Just the Beginning

I was the second of three children growing up in a suburb of New York City. My older sister was 5 years older than I, and my younger brother was 5 years younger. I was right in the middle. My parents had moved to New York City after my father completed his PhD in sociology. He was a professor at a large university in New York. The first memory I have of my father was when I was about 4 years old. I was excited about a new toy I had received from an aunt, and I wanted to show it to him. I remember running upstairs to his room in excitement, only to find my mother at the door of the bedroom standing guard. She told me to be quiet because my father was resting and too tired to play right now. Later that day our doctor came to the house to see my father (this was in the 1960s when doctors still made house calls). After our doctor saw him, he told us that my father would be going to the hospital for a while to get better, but that he would be home soon. My mother explained to us that my father needed to rest as he had been working too hard. A month later, my father came back home, looking tired, but ready to be with us as a family and to get back to work. I did not realize until I was a teenager, after my father was hospitalized for the third time, that there was something going on that made him chronically tired and sad.

No one in my family talked about what was happening with my father. My father never said much, only that he missed me when he was away. My mother told us that he got tired from his work because he was a busy professor. I asked my sister and she said she didn't know, and anyway she was too busy to answer my questions at the time. My brother was too little to know. So, I eventually stopped asking.

My father grew up in Alabama. He was the only one in his family who left to go to college outside of the state. He had two sisters and a brother who lived in Alabama. His sisters were teachers. His brother did odd jobs around the town, but never had steady work. We didn't see much of them, but heard about them from time to time. One of my aunts was very flamboyant and very excitable. Her nickname was Auntie Mame. She wore very flashy clothes and was always donating her money and clothes to charities, even when she didn't have the money or clothes to give. She never got married. She lived with my other aunt, who never married either. My uncle was basically a hermit. He lived in a log cabin in the woods. He liked hunting and fishing, but he didn't like people that much. He was a very quiet and reserved man from what I can recall. One winter, we heard from my aunt that he had died in a "hunting accident." My father went to his funeral, but we did not go.

When I got to college, I had a friend whose brother had what she called "bipolar" disorder, a condition where you have mood swings and can be elated or depressed. It was a cyclic thing for her brother, and he was taking medicine when he felt well. Then he would stop. I wondered then whether there was anything like that going on in my family. I asked my mother when I went home on school vacation. She said it would be best for me not to think too much about it.

I was 28 when I got married to my college boyfriend. We both worked at an insurance company in Connecticut. We had our first child, a boy, when I was 30 years old. Later that year my father died from a heart attack. He was 76 years old. This was so difficult for my mother, who had been with him for 45 years through thick and thin. At 74, I don't think she could bear the separation; she died the next year from a stroke. I had never been especially close to either of them. That was the way it was in my family. But I was so sad to lose them both so close together.

I became pregnant with my second child when I was 33, the year after my mother died. It was a difficult pregnancy from the start. At delivery, there were complications and I had to have an emergency cesarean section. Then my new daughter developed an infection and was in the hospital for a week. By the time we got home, I was exhausted. I had a three-year-old to take care of in addition to my daughter. I had quit my job so I could stay at home with both children. After a month, I began to feel sad all of the time. I could not get out of bed, take a shower, or comb my hair. Some part of me recognized that this is how my father used to get.

My husband was very concerned, and he took me to our doctor who then put me in touch with a psychiatrist. Part of the initial meeting with Dr. F. was reviewing my past medical and family history. When I told Dr. F. about my father's periods of exhaustion and hospitalization, my uncle's reclusive nature and accidental death, and my aunt's periods of excitement, he asked me whether depressive illness was inherited in my family. This was the first time I clearly put all of the pieces together to consider maybe that was so. Dr. F. said that his impression was that bipolar illness seemed to be present in my father's family. He said that he wanted me to take some medication to help me now so that I wouldn't fall into the same pattern as my father. With that in mind, he said that he wanted me to have a new test called the cytochrome P450 test, a genetic test that could help determine the proper medication and dose. I said I was willing to give it a try, as I wanted to be there for my kids; I wanted to be more present for them than my father was able to be.

After I had been on the medication for a while, my curiosity got to me and I called my father's sister—the one who had lived with Auntie Mame. I told her what was going on with me and asked her if she knew whether my father, his brother, and his sister had any mental health issues. My aunt told me that my father's reclusive brother had, in fact, committed suicide; it was not a hunting accident. The family did not want anyone to know that though. My other aunt was now in a nursing home with Alzheimer's disease and was being treated with medication to help with her excitability. My aunt told me that my father's father had also had a history of depression and would go off into the woods of Alabama for weeks at time to cope with his sadness. She herself had never had any problems with depression, but had seen the toll it had taken on her family.

When I learned all of this, I felt such sadness for my father and his family. They did not have any support. They were not able to talk about this illness in the family because of their concerns about rejection and stigmatization. I miss my father now, and I wish I could say to him "I understand." I am determined to take my medication and to work with my doctors to remain in good health. My husband has been very loving, understanding, and supportive. We have talked about having more children, but we know that depressive illness in my family is hereditary. We don't want to take any more chances. We will know what to look for in our children, the early signs, and we hope that newer treatments will become available to me and to others as more is learned. I know that I will be able to talk with my children about my history of depression as something that happened to me, just like a person can get diabetes or cancer. I will not need to hide it from them now.

I wish that there were more research monies being put into psychiatric research, and that people would and could talk more openly about this type of illness. I am glad that there are newer medications that are helpful and I hope that there will be more. I am mostly thankful to my husband for being so understanding and open to talking with me about how to take care of myself and our family.

Mary

Pharmacogenetics and Pharmacogenomics

The term "pharmacogenetics" was first introduced by Vogel (Vogel, 1979). Research in twin studies by Vessel and Page demonstrated interindividual variability in drug response and showed that drug response is much less variable in monozygotic twins who share all genes in common (Vogel, 1979; Vessel & Page, 1968). Adverse drug reactions were additional clinical events that revealed genetic variants of drug-metabolizing enzymes or drug targets. Incidental observations that some patients experienced unpleasant and unexpected adverse drug reactions when given the standard dose of a drug led to the discovery of polymorphisms (Meyer, 2000a). For example, adverse drug reactions such as nausea, diplopia, and blurred vision after the antiarrhythmic and oxytocic drug sparteine, or significant orthostatic hypotension after the antihypertensive agent debrisoquine have led to the discovery of the genetic polymorphism of the drug-metabolizing enzyme cytochrome P450 2D6 (CYP 2D6) (Meyer, 2000a).

The discipline of pharmacogenetics has since evolved to encompass the field of inquiry dealing with variability of responses to medications that is associated with genetic variation. Pharmacogenetics generally refers to the inherited variability in drug metabolism and disposition with a focus on the pharmacokinetic effects: drug absorption, distribution, metabolism, and elimination. As a discipline, pharmacogenetics takes into account a patient's genetic information of drug transporters, drug metabolizing enzymes, and drug receptors to create an individualized drug therapy that allows for optimal choice and dose of drugs (Tsai & Hoyme, 2002). Mary's story illustrates the initiation of this new way of prescribing medication based on human genome discoveries and their application—pharmacogenetics.

Pharmacogenomics is the research area directed toward the search for genetic variations that are associated with drug efficacy. Pharmacogenomics takes into account all aspects of drug behavior: absorption (active transport mechanisms), distribution (protein binding), metabolism (such as cytochrome P450), and receptor-target affinity. Its aim is to optimize patient management by customizing and synthesizing drugs based on genetic variations in drug response. Pharmacogenomics involves the search for genetic variation on a genomic scale with a focus on pharmacodynamic effects—target receptors or enzymes that determine cellular reactions to drugs (Mansour & Nimgaonkar, 2002; Tsai & Hoyme, 2002). The pharmaceutical industry is devoting intense efforts into pharmacogenomics for target characterization in drug development and research that aims at understanding the genetic basis for predicting drug responses in routine clinical care (Muller, 2003; Roses, 2000).

Individual variations in response to drugs, as well as adverse effects of such drugs, are influenced by many factors. Recognized factors include compliance, body weight, age, gender, general health, and nutritional status. Variability between different ethnic groups in drug response is also well-documented. As an example, primaquine-induced hemolytic anemia, due to an inherited deficiency in glucose-6-phosphate dehydrogenase (G-6-PD) is more common among black individuals and certain Mediterranean and Southeast Asian populations (Poolsup et al., 2000). As another example, individual differences in isoniazid-induced hepatitis and peripheral neuritis are related to ethnic differences or a polymorphism (small genetic change common in more than 1% of a given population) governing hepatic acetylation (Mansour & Nimgaonkar, 2002). Nurses have long recognized that patients show substantial variability in their response to drugs and that unusual responses may be clustered in families. These observations, in fact, indicate that at least some part of the variability of response to therapeutic interventions may be inherited.

Human genome research has revealed that there are about 30,000 genes in the human genome. New research strategies capitalize on databases generated from the human genome project to identify mutations or polymorphisms associated with drug response phenotype. Scientists can now localize genetic traits by statistical association to specific regions of the human genome and identify allelic variants—gene variations—that have subtle clinical significance and consequences. These are often characterized by point mutations that cannot be visualized by examining chromosomes. Identification of these mutations is crucial to better understanding of disease mechanisms and to developing and targeting treatments (Muller, 2003).

There are a total of 3.12 billion nucleotides in the human genome. Single-nucleotide polymorphisms (SNPs) occur at a frequency of 1/1250 base pairs between two individuals. This means that the total number of genetic variations between two individuals differs by about 2.5 million. Among these genetic variations, three major types have been identified: insertions or deletions, repetitive DNA, and SNPs. SNPs are the cause of about 40 to 60% of all cases of altered gene function and may affect protein structure or transcriptional regulation of normal proteins (Muller, 2003; Meyer, 2000b).

SNP databases are one important component of human genome research. The number of reported SNPs has increased rapidly *(http://www.ncbi.nlm.nih.gov/SNP)*. To date, approximately 4 million SNPs have been identified. Researchers believe that the majority of SNPs do not have any particular function and are located primarily between genes—intergenic SNPs. Another class of SNPs is located in non-coding gene regions, in introns, and may also consist of silent mutations—perigenic SNPs. Only about 100,000 SNPs are located in coding regions (cSNPs) that can cause alterations in amino acids. These discoveries have led researchers to understand the magnitude of interindividual differences and the importance of understanding the role of cSNPs in the genetic basis for disease, as well as for differential responses to drug treatment (Ingelman-Sundberg, 2001; Muller, 2003). Once a large number of these SNPs and their frequencies in different populations are known, they can be used to correlate an individual's genetic profile with likely individual drugs and drug response. The ability to predict interindividual differences in drug efficacy or toxicity based on genetic factors will be realized in future drug treatments (Meyer, 2000b).

Genetic Mechanisms and Variable Drug Response

Pharmacogenetic principles can be applied to two major therapeutic issues. First are the issues of adverse drug reaction (ADRs). Second is nonresponsiveness to drug treatment. ADRs create significant medical and economic challenges in routine drug therapy and development. It is estimated that the overall incidence of serious ADRs in the United States results in about 6% of hospitalizations and 100,000 of deaths per year, making these reactions a leading cause of death (Lazarou, Pomeranz, & Corey, 1998). In addition, ADRs are one of the most important reasons for attrition in the costly drug development process. Unpredictable reactions to medications appear to be strongly influenced by genetic factors and are categorized as dose-dependent and dose-independent ADRs. The association of polymorphisms in CYP2C9 with dose requirements and risk-of-bleeding complications with warfarin therapy is an example of dose-dependent ADR. Genetic variants can create an unexpected or non-dose-dependent drug effect, for example, hemolysis in glucose-6-phosphate dehydrogenase deficiency (Meyer, 2000b; Muller, 2003; Tsai & Hoyme, 2002).

Molecular studies in pharmacogenetics began with the cloning and characterization of CYP2D6 and have now been extended to many other human genes, including those coding for more than 20 drug-metabolizing enzymes and drug receptors, as well as several drug transport systems (Meyer, 2000b). The cytochrome P450 (CYP) system encompasses a family of enzymes, found primarily in hepatocytes, that metabolize and activate many drugs and synthesize compounds such as steroids. Drug metabolism occurs in two phases: Phase 1, in which functional groups are modified, and Phase 2, in which conjugation takes place. Polymorphisms affecting CYP function and expression in either phase can influence drug response. CYP2D6 and C2C19 are the most well-studied and well-understood (Tsai & Hoyme, 2002).

Nearly 100 variant alleles of the CYP2D6 locus have been reported and described *(http://www.imm.ki.se/CYPalleles/cyp2d6.htm)*. Some alleles code for nonfunctional products. These alleles differ from the normal gene (called wild type) by one or more point mutations, gene deletions, duplications, or multiduplications. The mutations can have no effect on enzyme activity, or they can code for enzymes with decreased or absent activity. Duplications lead to increased enzyme activity. Individuals who are homozygous (have two of the same gene copies) of the wild type or normal enzyme activity, or who are heterozygous for the normal enzyme activity (have two different gene copies) are referred to as *extensive metabolizers* (EMs) and make up approximately 75 to 85% of the population. *Intermediate metabolizers* (10 to 15% of the population) or *poor metabolizers* (PMs) (5 to 10% of the population) carry two decreased activity alleles or loss-of-function alleles. *Ultrarapid metabolizers* (UMs) (1 to 10% of the population) carry duplicated or multiduplicated active genes. The number and complexity of mutations in the CYP2D6 locus is likely quite large overall; however, only a few gene mutations—from 3 to 5 alleles—are common and account for more than 95% of the mutant alleles in a given population. Today, these alleles can be detected using modern DNA methods, including DNA chip microarrays, creating opportunities for many individuals to be assigned to a particular phenotype group and drug doses prescribed accordingly (Meyer, 2000b).

Pharmacokinetics

Pharmacokinetics (PK) is defined as "events which indicate what the body does to the drug." (Muller, 2003, p. 234). PK comprises drug absorption, distribution, metabolism, and elimination. The relevance of a particular genetic polymorphism to drug therapy depends on the characteristics of the drug being considered. Drug uptake mechanisms in the overall behavior of the drug and the particular agent's therapeutic range determine how much the dose has to be adjusted in PMs or UMs. The CYP2D6 polymorphism and its role in metabolizing nortriptyline provide an example. More than 90% of individuals require 75–150 mg/day of nortriptyline to reach a therapeutic plasma steady-state concentration. PMs, in contrast, need only 10–20 mg/day to reach the same concentration, while UMs may require as much as 300–500 mg/day or even more to reach the same plasma concentration. If the genotype of the individual patient is not known, PMs will be overdosed and at risk for drug toxicity. UMs, on the other hand, will be underdosed (Meyer, 2000b).

Another consideration regarding the specifics of a drug effect is when the therapeutic effect depends on the formation of an active metabolite. Such is the case with the formation of morphine from codeine. Poor metabolizers will experience no effect, and ultrarapid metabolizers may have exaggerated drug responses. Clinically relevant genetic polymorphisms for drug-related criteria are similar to those for drug-concentration monitoring, that is, a narrow therapeutic range or larger interindividual variation in kinetics, which is associated with overdose. Using new pharmacogenetics, a single DNA test performed only once in a lifetime can identify an individual's predisposition(s), especially for those individuals at the extreme ends of the spectrum (i.e., PMs and UMs) (Meyer, 2000b; Tsai & Hoyme, 2002).

Pharmacodynamics

Pharmacodynamics (PD) focuses on target receptors or enzymes that determine cellular reactions to drugs and has been described as "events indicating what the drug does to the body." (Muller, 2003, p. 236). The genetic basis of PD is not as well established as pharmacokinetics and results, to date, have a smaller impact on determining dosage for clinical use of drug therapy. The underlying genetic mutations often have subtle effects, affecting multiple proteins and leading to more subtle phenotypes. Examples of clinically relevant mutations in drug targets include the angiotensin-converting enzyme, apolipoprotein E, and cholesterol ester transfer proteins (Muller, 2003).

Clinical Applications of Pharmacogenetics

Practitioners are increasingly using a patient's genetic profile as determined by genetic testing to select the most appropriate medication for treatment. An individual's genetic profile is also used to select a medication that will produce minimal side effects by taking into account the differences in the production of the CYP450 enzymes (Roses, 2004). The following subsections provide common examples of the use of pharmacogenetics in clinical practice.

Psychiatry

Traditionally, therapeutic drug monitoring has been used as a tool to individualize drug dosage in the treatment of patients with psychiatric illness. The rapid advances in pharmacogenetic knowledge and genotyping methodology are increasingly used in clinical psychiatry. With most psychiatric medications, the CYP enzymes are essential for metabolism. Recent advances in the understanding of drug metabolism and the molecular biology of CYPs now allow for the use of pharmacogenetic testing as a complement or replacement for therapeutic drug monitoring when prescribing and monitoring psychiatric therapeutics. The pharmacogenetic tools currently used are phenotyping—the measurement of a specific enzyme activity by use of a probe drug, and genotyping—the analysis of functionally important mutations in gene coding for a specific enzyme (Dahl, 2002; Meyer, 2000b).

Studies of psychiatric patients treated with medications metabolized by CYP2D6 indicate that genotyping can improve efficacy, prevent ADRs, and decrease cost of therapy with these agents. Genotyping methods allow for identification of PMs, homozygous and heterozygous EMs and UMs with gene duplication and amplification. Several other alleles leading to decreased but not absent activity can also be identified with genotyping. Ethnicity is considered when choosing which alleles to analyze and in the interpretation of genotype-phenotype relationship, as genetic variation in drug metabolism has been observed between ethnic groups (Ansell et al., 2003).

Genotyping has several advantages for psychiatric drug prescription. First, it does not require drug intake followed by blood or urine sampling. Second, genotyping can be performed irrespective of the patient's drug treatment that may interfere with phenotyping by causing increased metabolic ratios, and, in some situations, the misclassification of genetically EM individuals as PM (called phenocopy). Genotyping is increasingly used in psychiatry and many other clinical situations as the preferable, if not the only, practically feasible tool for assessing patients' enzyme activity. Phenotyping, on the other hand, is used to identify UM individuals who do not carry duplicated or multiduplicated CYP2D6 genes (Dahl, 2002).

Antidepressants Many antidepressant medications are metabolized to a significant extent by the polymorphic cytochrome P450 (CYP) 2D6, which shows large, interindividual variations in activity. The metabolism of the tricyclic antidepressants—amitriptyline, clomipramine, desipramine, imipramine, and nortriptyline—is influenced by CYP2D6 polymorphisms to various degrees. Tricyclic antidepressants have a narrow therapeutic range, and considerable toxicity is associated with elevated concentrations. The PMs and UMs are two groups of patients who can have clinical problems with tricyclic drug dosage. PMs have increased plasma concentrations of tricyclic antidepressants, for example, when given the recommended doses of these drugs. The UMs, on the other hand, are prone to therapeutic failure because the drug concentrations at normal doses are much too low for therapeutic effect. In addition, adverse effects occur more frequently in PMs and may be misinterpreted as symptoms of depression, leading to further incorrect and adverse increases in the drug dose (Meyer, 2000b). Pretherapeutic genotyping for this class of antidepressants is now being piloted as a reasonable strategy for improving clinical outcome while minimizing adverse reactions and nonresponse to therapy (Mansour & Nimgaonkar, 2002).

The selective serotonin reuptake inhibitors (SSRI) are a group of antidepressants that interact with CYP2D6. Unlike the tricyclic antidepressants, SSRIs have a flat dose-response curve and thus a wide therapeutic index (Mansour & Nimgaonkar, 2002). Research data to date suggest that there is no clear correlation between serum concentrations and therapeutic effectiveness. Based on these assessments, there does not appear to be any significant benefit from pretherapeutic genotyping of metabolizing polymorphic enzymes. However, the ability of these agents to act as potent competitive inhibitors of CYP2D6 (paroxetine and fluoxetine) is of major importance. Inhibition means that the elimination of other CYP2D6 substrates, such as tricyclic antidepressants, is impaired. Other medications, such as fluvoxamine, are inhibitors and can thus cause important interactions with drugs that are partly metabolized by the cytochrome P450 enzyme, CYP1A2, for example, amitriptyline and imipramine (Mansour & Nimgaonkar, 2002).

Many of the classical as well as the newer antipsychotic drugs are also metabolized to a significant extent by the polymorphic CYP2D6. Classical antipsychotic medications have a relatively narrow therapeutic range, with concentration-dependent adverse effects occurring at concentrations similar to or only slightly higher than those required for therapeutic effect. The newer antipsychotic medications have a broader therapeutic index than the older drugs with respect to extrapyramidal effects. Pharmacokinetic interactions are common when patients receive more than one medication. Adverse effects may also resemble the symptoms of the disease being treated. Prospective trials are underway to examine the value of phenotyping or genotyping patients with depression and psychotic illnesses in selecting the correct starting dose to increase therapeutic efficacy and prevent toxicity (Weinshilboum, 2003).

Further information about drug metabolism for psychiatric and other medications is available at the research Web site of David Flockhart *(http://medicine.iupui.edu/flockhart)*. **Table 5-1** provides an overview of psychiatric medications, metabolizing enzymes, and their effects.

Table 5-1 Examples of Pharmacogenetic Applications in Psychiatric Medications

Gene	Medication	Clinical Consequence Linked to Polymorphism	Reference
CYP2C19	Diazepam	Prolonged Sedation	Ingelman-Sundberg et al., 1999
CYP2D6	Tricyclic antidepressants; Amitryptyline [Clomipramine] Desipramine Imipramine [Paroxetine]	Toxicity in PMs; inefficacy in UMs	Ingelman-Sundberg et al., 1999
	Antipsychotics: Haloperidol [Risperidone] thioridazine	Tardive dyskinesia, Parkinsonism	Meyer, 2000b; Krynetski & Evans, 1998
ApoE4	Tacrine	Less response in Alzheimer patients with a mutation in the 5′-upstream region	Poirier et al., 1995

For more detailed information go to *http://medicine.iupui.edu/flockhart*

Pain Management

Nurses assess patients for pain and provide therapeutic interventions for pain relief. Nurses are aware that some patients do not receive pain relief from the normal dose of codeine, for example, and that other patients become overly sedated with the usual dose of codeine. It is now known that the CYP2D6 polymorphism is associated with striking differences in adverse reactions to opioids. Codeine, hydrocodone, oxycodone, ethylmorphine, and dihydrocodeine are metabolized by CYP2D6. The metabolism of codeine is an important example of the use of pharmacogenetics in ensuring adequate pain management. PMs do not demethylate codeine to morphine, and therefore experience no analgesic effects. Similarly, respiratory, psychomotor, and pupillary effects of codeine are less in PMs in comparison to EMs. Codeine is often recommended as the drug of first choice for the treatment of chronic pain. Nurses and other health care providers need to know that no analgesic effect is to be expected in 5 to 10% of Caucasians who are PMs or who are EMs and receive concommitant treatment with a potent inhibitor of CYP2D6 such as quinidine (Meyer, 2000b).

Oncology

Most cancer chemotherapies used today are imprecise and extensively destructive. Clinicians have not had a way to distinguish which patients will respond to specific chemotherapeutic regimens. Because of this, chemotherapy has used a "one size fits all" approach (Meyer, 2000b). In addition, damage to healthy cells causes patients to experience devastating side effects, such as neutropenia, leading to severe infections, anemia, extreme weakness, and gastrointestinal symptoms and cause weight loss, nausea, and protracted vomiting. These side effects are often more debilitating than the cancer itself. Researchers have long searched for more selective methods of targeting cancer cells while leaving other body cells unharmed, and for ways to choose the correct dosage of treatment agents. Human genome discoveries and the expanding roles of pharmacogenetics and pharmacogenomics are leading to dramatic breakthroughs in cancer treatment. Some examples are described below.

Acute Lymphoblastic Leukemia (ALL) In cancer chemotherapy of acute lymphocytic leukemia, administration of drugs such as 6-mercaptopurine, 6-thioguanine, and azothioprine can cause severe hematologic toxicity and even death in 1 in 300 patients. The enzyme thiopurine S methyltransferase (TMPT) metabolizes 6-mercaptopurine and any medication with a purine base. The drug 6-mercaptopurine is also prescribed to individuals with irritable bowel disease. A pharmacogenetic test called genotyping and functional assays of TMPT in red blood cells can identify those patients who are homozygous for alleles encoding nonfunctional enzyme and therefore are unable to metabolize the drugs to their inactive methylated forms. Most people have normal to intermediate TMPT levels. However, about 10% of Caucasians and African-Americans have an inherited gene mutation that regulates TMPT activity, resulting in TMPT deficiency. These patients can be safely treated with doses 10 to 15 times less than is commonly prescribed, thereby avoiding severe and life-threatening hematologic and hepatic toxicity (Weinshilboum, 2003).

Knowing a person's genetic makeup for drug metabolism can help a health care provider select the appropriate drug dosage. Individuals with high or normal metabolism may need more of the recommended dose, while those who are TMPT-deficient or have slower metabolism will need less chemotherapy to achieve a therapeutic effect. Genotyping or functional enzyme analysis for TMPT has become standard practice in major cancer treatment centers. Event-free survival in childhood ALL has improved dramatically with this new knowledge (Muller, 2003; Tsai & Hoyme, 2002; Weinshilboum, 2002).

Chronic Myeloid Leukemia (CML) Chronic myeloid leukemia is a neoplastic proliferation of the hematopoietic stem cells made by the bone marrow. In CML, cancerous progenitor blood cells multiply uncontrollably within the bone marrow, enter circulation, and cause multiple systemic symptoms. Patients with CML have symptoms that include neutropenia, anemia, and thrombocytopenia. In most cases, CML is diagnosed in the chronic phase with presenting symptoms of fatigue, infection, weight loss, bleeding, and splenomegaly. Three to five years after onset, CML usually progresses to the more aggressive accelerated and blast phase (Capriotti, 2002).

A specific chromosomal abnormality called the Philadelphia (Ph) chromosome is found in 95% of patients with CML. The Ph chromosome is a shortened chromosome 22 resulting from a reciprocal translocation (rearrangement) between the long arms of chromosomes 9 and 22. The cause of the translocation is unknown. The reciprocal translocation causes neoplastic blood cells to produce a unique protein called BCR-ABL tyrosine kinase. The protein is an enzyme that is integral to the proliferation of CML cells (Ansell et al., 2003).

Testing patients with CML for the Ph chromosome identifies those individuals who can benefit from a new therapeutic agent, imatinib mesylate—Gleevec©. Gleevec© was recently approved by the Food and Drug Administration (FDA) for treatment of patients with CML. Gleevec© is a new class of anticancer agent called signal transduction inhibitors (STIs)—antiproliferative agents. STIs interfere with the specific pathways that signal growth of tumor cells. Gleevec© targets BCR-ABL tyrosine kinase and suppresses the synthesis of CML progenitor cells (Ansell et al., 2003; Goldman & Melo, 2001). Gleevec© has brought about dramatic results in many patients affected by CML. In many patients, the disappearance of the Ph chromosome and blood counts returning to normal have been reported (Drucker et al., 2001).

Gleevec© is still under investigation in terms of the required duration of treatment. However, its toxicity profile thus far appears very mild for a drug with such a high degree of potency. Gleevec© is also under investigation for treatment of other types of cancer, and has proven effective against some forms of ALL and gastrointestinal stromal tumors (Capriotti, 2002).

Breast cancer Overexpression of HER2, a proto-oncogene, is present in 25 to 30% of breast and ovarian tumors. This overexpression is associated with differences in prognosis, with those women who have amplification or overexpression of HER2 having a poorer prognosis and requiring other treatment alternatives (Ansell et al., 2003). Women with breast cancer can now be tested for HER2 before initiation of treatment. Those with HER2 benefit from treatment with trastuzumab (Herceptin©), a humanized monoclonal antibody against the HER2 receptor that is linked to HER2 overexpression. Herceptin©, produced by Genentech (South San Francisco, CA) is being used alone or in combination with chemotherapy, resulting in an improved response rate and an overall survival advantage for patients with HER2 overexpression. Patients with breast cancer who do not have HER2 overexpression are prescribed alternate therapy due to Herceptin's side effects and it is unlikely to improve treatment outcomes (Rusnak et al., 2001; Tsai & Hoyme, 2002).

Chemotherapy boosts survival rates in women with breast cancer after surgery. A recent study of women with breast cancer found patterns of gene expression that could predict how women responded to docetaxel, a new and widely-used chemotherapeutic agent. Using microarray technology (**Figure 5-1**), researchers

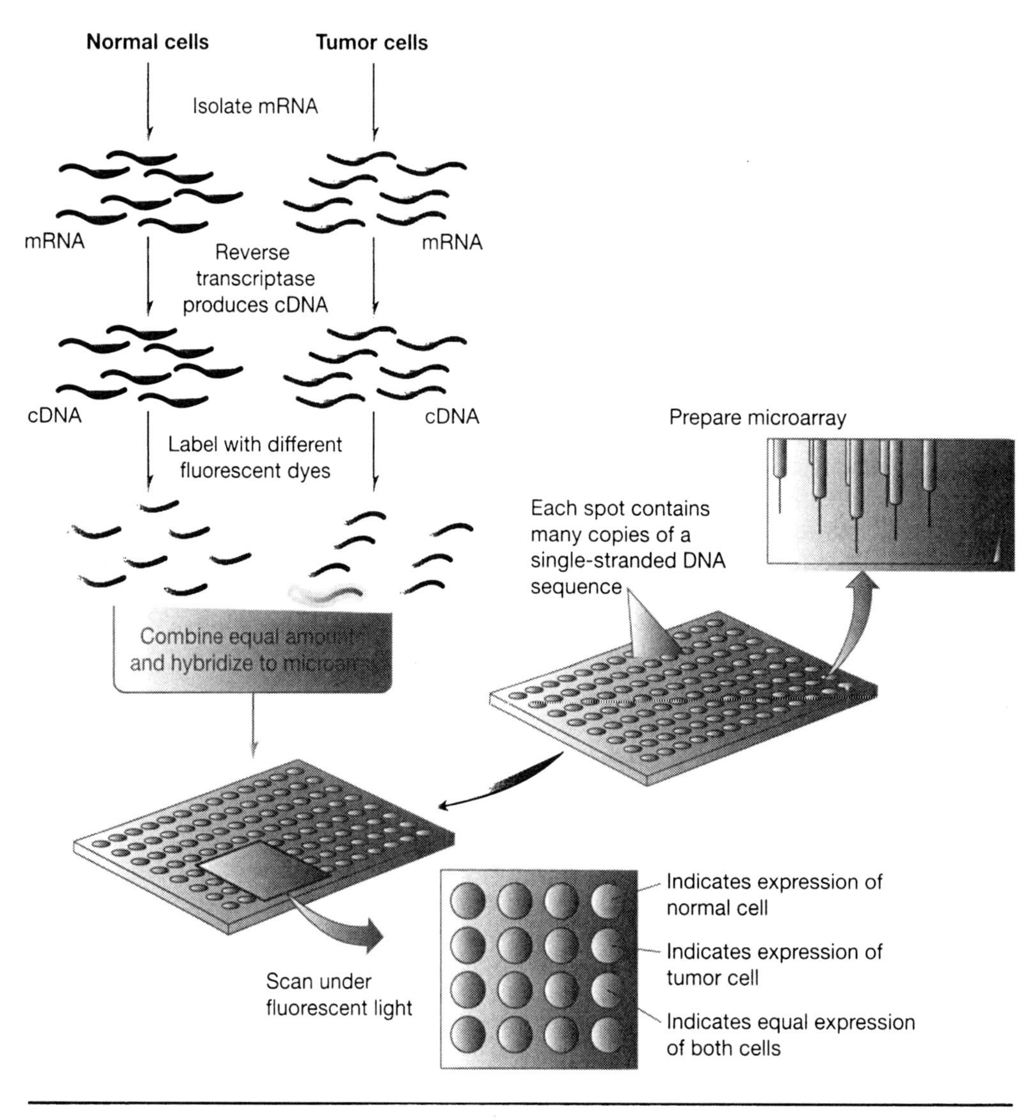

FIGURE 5-1 Microarray

conducted a gene analysis and found that women with different gene expression profiles had different responses to the drug. The tumors that responded to chemotherapy treatment had a higher expression of genes involved in the cell cycle, protein transport, and protein modification. This gene profile may be used to develop a test that could tell how a woman is going to respond to a particular drug.

In the near future, clinicians could have a test to tell which women will respond to specific chemotherapy (Chang et al., 2003).

Gastroenterology

Azothioprine and 6-mercaptopurine, used in treatment of children with ALL, is also used for treating inflammatory bowel disease. Azothioprine is metabolized by the enzyme thiopurine methyltransferase (TMPT). As noted earlier, about 1 in 300 individuals is unable to metabolize azotioprine due to decreased or absent TMPT activity. Some patients receiving azothioprine or 6-mercaptopurine medications experience leukopenia. Recent studies have shown that the leukopenia is related to decreased or absent TMPT. It has now become standard practice to measure TMPT activity before beginning treatment with azothioprine or 6-mercaptopurine for inflammatory bowel disease (Ansell et al., 2003; Dubinsky et al., 2000).

Cardiology

To date, a large number of genes and gene products that may contribute to the development of coronary artery disease (CAD) have been identified. These include genetic factors that encode proteins involved in lipid transport and metabolism, renin-angiotensin system, cytokines, adhesion molecules, and coagulation and fibrinolysis systems. Genetic factors are also involved in pathways that involve nitric oxide, homocystine metabolism, and inflammation. Polymorphisms in genes involved in lipid metabolism, such as apolipoptrotein E (Apo E) or rare variations in the low-density lipoprotein (LDL) receptor gene, have been shown to have a causal effect on both plasma LDL-cholesterol level and clinical phenotype of CAD (Winkelmann et al., 2000). The number of identified SNPs associated with CAD is also growing. Since CAD is a complex disease, a single gene or SNP may have only a limited contribution to the overall disease process. Therefore, the contribution of variations in multiple genes and their interactions with environmental factors—cholesterol-lowering drugs and diet, smoking and activity level—are under investigation (Wung, 2002).

There are several genetic polymorphisms that may play a role in variable responses to specific classes of cardiovascular agents (Humma & Terra, 2002). Polymorphisms in the genes encoding CETP and Apo E are associated with individual responses to the statins. A greater response to statins has been observed in individuals with CAD who have genotypes that are typically associated with a worse prognosis, such as CETP B1 allele and the ApoE4 allele (Winkelmann et al., 2000; Wung, 2002).

Gerontology

Aging is known to result in lower capacity for drug metabolism and in lower capacity to induce drug metabolizing enzymes. Polymorphisms involved in disease states related to aging also affect drug response. This is the case with apolipoprotein E mutations and tacrine efficiency in patients with Alzheimer's disease. Individuals with Alzheimer's disease who have the epsilon 4 subtype of the gene coding for apolipoprotein E (APOE 4) have been found to be less likely to benefit from the drug tacrine (Cognex). APOE epsilon 4 affects cholinergic function in the brain (Tsai & Hoyme, 2002).

As another example, increasing age is a strong risk factor for venous thrombosis. Health care providers may not always consider or attempt to identify genetic risk factors in elderly patients with venous thrombosis. Several studies suggest that individuals carrying the Factor V Leiden mutation—a common genetic risk factor for venous thrombosis—should be considered when the first episode of venous thrombosis occurs after age 50. This suggests that gene mutation testing for the Factor V Leiden mutation should not be limited to younger patients (Grody et al., 2001).

Infectious Diseases

Tuberculosis treatment The ability to metabolize and eliminate certain drugs depends on actylation in the liver by the enzyme N-acetyltransferase and on having two functional genes: NAT1 and NAT2. Drug efficacy and toxicity are linked to the functioning of this enzyme, which has genetically determined polymorphisms. Ethnic variation has been observed with 50 to 70% of the North American and European populations and up to 90% of some Mediterranean populations being slow acetylators. Individuals can be categorized as slow or poor, rapid, and ultrarapid acetylators (Lashley, 1998).

The importance of acetylator status in drug metabolism was first recognized during therapy for tuberculosis with isoniazid (INH). Slow acetylators are at significantly greater risk for developing INH-induced peripheral neuropathy, as they maintain higher serum levels of INH. Rapid acetylators, such as Japanese and Chinese persons, on the other hand, are at increased risk for developing INH-induced hepatitis (Lashley, 1998).

HIV treatment Discovery of the genetic basis of idiosyncratic drug reactions is increasing using new genome technologies (Muller, 2003). Abacavir hypersensitivity is one such example. Hypersensitivity to abacavir, a drug used in treatment of HIV, is a potentially life-threatening idiosyncratic ADR affecting approximately 4% of individuals treated. A research project was undertaken to identify a genetic pattern defined by SNPs that are associated with the ADR in patients. Results of the research revealed three HLA candidate genes that were highly associated with abacavir hypersensitivity. Follow-up clinical studies confirmed the significance of these loci. Withholding abacavir in individuals with HLA-B*5701, HLA-DR7, and HLA-DQ3 reduced the prevalence of hypersensitivity from 9 to 2.5% without the necessity of withholding abacavir from a patient who would benefit from the drug (Mallal et al., 2002; Muller, 2003).

Gene-Environment Interactions and Clinical Interventions

All human diseases result from interactions of genetic susceptibility factors and modifiable environmental factors including infectious, chemical, physical, nutritional, and behavioral factors. Common diseases such as cancer, diabetes, and heart disease are the result of complex interactions of such genetic and environmental influences.

Variations in a person's makeup are associated with nearly all diseases. Genetic variations do not cause disease, rather they influence a person's susceptibility to environmental factors. Contrary to popular beliefs, a person does not inherit a disease state. Instead, people inherit a set of genetic susceptibility factors to certain effects of environmental influences and therefore may inherit a higher risk of developing certain diseases. This concept helps explain why individuals are affected differently by the same environmental influences. As an example, some individuals will avoid smoking altogether, eat a low-fat diet, and still develop CAD at an early age; while other individuals who smoke and have a poor diet live to old age without developing CAD. Genetic variations are responsible, in part, for the differing responses to similar environmental factors *(http://www.cdc.gov/genomics/info/factshts/geneenviro.htm)*.

Cigarette Smoking and Coronary Artery Disease (CAD)

Dietary fat intake, smoking, alcohol consumption, and inactivity influence risk for CAD. For this reason, the interaction of genetic and environmental factors is considered in understanding the incidence of CAD. With smoking, for example, products of tobacco combustion directly damage vascular endothelium, leading to thrombosis and atherosclerosis. Smoking increases the risk for CAD in all individuals; however, some are especially vulnerable due to their genetic makeup. The interaction of the Apo E genotype, smoking, and the D9N polymorphism of the lipoprotein lipase gene on CAD risk have been studied. Individuals who carry the ApoE 4 genotype and those who carry the D9N polymorphism are especially vulnerable to the effects of cigarette smoking and are at a significantly increased risk for developing CAD. Preventive care and smoking cessation interventions should therefore be encouraged for these persons (Wung, 2002).

Cigarette Smoking and Bladder Cancer

The enzyme N-acetyltransferase (NAT2) is involved in detoxification of aromatic amines, known carcinogens found in tobacco and certain dyes. Individuals who carry a mutation in NAT2 have weakened enzymatic activity and are referred to as slow acetylators. The carcinogenic potential of aromatic amines depends on the balance of metabolic activation and elimination. Slow acetylators are inefficient in eliminating the active form of a chemical carcinogen. Studies of NAT2 slow acetylators and bladder cancer consistently show that those smokers who have increased exposure to aromatic amines have an increased risk of developing bladder cancer if they have the slow acetlyation genotype (Marcus, Vineis, & Rothman, 2000).

Thrombophilia and Oral Contraceptives

Venous thrombosis is a common, multifactorial disorder involving one or a combination of genetic risk factors and acquired or environmental conditions, such as pregnancy, oral contraceptive use, estrogen therapy, malignancy, stroke, trauma, surgery, or immobility. The risk increases with the number of genetic and/or environmental factors present. Known genetic factors are present in approximately 25% of isolated cases and up to 63% of familial cases (Grody et al., 2001).

The Factor V Lieden mutation, causing activated protein C resistance, is the most common genetic risk factor for venous thrombosis. It is present in 5% of Americans and is implicated in 60% of venous thrombosis cases in pregnant women. Factor V Leiden has also been associated with an increased risk for placental infarction, prematurity, and recurrent pregnancy loss (Grody et al., 2001).

As an example of gene-environment interaction, women who carry the blood clotting variant Factor V Leiden and who are taking oral contraceptives have been shown to have a dramatically increased risk of developing venous thrombosis. A simple DNA blood test can be used to identify those women with the Factor V Leiden variant and choices of more appropriate contraception can be offered (Grody et al., 2001).

Nursing Implications

The use of genetic testing—genotyping—to select the correct medication and correct dose for individual patients, while minimizing adverse drug reactions and therapeutic failures, is expanding. Nurses will be increasingly involved in caring for patients for whom genotyping is appropriate before the selection of medication and/or dosage is made. Key considerations for nurses and all health care providers are: (a) when to order a pharmacogenetic test, and (b) how to use the results in clinical decision making. When to order a genetic test depends on the availability of pharmacogenetic tests for a patient's specific condition. Nurses should be knowledgeable about whether or not genotyping tests are available that could affect the efficacy and safety of drugs that might be prescribed. In some cases, tests for general drug responsiveness, such as genotyping to determine whether a patient is a poor or rapid metabolizer, should be ordered. For other drugs, such as 6-mercaptopurine, the pharmacogenetic test should be specific to the drug and indicated on the drug's label (Robertson et al., 2002).

How to use pharmacogenetic test information depends upon the patient's condition, the therapeutic alternatives, and the positive predictive value of the test. Nurses should understand the clinical significance of the test results when they are administering a prescribed medication. In some instances, the genetic test will have significant relevance and determine the prescribing decision. This is the case with tricyclic antidepressants not being given to a particular patient or given only at a specific dosage. In other cases, even highly relevant tests will need to be weighed against other clinical facts to determine whether a particular medication is to be used. Such is the case with tacrine in the treatment of Alzheimer's patients with the Apo E4 allele. In other situations, although the genetic test strongly suggests against prescription, the patient may not have a better or safer therapeutic alternative. In this situation, having an awareness of potential risk will enable adjustment of dosage and close and appropriate monitoring (Robertson et al., 2002).

Nursing Roles

Nursing Assessments

Nurses have a number of important roles in the emerging era of pharmacogenetic medicine. Assessment of patients to determine whether a pharmacogenetic risk is present is essential. Nursing assessments include evaluation of past medical history to determine whether the individual has experienced any adverse drug responses, including response to pain medication and anesthesia. Assessment and recording of current medications, including naturopathic medications, are also important. A family history assessment is conducted to inquire about any other relatives who may have experienced an adverse drug or therapeutic response, for example, individuals who are at risk for malignant hyperthermia when certain anesthesia is administered. At present, inquiring about a person's ethnicity may help in assessing what type of genetic testing is appropriate or in determining a potential adverse drug response that may occur.

Health Teaching

Nurses will be involved in assisting with genotyping patients and in explaining the benefits and limitations of tailored therapeutics based on genotype. Health teaching is therefore a central nursing role. Patients will need to know that the safety or efficacy of a prescribed medication may depend on their genotype; thus, testing for those variations before prescribing the medication may be needed for proper clinical decision making. Nurses will also need to explain to patients that certain medications may not be prescribed if the patient has a particular genotype. Before pharmacogenetic technology, newly developed drugs were available and marketed to all patients with the same clinical diagnosis (Robertson et al., 2002). Patients may know of others with the same diagnosis receiving a particular medication and be concerned that they are not receiving the same treatment. Without understanding the genetic component and link to the treatment decision, they may feel that they are being discriminated against when a particular medication is not prescribed. Nurses need to be prepared to explain the rationale behind such clinical decisions and how medication administration based on genotype differs from therapies based on clinical symptoms or diagnosis in ways that are understandable to the patient (Prows & Prows, 2004).

Discrimination Concerns

Patients may also express fears about the use of genetic tests for treatment decisions and the potential for discrimination. Nurses can explain that all attempts will be made to keep the individual's genetic information private and confidential. It can also be explained that genetic knowledge is increasingly used to help guide clinical management and offer examples of such knowledge as family history of heart disease or blood type to facilitate acceptance of the routine use of genotype information in medical care (Robertson et al., 2002).

At present, national dialogue and discourse continues regarding the development of a set of criteria by which the use of genetic and all personal medical information should occur. Issues for consideration include discrimination against those who are less likely to respond to treatment or who are at higher risk for developing disease. There are, at present, no specific rules or regulations that guarantee the beneficial and legitimate use of genetic information in the patient's interest while prohibiting their use in ways that may be harmful to the individual—personally, financially, or otherwise. Protection of personal genetic information and patient–subject protection are central components of the mandates included in the World Health Organization's "Proposed International Guidelines on Ethical Issues in Medical Genetics and Genetic Services," which mandate autonomy, beneficence, nonmaleficence, and justice (Lindpaintner, 2003).

Informed Consent

Informed consent is another central nursing role. Nurses will need to consider the nature of the test, the information that will be gained, and the level of informed consent required for a particular pharmacogenetic test and treatment. Some consider pharmacogenetics to be similar to other diagnostic or clinical laboratory tests that are now routinely administered with minimal informed consent. The risk for psychosocial harm may to be lower than that associated with genetic tests that confirm the diagnosis of a genetic condition such as Huntington Disease; predict disease, such as hereditary breast–ovarian cancer testing; or are for carrier testing, such as cystic fibrosis (Robertson et al., 2002). The pharmacogenetics working group has developed a list of specific elements of informed consent for pharmacogenetic research that will help guide the informed consent process when pharmacogenetic interventions are applied clinically. **Table 5-2** provides a summary of these recommendations (Anderson et al., 2002).

TABLE 5-2 Elements of Informed Consent for Pharmacogenetic Research

- *Purpose and intent of the pharmacogenetic study.* Readily understandable endpoints should be conveyed such as "to identify variations of genes, which may cause or modify disease" (p. 285).
- *Trial procedures.* Describe the procedures involved in collecting and handling samples, the options available to the patient once a sample has been acquired, and the genetic information derived.
- *Voluntary participation.* A separate informed consent document for pharmacogenetic sampling must provide sufficient information for the individual to make an informed decision to donate genetic materials based on the risks and benefits of the pharmacogenetic research.
- *Pharmacogenetic sample collection, storage, and distribution.* A description of collection procedures including an indication of which procedures are part of routine care and which are specifically applicable to the pharmacogenetic research objectives; a discussion of where and how long the samples will be stored, and an explanation that a detailed family history of disease or other genetic traits may be required.
- *Withdrawal options and timelines.* The informed consent should indicate when sample destruction will not be possible, for example, because of pooling of individual samples in the laboratory, and the circumstances under which individual genetic results cannot be retrieved.
- *Contact information.* Clearly indicate the information necessary to enable the research subject to contact the researcher if questions or concerns arise.
- *Options for communication and sharing of genetic results with research subjects.* Describe the intended types of pharmacogenetic results to be gained from a study, and inform the research subject about the realistic expectations and health implications, if any, of the study.
- *Sharing of unintended genetic results.* Research subjects should know that sensitive genetic information may inadvertently be revealed as a result of pharmacogenetic research. It may be determined that a research subject is genetically at risk for a serious condition or is unrelated to someone previously assumed to be a family member, even though such results were not intended. Circumstances requiring disclosure of unintended genetic results, including respect for the person's right not to know, must be carefully considered by weighing the health benefits to the research subject and/or family members against the risks related to the genetic information.

TABLE 5-2 continued

- *Risks and benefits.* Convey the relative risks (i.e., potential risk for genetic discrimination based on test results) and benefits (opportunities to define the molecular-genetic basis of variable drug responses) as well as scientific design of the study.
- *Privacy and confidentiality.* Describe procedures to protect the research subject's privacy including anonymizing samples.
- *Commercialization.* When clinical research is sponsored by pharmaceutical or private companies, inform the research subject that the contribution of his or her pharmacogenetic sample may result in commercial gains or intellectual property for the sponsoring pharmaceutical company.
- *Compensation.* Discussion of issues of compensation associated with the reimbursement of expenses incurred by research subjects who participate in a study as distinguished from financial compensation that would be an inducement for individuals to participate in pharmacogenetic research.

(Anderson et al., 2002).

As pharmacogenetics moves forward into the clinical realm, the informed consent process may be simplified. For example, the nurse may explain that the provider wishes to "test a sample of the patient's DNA to see if a drug will be safe or whether it will work" (Robertson et al., 2002, p. 163) and offer assurance that this will be the only use of the sample. The Mayo Clinics are using this approach and have developed a simple brochure regarding pharmacogenetic testing for drug metabolism that is given to patients. In the future, nurses may be involved in designing additional culturally and educationally relevant materials to help educate patients about the purposes and benefits of pharmacogenetics.

Nurses should be aware that pharmacogenetic tests may carry secondary information about a patient's genetic health status that should not be communicated to others without the patient's written consent. As an example, to assess cardiovascular risk, a patient has genetic testing to determine which of the four ApoE alleles they carry. Researchers have learned that there is an association of a particular ApoE allele with early-onset sporadic Alzheimer's disease. A patient's desire to know his or her genotype for a treatable disorder (cardiovascular disease) may be very different from their desire to know their genotype for an untreatable disease (Alzheimer's disease). It is possible that a genotype result may inadvertently reveal information about future health risks that a patient may not want to know (Prows & Prows, 2004).

As another example, a patient may be tested and found to have a mutation in a gene that indicates an enzyme essential for metabolizing one drug is absent. This result indicates that the patient is not likely to respond to other similar drugs, as is the situation with cytochrome P450 gene mutations. Such information could be of interest to insurers and employers who want to minimize health costs. In this case, a more extensive informed consent discussion would be indicated. The nurse should be aware that the nature of the secondary information gained by the pharmacogenetic test result and the social, economic, and cultural context within which the test occurs has the potential to carry risk of financial harm (Robertson et al., 2002).

The privacy regulations now in effect under the Health Insurance Portability and Accountability Act of 1996 are an important step forward in safeguarding patients' rights to privacy and confidentiality of genetic information. State laws protecting the privacy of DNA samples and test results may also be needed. Nurses should be aware of the specific privacy laws in their own state about the use of genetic information *(http://www.genome.gov; http://www.ncsl.org/programs/health/genetics.htm)*.

Monitoring Patients for Therapeutic Benefit and Effects

Monitoring patients for therapeutic benefit and adverse drug responses based on pharmacogenetic testing will become an integral nursing activity. Patient education before drug administration and throughout monitoring includes the benefits, limitations, and risks of genotype-based drug therapy. Nurses will be involved in preventing ADRs and therapeutic failures by teaching patients to avoid certain drugs based on their genotype and by adjusting initial doses based on genotype. Monitoring for specific side effects, compliance issues, and psychosocial responses to the condition and interventions are ongoing nursing activities.

Gene-Environment Monitoring

Genetic information gained from pharmacogenetics can be used to target interventions. All individuals carry genetic variants that increase susceptibility to some diseases and that cause variation in response to medications and other therapeutic interventions. Identifying gene-environment interactions, such as smoking and cardiovascular disease, provides opportunities to effectively target intervention strategies.

Nurses will become increasingly involved in monitoring environmental factors that may interact with and influence a patient's response to treatments and prevention efforts. For example, some recent research has focused on genetic testing for lung cancer susceptibility. The CYP2D6 gene is associated with metabolism of carcinogens in tobacco. Individuals with the genotype associated with extensive metabolizing may be more susceptible to lung cancer. In a randomized study, Lerman et al. (1997) studied individuals who received both smoking cessation counseling and CYP2D6 genotyping. Smokers who received the genetic testing feedback reported increased levels of perceived risk and perceived benefits with smoking cessation when compared to two other groups who had received the counseling and carbon monoxide feedback only. At the 12-month follow-up, however, there were no significant differences between the groups (Lerman et al., 1997). Future nursing research will need to focus on the impact of genetic testing on decisions about lifestyle changes such as smoking.

Many of the genetic risk factors for diseases and the ways in which individuals metabolize and respond to medications and therapies have not been identified. The complex interaction of genes with other genes and genes with environmental factors is not yet fully understood *(http://www.cdc.gov/genomics/info/facshts/geneenviro.htm)*. Nurses can facilitate clinical and epidemiologic studies to further describe and understand these factors and their interactions. Such studies will lead to the identification of genetic information that can be used to plan appropriate intervention strategies for high-risk individuals.

Ethical and Societal Issues

Access to Pharmacogenetic Testing

The application of pharmacogenetics to the delivery of medications has the potential to improve health care outcomes and impact medical care at multiple levels. For example, in the future medications would be prescribed to only those patients where there is a high probability of success without significant adverse effects. Selection of predicted responders offers a more efficient and economical solution to the growing problem that is leading insurers and health care providers to deny effective medicines to the few because a proportion of patients do not respond to treatment (Roses, 2000). However, a major determinant of improved health using new genetic technologies, including pharmacogenetics, is adequacy of access to testing and to the medical interventions that may consequently be warranted. Because of our country's current multipayer system of health care, individuals will have widely different levels of access to pharmacogenetics and other genetic technologies (Clayton, 2003). Nurses and other health care providers cannot assume that all people will have ready access to new pharmacogenetic and pharmacogenomic advances. Nurses will need to become involved in dialogue with other health care providers and policy makers to develop policy solutions that make health care options more accessible and affordable.

Another consideration in terms of access is the rapid entry of pharmacogenetic and other genetic tests into the marketplace, some with few or no long-term data to support their use. Efforts to determine when genetic tests are reliable enough for routine clinical use are imperative for the general health and safety of the public. The Secretary's Advisory Committee on Genetics, Health, and Society was created to provide such guidance. Nursing is represented on this Committee, whose goal is to ensure that public safety and appropriate public education about genetic testing are adequately addressed *(http://www.sacghs.org)*.

Pharmacogenetics and Managed Health Care

Managed care organizations will face reimbursement decisions regarding pharmacogenetic tests on a growing scale. It is anticipated that as well-tolerated and effective medications that treat, cure, or prevent common diseases become a greater proportion of the medical care bill, the costs of chronic debilitating illnesses will be significantly reduced (Roses, 2000). How individuals and society will view pharmacogenetic information will be an important challenge. For example, individuals undergoing pharmacogenetic testing who are labeled as "non-responders" or "responders," terms that describe a SNP profile result, may be thought of as having a disease state. Furthermore, the label meaning may vary depending on whether or not a patient ever needs a specific medication for which he or she is a "non-responder" and whether or not alternative medications exist and can be used effectively (Prows & Prows, 2004). Criteria need to be developed that can assist decision makers in evaluating the cost-effectiveness of pharmacogenetic testing and inform the decision-making process in the context of these ethical, legal, and social issues related to genetic testing (Higashi & Veenstra, 2003). Nurses can become involved in creating health policy solutions that will help to support innovations in managed care and coverage of pharmacogenetic tests.

Dubious Pharmacogenetic Testing

Increasingly, patients and families are turning to the Internet to obtain genetic information and to order medical tests and treatments without a prescription. Direct-to-consumer marketing for genetic testing is taking place. Human genome discoveries offer the opportunity for personalized genetic medicine, with genetic testing becoming a standard component of clinical care. These discoveries also bring the risk of misuse of genetic information due to the lack of accompanying counseling services, adequate provider education, and questions about test oversight (Barret & Hall, 2003). As an example, the Internet can now be used to purchase directly medical products, such as pharmaceuticals, dietary supplements, and diagnostic tests, without the supervision of the patient's primary care physician. Test results received may influence patients to change their behaviors adversely, such as inappropriate changes in medication or diet. Patients may use the Internet instead of pursuing appropriate interventions to help guide in decision making (Gollust, Wilfond, & Hull, 2003). **Box 5-2** presents a summary of some of the questions and concerns nurses should consider about dubious genetic testing.

Nurses are often the first health professionals to whom patients and families may turn with information that they have gathered from the Internet or media

Box 5-2 Genetic Testing: When and What to Question

Increasingly, private companies are offering genetic testing combined with guidance on drug usage, lifestyle changes, diet, and supplement strategies that they claim will improve health outcomes. They claim that expensive genetic tests will "take the guesswork" out of prescribing supplements and medications and making clinical recommendations. Nurses need to be aware of such companies and claims and help the public understand that testing a person's genetic makeup may not really be the answer for their health needs. Following are some basic concepts nurses can share with patients.

- *Many genetic test results are still not well understood.* What genes do and how they influence health is not well understood. Most common diseases are complex and caused by multiple genes interacting with environmental factors. The genetic and environmental factors have not been completely sorted out. Much more research is still needed for a better understanding of how genetic testing can be used to prevent and treat complex diseases such as diabetes, heart disease, and cancer.
- *Genes do not determine who you are or when you will develop a disease or condition.* People vary in their susceptibility to different diseases, illnesses, and pollutants. Some of this variation lies in their genes, but much of it does not. Age, gender, diet, and overall health are important factors as well. People who have a gene linked with a common disease may not get the disease, while people who do not have the gene will.
- *Genetic tests do not take the guesswork out of treatment.* Companies selling genetic tests imply that provider experience and knowledge can be replaced with tests of just a few of one's genes. This is the antithesis to a holistic approach involved in providing therapeutic interventions. Testing a person's genetic makeup that they were born with may be less, not more, precise than testing changes actually happening in a person's body, and it does not replace provider care and experience.
- *Genetic tests do not help you change your life or environment.* The limited research that has been conducted to date suggests that genetic tests do not help people change unhealthy behaviors, such as smoking and obesity. The current major increase in obesity in the United States is not due to an increase in genes for obesity. Tailoring treatments to a person's genetic makeup may not address the underlying problems that cause most common diseases: smoking, alcohol, fast food, and lack of physical activity. Furthermore, genetic tests do not help people escape from poverty or pollutants.

(http://www.genewatch.org)

advertising. Nurses need to be prepared to respond to patients' questions, concerns, and activities and to have knowledge of reliable resources to provide information to their patients (**Box 5-3**).

Public and Professional Education: What Are the Differences in Genetic Testing?

The term "genetic testing" is now used in common vocabulary and with little specificity, often referring to very different applications of the testing. The class of genetic tests that has received most of the attention is those for testing single-gene genetic disorders. Pharmacogenetics is another class of "genetic test" used to identify polymorphic detoxifying enzymes, drug receptor variants, and other inherited polymorphic traits that are not diagnostic of a disease (Roses, 2000). The use of SNP genetic profiles to choose the right drug at the right dose for the right person will predict patients' responses to medications but will not specifically test a person for the

Box 5-3 Case Example: *"Finding What I Need"*

Mr. and Mrs. N. have a 4-year-old son. They are concerned about their son who is "very active and doesn't seem to need much sleep." Mr. N. had a cousin who had attention-deficit hyperactivity disorder. He is worried that his son has the same thing. One night he goes on the Internet to search for more information about how to diagnose hyperactivity. He finds a site that offers DNA testing for a variety of health conditions including "DNA testing that can help you to manage your child's behavior." The cost of the testing is $400 and the sample required is merely a brushing of the inside of the cheek "just like a tooth brush" to obtain cells for analysis. The Internet site offers information about specific dietary changes that can be put in place if a child test positive, including megavitamin therapy. Mr. and Mrs. N. are very excited about this information and bring it to the nurse practitioner at the pediatric clinic, requesting a referral for the testing.

Nursing Interventions: In this situation, the nurse practitioner would advocate for the safety of the child and engage in an educational dialogue with the parents. He or she could explore with the parents the nature and legitimacy of the Web site. He or she could inquire about whether the genetic test has been approved by the national organization CLIA or whether it is experimental and would yield no or useless information (e.g., the clinical specificity and utility of the genetic test). The nurse practitioner could also explain that megavitamin therapy is not yet a proven intervention, and could in fact cause harm to their child. A recommendation to a national specialist and to the Genetic Alliance *(http://www.geneticalliance.org)* for more reliable diagnostic and treatment information could be made.

presence or absence of a disease gene-specific mutation. Furthermore, this type of genetic test may not provide any other significant disease-specific predictive information about the individual or family members. Medicine selection and response profiles will simply predict a patient's response to medications based on a pattern of inherited factors detected as small genetic changes. Genetic testing methods in this application would be used to differentiate those patients who experience good efficacy and lower significant adverse events in response to a medication from other patients who fail to respond or who develop significant ADRs (Roses, 2000).

The genetics of response to or selection of medication is the only data generated using SNP profiling and could be designed and safeguarded to have no meaning or relationship to any known disease-specific gene information. The application of genetic testing for pharmacogenetics, therefore, would not warrant traditional genetic counseling and education about disease inheritance because the SNP profile would not yield predictive information about disease. Thus, the ethical, social, and legal considerations of drug-target or metabolic gene profiles and SNP profiles for the purpose of medication prescription need to be given separate consideration from the concerns raised by prenatal, predictive, and carrier genetic testing (Roses, 2000). Nurses can participate in educating the public about the differences between the different types of genetic testing and the role of pharmacogenetic testing for the purpose of effective and safe medication selection. Resources for nurses and their patients are listed in **Table 5-3.**

TABLE 5-3 Pharmacogenomics Resources on the Web

Genomics and Its Impact on Medicine and Society *http://www.ornl.gov/hgmis/pubilcat/primer2001/index.html*
Medicine and the New Genetics: Gene Testing, Pharmacogenomics and Gene Therapy *http://www.ornl.gov/hgmis/publicat/primer2001/6html*
Human Genome Project Education Resources *http://www.ornl.gov/hgmis/education/education.html*
TIGR Genome News Network *http://gnn.tigr.org/main.shtml*
NHGRI Educational Resources *http://www.genome.gov/page.cfm?pageID* = 10000002
Single Nucleotide Polymorphism Database *http://www.ncbi.nlm.nih.gov/SNP/*

Pharmacogenetics, Race, and Ethnicity

Social categories of race and ethnicity have long been used in predicting therapeutic response, and are increasingly used in the clinical application of pharmacogenetics and pharmacogenomics. The FDA has been concerned about variable drug effects among population subgroups and has issued guidelines that encourage the collection and analysis of clinical trial data from subgroups. A recent draft guidance issued by the FDA recommends a standardized approach for the collection of race and ethnicity in data in clinical trials to "enhance the early identification of differences in physiological response among ethnic subgroups" (FDA, January 2003).

Pharmacogenetic tests could make this use of ethnicity and race largely irrelevant by allowing clinicians to base diagnosis and treatment decisions on the unique genetic profile of individual patients. However, current uses of ethnic and racial categories in pharmacogenetic research, as an example, show how drug development and marketing may perpetuate the use of social categories of race and ethnicity (Haga & Venter, 2003).

Nurses should be aware of the considerable controversy that exists about the use of race and ethnicity in pharmacogenetic and other medical research (Cooper, Kaufman, & Ward, 2003; Foster, Sharp, & Mulvihill, 2001; Haga & Venter, 2003). Specifically, those uses may unintentionally diminish the precision of genetic technologies and create new threats to socially identifiable populations. The population in the United States is becoming increasingly diverse and complex, with more than 6 million Americans reporting two or more races (U.S. Census, 2000). The meaning of such ethnic categories as "Black" has changed significantly as well. This makes it very difficult to classify individuals accurately and consistently for the purposes of identifying the right drug and evaluating differences in drug effects (Haga & Venter, 2003). For these reasons, some researchers recommend that greater caution be taken in using social categories as indicators for specific genetic tests or therapies. Such steps may allow the full potential of pharmacogenetic testing to be realized while minimizing social disparities in health care (Foster, Sharp, & Mulvihill, 2001; Haga & Venter, 2003).

Summary

Pharmacogenetics and pharmacogenomics are two new fields related to basic drug discovery research, the genetic basis of pharmacokinetics and pharmacodynamics, new drug development, patient genetic testing for the purpose of selecting the appropriate drug and dosage, and clinical patient management. The ultimate goal of pharmacogenetics is to be able to predict a patient's genetic response to a specific drug as a means of delivering the best possible medical treatment. Pharmacogenetics and pharmacogenomics hold the promise of eliminating adverse drug responses and increasing tailored treatments that improve responsiveness to medications for all populations. Nurses caring for patients receiving drug therapies need to be aware of the genetic differences that can manifest as variations in the response of different individuals to the same medication or drug treatment. In most acute and chronic health care settings, it is nurses who have the greatest contact with patients. Nurses administer medications and have the obligation to detect changes in a patient's condition that may arise as a consequence of any prescribed medications (Nicol, 2003).

Nurses will increasingly care for patients from diverse ethnic backgrounds and cultures. Culturally safe care is a type of nursing care and an approach that reflects the priorities and health needs of those receiving care. Pharmacogenetics is an additional component in the delivery of nursing care that is culturally safe. When ethnic differences are known as influential to the responses to a particular medication, the nurse needs to be observant and vigilant (Nicol, 2003). Nurses also need to realize that, as the technologies of pharmacogenetics develop, new and more individualized tools will become available that are more precise than ethnicity and will allow for culturally safe care for all populations.

Nurses are also concerned with equal access to and the affordability of health care, including reimbursement for pharmacogenetic technologies for targeted treatment. Nurses are represented at the policy level on the Secretary's Advisory Committee on Genetic Health Care to address these issues. All nurses share in the responsibility for helping to shape new health care delivery options that include pharmacogenetics and that are made available to all populations.

Chapter Activities

1. Mary's story illustrates the initiation of a new way of prescribing medication based on human genome discoveries and their application—pharmacogenetics. Some commercial companies are advertising genetic testing on the Internet to determine drug prescribing, diet interventions, and disease predisposition. Go to the Web site *http://www.quackwatch.org/01QuackeryRelatedTopics/Tests/genomics.html* and read the article on *Dubious Genetic Testing.* From your reading, list three Web sites that are reliable information sources regarding genetic testing and pharmacogenetics.
2. Richard is a 25-year-old man who has a history of depression. He has a family history of depression as well as panic attacks in his 30-year-old brother. His mother has a history of depression as do her twin brothers. The twin brothers

each have two children who have histories of depression. Richard has been on several medications to treat his depression. He tells you that he has been looking on the Internet for a better treatment and learned that there are some people with depression who do not respond to the usual dose of antidepressants and that there are now some tests for this. He wonders whether he might be one of those people who does not respond to the usual dose of antidepressant medication and wants to have testing to find out. Construct Richard's family tree. What inheritance pattern for depression seems to be present in Richard's family? How would you respond to Richard's request for testing?

References

Anderson, D. C., Gomez-Mancilla, B., Spear, B. B., Barns, D. M., et al. (2002). Elements of informed consent for pharmacogenetic research; perspective of the pharmacogenetics working group. *Pharmacogenomics, 2*(5), 284–292.

Ansell, S. M., Ackerman, M. J., Black, J. L., Roberts, L. R., & Tefferi, A. (2003). Primer on medical genomics. Part VI: Genomics and molecular genetics in clinical practice. *Mayo Clinic Proceedings 78*(3), 307–317.

Barrett, S. & Hall, H. (2003). Dubious genetic testing. From *http://www.quackwatch.org/01QuakeryRelatedTopics/Tests/genomics.html*

Capriotti, T. (2002). Gleevac: Zeroing in on cancer. *Medsurg Nursing, 11*(6), 301–304.

Chang, J. C., Wooten, E. C., Tsimelzon, A., Hilsenback, S. G., Gutierrez, M. C., et al. (2003). Gene expression profiling for the prediction of therapeutic response to docetaxel in patients with breast cancer. *The Lancet, 362,* 362–369.

Clayton, E. W. (2003). Ethical, legal and social implications of genomic medicine. *New England Journal of Medicine, 349*(6), 562–569.

Collins, F. (2003). The genome era and mental illness. Accessed Septemeber 15, 2004, from *http://www.genome.gov/Pages/News/Documents/GenomeMental.pdf.*

Cooper, R. S., Kaufman, J. S., & Ward, R. (2003). Race and genomics. *New England Journal of Medicine, 348*(12), 1166–1170.

Dahl, M. L. (2002). Cytochrome P450 phenotyping and genotyping in patients receiving antipsychotics. *Clinical Pharmacokinetics, 41*(7), 453–470.

Drucker, B. J., Talpaz, M., Resta, D., Peng, B., Buchdunger, E., Ford, J., Lydon, N., Kantarjian, H., Capdeville, R., Ohno-Jones, S., & Sawyers, C. (2001). Efficacy and safety of a specific inhibitor of the BCR-ABL tyrosine kinase in chronic myeloid leukemia. *New England Journal of Medicine, 344*(14), 1031–1037.

Dubinsky, M. C., Lamothe, S., Yang, H. Y., Targan, S. R., Sinnett, D., Theoret, Y., & Seidman, E. G. (2000). Pharmacogenomics and metabolite measurement for 6-mercaptopurine therapy in inflammatory bowel disease. *Gastroenterology,118*(4), 705–713.

FDA (30 January 2003). Guidance for industry: Collection of race and ethnicity data in clinical trials, Draft Guidance, January 2003 cited at 68 FR 4788.

Foster, M. W., Sharp, R. R., & Mulvihill, J. J. (2001). Pharmacogenetics, race and ethnicity: Social identities and individualized medical care. *Therapeutic Drug Monitoring, 23,* 232–238.

Goldman, J. M., & Melo, J. V. (2001). Targeting the BCR-ABL tyrosine kinase in chronic myeloid leukemia. *New England Journal of Medicine, 344*(14), 1084–1086.

Gollust, S. E., Wilfond, B. S., & Hull, S. C. (2003). Direct-to-consumer sales of genetic services in the internet. *Genetics in Medicine, 5*(4), 332–337.

Grody, W. W., Griffin, J. H., Taylor, A. K., Korf, B. R., & Heit, J. A. (ACMG Factor V Leiden Working Group) (2001). American College of Medical Genetics Consensus Statement on Factor V Leiden Mutation Testing. *Genetics in Medicine, 3*(2), 139–148.

Haga, S. B., & Venter, C. J. (2003). Policy Forum: FDA races in wrong direction. *Science, 301*(5632), 466.

Higashi, M. K., & Veenstra, D. L. (2003). Managed care in the genomics era: Assessing the cost effectiveness of genetic tests. *American Journal of Managed Care, 9,* 493–500.

Humma, L. M., & Terra, S. G. (2002). Pharmacogenetics and cardiovascular disease: Impact on drug response and applications to disease management. *American Journal of Health Systems Pharmacy, 59*(13), 1241–1252.

Ingelman-Sundberg, M. (2001). Pharmacogenetics: An opportunity for a safer and more efficient pharmacotherapy. *Journal of Internal Medicine, 250,* 186–200.

Ingelman-Sundberg, M., Oscarson, M., & McLellan, R. A. (1999). Polymorphic human cytochrome P450 enzymes: An opportunity for individualized drug treatment. *Trends in Pharmacological Sciences, 20,* 342–349.

Krynetski, E. Y., & Evans, W. E. (1998). *American Journal of Human Genetics, 63,* 11.

Lashley, F. (1998). *Clinical genetics in nursing practice* (2nd ed.). New York, NY: Springer Publishing Company.

Lazarou, J., Pomeranz, B. H., & Corey, P. N. (1998). Incidence of adverse drug reactions in hospitalized patients: A meta-analysis of prospective studies. *Journal of American Medical Association, 279,* 1200–1205.

Lerman, C., Gold, K., Audrain, J, Lin, T. H., Boyd, N. R., Orleans, C. T., Wilfond, B., Louben, G., & Caporaso, N. (1997). Incorporating biomarkers of exposure and genetic susceptibility into smoking cessation treatment: Effects of smoking-related cognitions, emotions, and behavior change. *Health Psychology, 16*(1), 87–99.

Lindpaintner, K. (2003). Pharmacogenetisc and pharmacogenomics in drug discovery. *Clinical Chemistry and Laboratory Medicine, 41*(4), 398–410.

Mallal, S., Nolan, D., Witt, C., Masel, G., Martin, A. M., Moore, C., Sayer, D., Castley, A., Mamotte, C., Maxwell, D., James, I., & Christianson, F. T. (2002). Association between presence of HLA-B*5701, HLA-DR7 and HLS-DQ3 and hypersensitivity to HIV-1 reverse-transcriptase inhibitor abacavir. *Lancet, 359,* 727–732.

Mansour, H. A., & Nimgaonkar, A. M. (2002). Pharmacogenetics of bipolar disorder. *Curr Psychiatry Rep, 4*(2), 117–123.

Marcus, P. M., Vineis, P., & Rothman, N. (2000). NAT2 slow acetylation and bladder cancer risk: A meta-analysis of 22 case-control studies conducted in the general population. *Pharmacogenetics, 10,* 115–122.

McInnis, M. G., & DePaulo, J. R. (2002). Chapter 109. Major mood disorders. In Rimoin, D. L., Connor, J. M., Pyeritz, R. E., & Korf, B. R. *Principles and Practice of Medical Genetics* (4th ed.) London, England: Churchill Livingstone.

Merikangas, K. et al. (2002). Workgroup reports: NIMH strategic plan for mood disorders research. Future of genetics of mood disorders research. *Society of Biological Psychiatry, 52,* 457–477.

Meyer, U. A. (2000a). Drugs in special patient groups: clinical importance of genomics in drug effects. In: Carruthers G. S., Hoffman, B. B., Melmon, K. L., & Nierenberg, D. W. eds. New York: McGrawHill, 1179–1205.

Meyer, U. A. (2000b). Pharmacogenetics and adverse drug reactions. *The Lancet, 356,* 1667–1671.

Muller, M. (2003). Pharmacogenomics and drug response. In *International Journal of Clinical Pharmacology and Therapeutics, 41*(6), 231–240.

Nicol, M. J. (2003). The variation of response to pharmacotherapy: Pharmacogenetics—A new perspective to 'The right drug for the right person.' *MEDSURG Nursing, 12*(4), 242–249.

Poirier, J., Delisle, M. C., Quirion, R. et al. (1995). Apolipiprotein E4 allele as a predictor of cholinergic deficits and treatment outcome in Alzheimer disease. *Proceedings of the National Academy of Sciences of United States of America, 92*(12), 260–264.

Poolsup, N., Li Wan Po A., & Knight, T. L. (2000). Pharmacogenetics and psychopharmacotherapy. *Journal of Clinical Pharmacy and Therapeutics, 25,* 197–220.

Prows, C. A., & Prows, D. R. (2004). Medication selection by genotype. *American Journal of Nursing, 104*(5), 60–70.

Robertson, J. A., Brody, B., Buchanan, A., Kahn, J., & McPherson, E. (2002). Pharmacogenetic challenges for the health care system. *Health Affairs, 21*(4), 155–167.

Roses, A. (2004). Pharmacogenetics and drug development: The path to safer and more effective drugs. *Nature Reviews Genetics, 5,* 645–656.

Roses, A. D. (2000). Pharmacogenetics and the practice of medicine. *Nature, 405,* 857–865.

Rusnak, J. M., Kisabeth, R. M., Herbert, D. P., & McNeil, D. M. (2001). Pharmacogenomics: A clinician's primer on emerging technologies for improved patient care. *Mayo Clinic Proceedings, 76,* 299–309.

Tsai, Y. J., & Hoyme, H. E. (2002). Pharmacogenomics: The future of drug therapy. *Clinical Genetics, 62,* 257–264.

U.S. Census 2000. *(http://factfinder.census.gov/servlet/BasicFactsServlet)*

Vessel, E. S., & Page, J. G. (1968). Genetic control of drug levels in man: Phenylbutazone. *Science, 159,* 1479–1480.

Vogel, M. J. (1979). The specter of progressive medicine. *Reviews in American History, 7*(2), 229–235.

Weinshilboum, R. (2003). Inheritance and drug response. *New England Journal of Medicine, 348*(6), 529–537.

Winkelmann, R. R., Hager, J., & Kraus, W. E. et al. (2000). Genetics of coronary disease: Current knowledge and research principles. *American Heart Journal, 140*(4), S11–26.

Wung, S. F. (2002). Genetic advances in coronary artery disease. *MEDSURG Nursing, 11*(6), 296–300.

CHAPTER 6

Putting the Pieces Together: How I Learned of the Genetic Condition in My Family

Being able to identify individuals and families who would benefit from genetics services, and knowing when to seek assistance from and make referrals to appropriate genetics experts and peer support resources will greatly enhance the nurse's ability to provide support to and coordinate the care of patients with genetic concerns.

Our dreams are different . . . "Welcome to Holland."

Imagine you have planned a wonderful trip to Italy, where you will meet other friends who also plan to go there! You go on your way, but when you arrive, the steward says, "Welcome to Holland!" You find that you must learn a new language, buy new guide books, and you wonder what it would have been like had you gone to Italy with your friends as you had planned. But, the author reminds us, ". . . if you spend your life mourning the fact that you didn't get to Italy, you may never be free to enjoy the very special, the lovely things . . . about Holland. (Emily Perl Kingsley. © 1987) *http://www.angelfire.com/ky/touristinfo/holland.html*

Introduction

The nature of a genetic condition, diagnostic resources available, and family sharing of genetic health and family history information are just a few factors that affect how individuals and families learn about a genetic condition(s) in themselves or family members at various points throughout the life cycle. Some individuals, for example, may be aware of a genetic condition in the family because it has been passed down from generation to generation. Huntington Disease is an example of such a condition; it is inherited through families in an autosomal dominant manner. Prenatal diagnosis of a condition such as Down syndrome or neural tube defect may be how a woman and her family learn of the genetic condition for the first time. For other families, a positive newborn screening test for phenylketonuria (PKU) or sickle cell anemia may be the first occurrence of a genetic condition in a family. Others may have heard about the availability of a new genetic test, such as breast cancer testing for hereditary breast–ovarian cancer, and seek information and counseling before making a decision to have testing. In each of these situations, individuals and families will have questions and concerns about how the condition occurred, why it occurred in their family, whether it will affect other family members, and what treatments or interventions are available. To answer these questions, individuals and families who are concerned about an inherited or genetic condition may come to you for guidance, and when appropriate should be referred to a specialty center for a genetics consultation.

The focus of clinical genetics is on the diagnosis and management of the medical, social, and psychological aspects of genetic or hereditary conditions. As with all other areas of medicine, a correct diagnosis, or at least an appropriate differential diagnosis, is essential for the provision of appropriate testing and treatment. This includes helping the person with the condition and their family members understand and come to terms with the nature and consequences of the specific condition. In clinical genetics there is an added dimension when the condition is inherited. This is the need to inform other family members of their chance of having the condition or having children with the condition, and about the resources, including genetic consultation, available to them to learn more about their chances and available interventions. The focus in clinical genetics, not only on the individual but also on members of the individual's family, both present and future, is a unique component of clinical genetics (Nussbaum, McInnes, & Willard, 2001).

Clinical genetics services are provided by medical geneticists (physicians with specialty training in genetics), genetic counselors, advanced practice nurses with a credential in genetics, and nurses credentialed in genetics at the baccalaureate level (ISONG, 2003; NSGC, ACMG). These clinical services are usually provided in a pediatrics genetics division at academic medical centers or community hospitals and through state-sponsored genetics services programs. Children with genetic conditions may be followed in disease-specific clinics such as cystic fibrosis, spina bifida, craniofacial, or metabolic clinics. Prenatal genetics services are often provided through a maternal–fetal medicine program where prenatal diagnostic procedures (high-resolution ultrasound, amniocentesis, chorionic villus sampling) are also available. Adult genetics services, such as neurology, psychiatry, and oncology, are offered through specialty clinics.

As more is learned about the genetic contribution to the many different rare and common conditions and to all health and disease, the need for clinical genetics services will increase. It is anticipated that these services will be integrated into all areas of medicine and public health. Clinical genetics services include genetic family and medical history risk assessments, genetic evaluation including physical examination and genetic testing, genetic counseling and supportive counseling, and follow-up. Nurses in all practice settings will need to become knowledgeable and develop the skills to identify persons and families who may benefit from a genetics referral. The American Academy of Nursing has created a position statement that outlines strategies for incorporating genetic knowledge and skills, including indications for a genetic referral, into undergraduate and graduate education to enhance clinical practice (Lea, 2002).

Alec's story provides a starting point for introducing the common indications for a referral for genetics consultation. Through Alec's story, we learn of one indication for a genetics referral—Fragile X syndrome—and the long journey some individuals and families may have before a genetic diagnosis is made. Alec's story illustrates how other health professionals may be involved in making a genetic diagnosis before the family becomes involved with clinical genetics. In telling Alec's story, his mother, Julie, shares with us challenges that families may face during the diagnostic process and the impact on family. As she points out in quoting "Welcome to Holland," learning of a genetic condition in a family for the first time involves a change in expectations and lifestyle. Although different from what was expected, with support from genetics professionals and equally importantly peer support, Julie's life with Alec has opened other doors for her and her family and new possibilities for Alec.

Alec's Story: A Change in Plans

"Why not you?" is what my friend replied frankly when I asked, "Why me?" The impact of discovering that I was a carrier of Fragile X syndrome and had passed it on to my son, Alec, had finally sunk in. I was angry, afraid, and sad. I was angry that, out of five siblings, I appeared to be the only one in the family who had passed it on to my child (my sister chose not to have children when she found that she, too, was a carrier). I was afraid of what the future would hold for my son. I was sad for myself, for my husband, and for Alec. It is hard to believe it has been 11 years since Alec was diagnosed with Fragile X syndrome. Eleven years since my husband and I *thought* our dreams and plans were gone forever. Over the years, we have found our dreams and plans were not gone forever; they have just changed. Has it been easy getting to this point of acceptance of different dreams and plans? No. Am I still angry? No. Am I still afraid and sad? Yes, sometimes I am. This is not what I had planned or how I thought things would be, but it is okay. I am glad the years around and following Alec's diagnosis are behind us, as they were the most challenging.

How we found out . . . Alec was a very irritable and seemingly anxious baby and not easily consoled. In addition, he didn't sleep or eat much, and that's what I thought babies were supposed to do. He was hospitalized when he was 4 months old because he was not gaining weight. Following his hospitalization, he continued to be irritable, and it became increasingly difficult to take Alec to large family gatherings or crowded, noisy stores. My husband and I sensed that something was not quite right.

Then, Alec began to miss some normal milestones of babbling, crawling, and walking. He continued to be irritable, cranky, and just not much fun. My father felt that what he needed was a spanking, and my mother-in-law was convinced his behaviors were due to being bottle-fed instead of breast-fed! In our quest for answers, one group of doctors questioned whether my husband and I were, in fact, related. After we assured them we were not, we were informed that Alec most likely had a brain tumor. He didn't. Another doctor told us that he would be globally retarded and would never walk or talk. Well, he is not, and he does!

Following a series of blood tests, Alec's neurologist diagnosed Alec with Fragile X syndrome (**Box 6-1**). Alec was almost 2 years old. We were relieved, because now we had a diagnosis for Alec's problems.

Box 6-1 Fragile X Syndrome

General Information

Fragile X syndrome is the most common inheritable form of mental retardation. It is seen in 1 in 4,000 males and 1 in 8,000 females, and is second only to Down syndrome in causing mild to moderate mental retardation.

The designation "Fragile X syndrome" comes from the presence of a "fragile site" on the long arm of the X chromosome. The fragile X gene—FMR-1, located on the long arm of the X chromosome—codes for the protein FMRP. Within a normal FMR-1 gene there is a segment in which a sequence of three nucleotides (CGG) occurs repeatedly, but in fewer than 50 repeats. Individuals with Fragile X syndrome have 230 to 1,000 or more repeats. An intermediate number of repeats, ranging from 50 to 230 is seen in normal transmitting males and their female offspring.

Clinical Symptoms

Males with Fragile X syndrome usually have mild to severe mental retardation. In a small number of families, individuals with Fragile X syndrome have significant autism. The degree of mental retardation seems to be milder and more variable in females, and they may not have the characteristic physical appearance.

Individuals with Fragile X syndrome may have such developmental problems as learning disabilities, mental retardation of variable severity; autistic-like behaviors; attention deficit disorder; poor eye contact; and unusual speech patterns. In addition, individuals with Fragile X syndrome have a characteristic facial appearance, large ears and a long face; hypermobile joints; and in postpubertal males, larger than average testes.

Box 6-1 continued

Prognosis

Males with Fragile X syndrome (e.g., the full mutation) have mild to severe mental retardation. In a small percentage of families, affected individuals are autistic, have little or no speech, and are capable of only limited social interaction. Females who have the full Fragile X mutation may not have the characteristic physical appearance but approximately one half to two thirds have mental impairment. Female carriers (females who have the premutation) may have premature ovarian failure. The variability of physical and intellectual features among females is believed to result from the ratio in the brain of the active X chromosomes with the FMR-1 full mutation to inactive X chromosomes with the normal FMR gene.

Inheritance

Fragile X syndrome is inherited in families in X-linked dominant manner. A female who has the Fragile X gene mutation, therefore, has a 50% chance to pass on the Fragile X gene to a male and a 50% chance to pass on the Fragile X gene to a female. Eighty percent of males but only 30% of females with the Fragile X gene mutation exhibit symptoms. The relative lack of symptom exhibition in females shows that they have a lower degree of penetrance as well as variability in expression, probably because of X-inactivation. Males who carry the Fragile X gene mutation but who do not express the disease are said to be "normal transmitting males." Normal transmitting males will pass the Fragile X gene mutation on to all of their daughters who will be mentally normal but who, in turn, will pass the gene on to half of their sons, all of whom will be affected. When unaffected females transmit the gene to their offspring, there is often an expansion of the Fragile X gene from this "premutation" of 50 to 230 repeats to the full mutation of more than 230 repeats. The repeats tend to become larger in successive generations, and larger premutations are more likely to expand to a full mutation.

Because Fragile X syndrome is common, it requires consideration in any differential diagnosis of developmental delays or mental retardation in either males or females.

Relevant Testing

DNA analysis of the gene mutation on the X chromosome is the key diagnostic test for the syndrome. Historically, chromosome analysis was the diagnostic test of choice for Fragile X syndrome. With the discovery of the FMR-1 gene, DNA analysis for the CGG repeat became, and remains, the currently recommended test for diagnosis, carrier testing, and prenatal diagnosis. Pre-implantation diagnosis using in vitro fertilization is also available for couples with increased risk for Fragile X syndrome.

Box 6-1 continued

Treatment

There is no specific cure for Fragile X syndrome. Identification of Fragile X syndrome is helpful, however, to a family. The diagnosis can also benefit affected individuals by serving to guide their educational and health programs. Medication to control symptoms, treatment programs, and special educational strategies supported by related services such as speech, language, and occupational therapies are all beneficial in helping individuals with Fragile X syndrome improve their quality of life.

Prevention

Identification of carrier status in prospective parents before conception will help define the risk for having a child with Fragile X syndrome. If there is a family history of diagnosed Fragile X syndrome or of undiagnosed but prevalent mental retardation or even mild mental retardation, diagnostic evaluation, counseling, and testing is recommended.

Patient Resources

National Fragile X Foundation
P.O. Box 190488
San Francisco, CA 94119
Phone: 800-688-8765
Fax: 510-763-6223
E-mail: natlfx@fragilec.org
http://www.fragilex.org

Source: Adapted with permission Lea & Smith, 2003. The Genetics Resource Guide, pp. 83–85. Scarborough, ME: Foundation for Blood Research Publisher.

My husband and I were referred for genetic counseling at a regional genetics services program. It was there that we received the most valuable information—the name and phone number of a parent who had two children with Fragile X syndrome and who was available to talk with us. I called her as soon as I got home. She was honest, yet positive, supportive, and empathetic and she had a wonderful sense of humor. We have kept in touch over the years, and I recently called her for words of wisdom she might have to help us make it through Alec's adolescence.

She laughed when I told her I thought, given everything else, we should be entitled to bypass this stage. "Why should you?" she said.

Alec is 13 years old now and in the seventh grade. The academic arena is still a struggle for him, and we have had to be realistic about what is important for him to learn in order to help him be as happy and independent as possible. Alec is a charming, sensitive young man with a fantastic sense of humor. Alec has had the opportunity of being the manager of several athletic teams at school. He loves to go fishing, canoeing, and snowshoeing, and loves to cook. He is happy and fun to be with.

No, our dreams are not gone, they are just different. ***And that's okay.***

Julie, Alec's mom

Indications for Making a Genetics Referral: Putting the Pieces Together

Knowledge of the indications for making an appropriate genetics referral will enable the nurse to identify individuals and families who may benefit from a genetics consultation. As Julie points out in Alec's story, the genetics consultation helped them to find an answer to their long search for a diagnosis. Julie's story illustrates another important component of the genetics consultation—making peer support resources available. Often the individuals seeking genetics consultation are the parents of a child with a potential or known genetic condition. Others seeking genetic consultation may have a family history of a genetic condition and have concerns about the implications for their reproduction decisions or their own health. In this situation, the person with the genetic condition may be a sibling, parent, or other relative in the family. Genetic consultation is also an integral part of prenatal testing and screening programs, and of cancer risk assessment and evaluation programs. Some general reasons that people seek clinical genetics services and genetic consultation are listed in **Table 6-1.**

In this section, we outline a general overview of the reasons identified through nursing assessments that will help trigger a genetics referral. Nurses need to be familiar with available and appropriate genetics experts and peer support resources for patients and families who have or who are at risk for genetic conditions as an important component of comprehensive genetics services. For Julie, for example, introduction to another parent who had two children with Fragile X syndrome has given her ongoing support and friendship over the years. A listing of such resources is found in **Table 6-2.**

Table 6-1 General Reasons People Seek Genetics Services

- Family or personal history of mental retardation or developmental delay
- Presence of a hereditary condition in a person or family such as cystic fibrosis or Fragile X syndrome
- Previous child with multiple congenital anomalies or an isolated birth defect such as cleft lip–palate, congenital heart defect, or neural tube defect
- Genetic relatedness (consanguinity) in a couple, usually a first cousin or closer relationship.
- Person or family with questions about the genetic aspects of a medical condition, including those who are considering presymptomatic or susceptibility testing and those who have received test results for a neurologic condition or cancer
- Preconception or prenatal risk assessment and counseling for advanced maternal age or other indications for prenatal diagnosis
- Couples with a history of infertility or recurrent miscarriage
- Exposure to teratogenic agents such as alcohol, prescribed medications, or occupational chemicals
- In the newborn, follow-up for a positive newborn screening test such as PKU
- Person with a newly diagnosed genetic condition or syndrome such as Down syndrome

(Nussbaum, McInnes, & Willard, 2001; Jorde et al., 2003)

A Nursing Framework for Making a Genetics Referral

Helping people be appropriately referred for a clinical genetics consultation, whether that be for preconception or prenatal concerns, or pediatric or adult diagnoses, is a required nursing competency in all health care settings. This competency requires being an active and informed member of a health care team and developing a philosophy of care that has genetics at its core (Lea & Smith, 2003). This means always including in a nursing assessment how genetic factors are influencing an individual's health. It involves effective family history gathering and interpretation by use of genetic screening questionnaires and creating a family pedigree, because the family history continues to be an important predictor of risk for many conditions. It includes assessments of other categories of symptoms that can initiate recognition of need for a genetics referral including reproductive issues, ethnicity, and developmental and physical assessments.

A nursing framework for making a genetics referral also requires consideration and appreciation of the relationship of genetic information and interventions to the individual's concept of self. A person's concept of self develops over time through complex psychological processes that include the biological and genetic aspects,

TABLE 6-2 Selected Support Resources for Families with a Genetic Condition

Support Organizations with Listings of Many Genetic Conditions	
• Genetic Alliance	*http://ww.geneticalliance.org*
• Genetic and Rare Diseases Information Center	*http://raredisease.info.nih.gov/info_center.html* or *http://www.genome.gov/Health/GARD*
• National Organization for Rare Disorders (NORD)	*http://www.rarediseases.org*
• Birth Defects Research for Children, Inc. (BDRC)	*http://www.birthdefects.org*
• March of Dimes Health Library	*http://www.marchofdimes.com/HealthLibrary*
Condition-Specific Support Groups (partial listing)	
• American Heart Association, National Center	*http://ww.americanheart.org*
• American Hemochromatosis Society, Inc.	*http://www.americanhs.org*
• Cystic Fibrosis Foundation	*http://www.cff.org*
• FRAXA, Research Foundation, Fragile X	*http://www.fraxa.org*
• Huntington Disease Society of America	*http://www.hdsa.org*
• Muscular Dystrophy Foundation	*http://www.mdusa.org*
• National Breast Cancer Coalition	*http://www.natlbcc.org*
• National Foundation for Jewish Genetic Diseases, Inc.	*http://www.nfjd.org*
• National Fragile X Foundation	*http://www.fragileX.org/home/hgm*
• National Down Syndrome Society (NDSS)	*http://www.ndss.org*

cognitive development, and experiential events within relationships. If an individual's perception of self is adversely affected, or if an individual feels that how others view him or her will be adversely affected, that individual may feel guilty or stigmatized (Peters, Djurdjinovic, & Baker, 1999). As Julie asks her friend "Why me?" following Alec's diagnosis of Fragile X syndrome, she expresses feelings of anger that she might be the only one of her five siblings to have passed the gene on to her children. Julie expressed concern and fear about what the future would hold for Alec.

For many, genetic information is considered private information, and there may be a sense of embarrassment or shame. People may be reluctant to share family history or genetic information for fear of being blamed for familial conditions (Bennett, 1999). Throughout the genetic assessment and referral process, nurses and other health professionals have a responsibility to keep in mind the potential personal and societal implications when assessing and referring individuals and families for genetics evaluation and counseling services.

Nursing Assessments and Genetic Referrals

Nursing involvement with clinical genetics begins with knowledge of when to consider a genetics referral. The ability to identify genetic factors in a family is an essential nursing skill (Williams & Lea, 2003). Family history assessment is a first step. When taking a family history, the nurse is mindful that he or she is inquiring about very sensitive information for an individual. Family history assessment involves collecting and interpreting relevant family and medical histories. The nurse asks about an individual's personal health, and also about his or her biological family relationships and the health of family members. Prior to taking the family history, the nurse can prepare individuals for the inquiry by informing them that the questions to be asked, although personal, are an important part of providing appropriate medical care (Bennett, 1999).

A person's particular cultural and health beliefs are considered when taking a family history. Culture consists of shared knowledge, meanings, patterns, and behaviors of a particular social group (Fisher, 1996). When conducting a family history assessment, the nurse needs to be aware that individual beliefs or cultural values may not be the same as traditional mainstream medical explanations for a specific health condition. For example, individuals from a traditional Southeast Asian culture may hold strong beliefs in karma and fate as causes of health conditions. An individual from a traditional Latino culture may believe that his or her child's cleft lip and palate are caused by supernatural forces during a lunar eclipse (Cohen, Fine, & Pergament, 1998). In other situations, individuals may not know about health conditions in relatives. They may need to contact family members for clarification and to determine whether family members are willing to share such information. Thus, the process of genetic family history risk assessment and genetics referral may take time.

Genetic Family History Assessment

When taking the family history, the nurse uses a combination of genetic interviewing techniques and a genetic health history questionnaire or family pedigree. The purpose and focus of family history assessment depend upon the reason the individual or family is seeking genetic information. Individuals and families may be referred for genetic consultation for preconception or prenatal evaluation, diagnosis and counseling, pediatric diagnosis, and adult-onset conditions. In each of these settings, the nurse inquires about the present and past health of three generations: the individual patient, his or her offspring, siblings, nieces and nephews, parents, aunts, uncles, cousins, and grandparents. The purpose is to identify specific genetic conditions or patterns of inheritance that suggest a genetic factor or condition for which further genetic consultation would be beneficial. See Chapter 2 for more details on family history and pedigree construction.

Prenatal Genetic Family History Assessment

In the prenatal or preconception setting, the focus is on the woman–couple's concern for a pregnancy. Therefore, the purpose of the family history assessment is to identify actual or potential genetic conditions or factors that may be inherited in the family or affect the pregnancy, especially those for which genetic testing and prenatal diagnosis may be available. Common indications for a genetic referral in this setting are listed in **Table 6-3.** Indications include couples who have had a previous child with multiple congenital anomalies, mental retardation, or an isolated birth defect such as a neural tube defect or cleft lip and cleft palate. When assessing family history, the nurse inquires about the woman's age at the expected time of delivery. Women who are 35 years or older have a higher chance for having a baby with a chromosome difference such as Down syndrome. Such women are routinely offered genetic counseling to discuss their age-related risks and the available prenatal testing and diagnostic options, including amniocentesis and multiple marker prenatal screening available to them. Ethnicity of maternal and paternal grandparents is also collected because certain genetic conditions are more common in particular ethnic groups for which carrier testing and prenatal diagnosis are now

Table 6-3 Common Indications for a Prenatal Genetic Referral

- Preconception or prenatal risk assessment and counseling for advanced maternal age and other indications for prenatal diagnosis
- Either parent has a child with a chromosome problem
- Either parent has a child with single or multiple congenital anomalies or genetic syndrome (e.g., neural tube defect, Down syndrome, velo-cardio-facial syndrome)
- Previous child with an inherited condition such as sickle cell anemia, cystic fibrosis
- Family history of mental disability
- Couples with a history of infertility or multiple miscarriage
- Maternal insulin-dependent diabetes, epilepsy, alcoholism
- Maternal exposure to certain medications or drugs (e.g., antiseizure medications) during pregnancy
- Preconception or prenatal counseling for carrier screening for genetic conditions such as cystic fibrosis, Tay-Sachs disease, sickle cell anemia
- Couples who are genetically related (consanguinity), usually a first cousin or closer relationship.

(Jorde et al., 2003; Nussbaum, McInnes, & Willard, 2001)

available. For example, couples who are of Ashkenazi Jewish ancestry have a higher chance for having children with Tay-Sachs disease and other autosomal recessive inherited conditions such as Canavan disease. **Table 6-4** provides a listing of common ethnic origins and testing indications that should be made available to couples considering a pregnancy or who are already pregnant.

The American College of Obstetrics and Gynecology and the American College of Medical Genetics recently developed guidelines for offering cystic fibrosis carrier testing. These organizations recommend that cystic fibrosis carrier testing be offered to all couples planning pregnancy or seeking prenatal care. The nurse can use the family history assessment to identify other women–couples who should be offered cystic fibrosis carrier testing. The following is a list of possible candidates:

1. Individuals with a family history of cystic fibrosis
2. Reproductive partners of individuals who have cystic fibrosis
3. Couples in which one member is a Caucasian (ACOG & ACMG, 2001)

Nurses in prenatal and obstetric settings are now involved in offering carrier testing for cystic fibrosis and in referring couples for further genetic consultation when one or both are identified to be cystic fibrosis carriers. For more information about cystic fibrosis, see Chapter 1.

The prenatal history assessment includes inquiring about exposure to known teratogenic agents during pregnancy that place the developing baby at risk for preventable, but irreversible birth defects (Williams & Lea, 2003). While several environmental agents are thought to affect fetal development, there are only a few that are known to cause birth defects. **Table 6-5** lists the more common exposures

TABLE 6-4 Ethnic Origins and Genetic Testing

Ancestry	Genetic Condition	Carrier Test
Ashkenazi Jewish	Tay-Sachs Disease	Serum Hex-A or Leukocyte testing
	Canavan Disease	DNA-based mutation analysis
Northern European Caucasian and Ashkenazi Jewish	Cystic Fibrosis	DNA-based mutation analysis
Southeast Asian	alpha-thalassemia	CBC (MCV) and Hgb electrophoresis
Greek, Italian, Middle Eastern	beta-thalassemia	CBC (MCV) and Hgb electrophoresis
African-American (Black)	sickle cell anemia	Hgb electrophoresis

(Jorde et al., 2001; Nussbaum, McInnes, & Willard, 2000; Williams & Lea, 2003)

TABLE 6-5 Common Maternal Exposures and Conditions That Affect Fetal Development

While many environmental agents (teratogens) are believed to adversely affect fetal development, there are only a few well-established human teratogens known to cause birth defects. Exposure to the following agents during a pregnancy has proven to be responsible for harmful effects in unborn babies.

Infections

- Cytomegalovirus (CMV)—increased fetal loss, congenital malformations, prematurity, and growth retardation
- Rubella (German Measles)—cardiac and fetal eye malformations
- Toxoplasmosis—increased fetal loss, congenital malformations, growth retardation, and prematurity.

Commonly Used Drugs

- Alcohol—fetal alcohol syndrome, intrauterine growth retardation, microcephaly, cardiac defects, mental retardation
- Nicotine—decreased birth weight; increased perinatal complications

Maternal Conditions

- Maternal insulin-dependent diabetes—increased risk for cardiovascular, skeletal, and renal malformations
- Maternal PKU (phenylketonuria)—microcephaly, cardiovascular malformation, mental retardation

Drugs and Therapeutic Agents

- Androgens/norprogesterones—masculinization of female genitalia
- Carbamazapine (antiseizure medication)—spina bifida
- Diethylstilbesterol (DES)—uterine abnormalities, vaginal adenocarcinoma; male infertility
- Isotretenoin (retinoic acid or Accutane)—fetal death, CNS defects; microtia/anotia; conotruncal cardiac defects
- Lithium—Ebstein abnormality (cardiac defect)
- Phenytoin (antiseizure medication)—craniofacial dysmorphology, hypoplastic nails
- Solvents, abuse (entire pregnancy)—small size for gestational age; developmental delay
- Streptomycin—hearing loss
- Tetracycline—stained teeth and bone
- Thalidomide—limb reduction defects; limb hypoplasia; ear anomalies
- Valproic acid (Depakote, antiseizure medication)—spina bifida
- Warfarin—nasal hypoplasia; CNS defects secondary to cerebral hemorrhage

(Jorde et al., 2003; Lea & Smith, 2003; Williams & Lea, 2003)

known to cause birth defects that nurses may encounter in women who are pregnant or who may become pregnant and for whom a genetics referral would be appropriate. Maternal conditions that may also influence fetal development and lead to certain developmental and physical differences include maternal seizure disorder and maternal insulin-dependent diabetes (see Table 6-5). Regardless of the indication, genetic counseling in the prenatal setting is offered so that women and couples will have the opportunity for further evaluation of genetic and environmental factors that may influence a pregnancy and for discussion and informed decision making with regard to prenatal testing options.

Pediatric Genetic Family History Assessment

In the pediatric setting, the nurse would consider a genetic referral in an infant or child with

1. One or more major birth defects such as a neural tube defect or cleft lip–palate
2. Developmental delays
3. Physical findings suggestive of a chromosomal condition such as Down syndrome

A positive newborn screening result, such as PKU or failure of newborn hearing screening, is another indication for a pediatric genetics referral. Like Julie in Alec's story, parents of children with physical or developmental health issues are often seeking a diagnosis and answers to questions such as "Why and how did this happen? Will it happen again in other children? and What does this mean for my family?"

The family history assessment is focused on genetic factors that may have caused or contributed to the birth defect or condition in the infant or child. For example, the nurse inquires about and documents whether there are other family members with a similar condition or if the condition has appeared in the family for the first time. As Julie points out in her story, the family may be asked about genetic relatedness (consanguinity). For Julie, this seemed odd and out of context with the diagnostic process. The reason for asking this question when a genetic condition is suspected in a child is that individuals who are blood relatives share more genes in common. Therefore, they may have a higher chance for having children with autosomal recessive inherited conditions or birth defects, depending on the degree of relatedness. The concern for increased risk is usually with first cousin or closer relationships. **Table 6-6** outlines pediatric indications for making a genetics referral.

Past medical history is also obtained during assessment, including any genetic testing that may have been done. Any child or person with unexplained develop-

Table 6-6 Common Indications for a Pediatric Genetics Referral

- Evaluation of a child with developmental delay or mental retardation
- Evaluation of an infant or child with multiple congenital anomalies or a single anomaly such as cleft lip or neural tube defect
- Evaluation of an infant or child for possible fetal alcohol syndrome or another syndrome associated with a prenatal exposure
- Follow-up evaluation and counseling when there is a positive newborn screening result such as PKU, sickle cell anemia, or cystic fibrosis
- Presence of a single-gene condition such as Fragile X syndrome or hemophilia in an infant or child
- Presence of a chromosomal abnormality such as Down syndrome or velo-cardio-facial syndrome (VCFS) in an infant or child
- Suspicion of a metabolic disorder
- Unusually tall or short stature or growth delays

(Jorde et al., 2003; Nussbaum, McInnes, & Willard, 2001)

mental delays or mental retardation, or with unusual facial features and/or more than one major birth defect, often has a chromosome study as a first step in genetic evaluation, particularly if there is a family history of mental retardation and/or multiple miscarriages. Mental retardation and multiple miscarriages may be caused by the presence of a chromosomal rearrangement, called a translocation, that may be inherited in a family. In an older child or adult, a normal chromosome study, which was frequently the option before newer laboratory techniques were developed, should be repeated. When a current chromosome study is normal, further testing and evaluation may be recommended because a normal chromosome study does not rule out a genetic diagnosis.

Pediatric genetic testing assessment In preparation for the genetics referral, the nurse also gathers information about whether other types of studies may have been done, such as metabolic studies (amino or organic acid studies) and neurological studies (brain imaging). In Alec's situation, he had multiple studies, including evaluation for a possible brain tumor, before he eventually had the genetic diagnostic test for Fragile X syndrome. For Alec and his family it was a pediatric neurologist who ordered the test and made the genetics referral. This is not unusual; children with developmental delays and disabilities may be referred to a pediatric neurologist as a first step in evaluation and diagnosis.

Other pediatric genetic nursing assessments Other nursing inquiries and assessments in the pediatric setting in preparation for a genetics referral include:

- When the condition was first noted to be present
- Are developmental delays static or progressing? If progressing, at what age did the changes begin?
- Whether there were problems with the pregnancy or with birth
- Exposures, such as medications during pregnancy

Answers to these questions provide important information for a genetics referral and will help the genetics team sort out whether the condition in the child is due to genetic or environmental causes. For example, maternal factors such as maternal PKU, maternal myotonic dystrophy, and teratogens (e.g., alcohol, fetal hydantoin syndrome if the mother was taking the antiseizure medication, hydantoin) need to be considered as possible causes of the condition. In addition, prematurity, prenatal or perinatal infections or illnesses, birth trauma (e.g., cerebral hemorrhage), and asphyxia (e.g., abruptio placenta, cord prolapse, meconium aspiration) are also considered (Bennett, 1999). Family beliefs about why and how a birth defect or genetic condition occurred often focus on car accidents, environmental exposures, or other factors as the cause. In Alec's family, Julie's mother-in-law attributed his irritability and behaviors to having been bottle-fed instead of breast-fed, while her father felt what he needed was a good spanking! These beliefs need to be elicited and explored so that they can be clarified during the genetics consultation.

Pediatric genetic physical assessment Physical assessment is another way that nurses may identify children and adults who may have a genetic condition. When conducting a physical assessment, the nurse looks for variations from normal. For example, clinical features that may suggest the presence of Down syndrome in a newborn include increased nuchal fold, hypotonia, and up-slanting palpebral fissures of the eye. The diagnosis of velo-cardio-facial syndrome is considered in an infant who has a cleft palate and a congenital heart defect. Children with Fragile X syndrome may have a high forehead, protruding ears and, in adolescent males, macroorchidism. When describing clinical findings, dysmorphology terms are used, as these are more descriptive and appropriate than terms such as "funny looking kid" or "mongoloid" (Williams & Lea, 2003).

The American College of Medical Genetics Foundation has published a guide for evaluation of the newborn with congenital anomalies that is useful for nurses and other primary care providers. The guide outlines a detailed, step-by-step description of the type of data to be collected, the diagnostic decision-making process, and communicating findings, concerns, and recommendations to families (ACMG Foundation, 1999). Other references and videos on physical assessment and evaluation of newborns and children are listed in **Table 6-7.**

TABLE 6-7 Resources for Nurses to Guide Genetic Physical Assessment

- Aase, J. (1992). Dysmorphologic diagnoses for the pediatric practitioner. *Pediatric Clinics of North America, 39*(10), 135–156.
- American College of Medical Genetics Foundation Clinical Guidelines Project (1999). Evaluation of the newborn with single or multiple congenital anomalies: A clinical guideline. Published by the American College of Medical Genetics under the auspices of a grant from the New York State Department of Health to the American College of Medical Genetics Foundation.
- Pacific Northwest Regional Genetics Group (1994). A nursing assessment of children for detection of genetic disorders and birth defects. Presented at Child Development and Rehabilitation Center, Regional Services Center, Oregon Health Sciences University, Eugene, OR.
- Wardinsky, T. (1994). Visual clues of diagnosis of birth defects and genetic disease. *Journal of Pediatric Health Care, 8*(2), 63–73.

Adult Genetic Family History Assessment

Indications for referring adults for genetics evaluation are increasing as more is learned about the genetic contribution to adult-onset conditions such as cancer, heart disease, thrombosis, and dementia. Until recently, presymptomatic testing was available for only a few rare adult-onset genetic conditions such as Huntington Disease and myotonic dystrophy. Results of human genome research have led to increased knowledge about and testing for more common health conditions that may be inherited in families such as breast–ovarian and colon cancer, familial hyperlipidemias, hereditary hemochromatosis, and Factor V Leiden. When assessing family and personal history, for example, the nurse should be aware of multiple family members in more than one generation with early-onset or rare cancers as an indication of the need for referral for further genetic evaluation. A personal and family history of early-onset heart disease or hypercholesterolemia is another indicator for further genetic evaluation. Individuals who have abnormal iron levels and a family history of liver disease, diabetes, and heart disease may be at risk for hereditary hemochromatosis and require more in-depth genetic testing for clarification. As well, a person with a personal or family history of early-onset deep vein thrombosis, premature delivery, or stroke may have a genetic predisposition caused by factor V Leiden that needs further exploration. Adult clients may also seek genetic consultation to learn more about the genetic aspects of their particular condition so that they can share this genetic information with their children and other family members. Common reasons for making an adult genetics referral are found in **Table 6-8.**

TABLE 6-8 Common Indications for an Adult Genetics Referral

- Adult with undiagnosed mental retardation
- Personal or family history of thrombotic events
- Adult-onset conditions such as hemochromatosis, hearing loss, or visual impairment
- Family history of adult-onset neurologic conditions such as Huntington Disease or myotonic dystrophy for which presymptomatic testing is available
- Personal history of early-onset breast–ovarian, colon, or other cancer for consideration of genetic testing for clinical management purposes
- Family history of cancer before undertaking genetic testing and after receiving results
- Features of a genetic condition such as neurofibromatosis (café-au-lait spots, neurofibromas of the skin) or Marfan syndrome (unusual tallness, dilation of the aortic root)

(Jorde et al., 2003; Lea, 2002; Nussbaum, McInnes, & Willard, 2001)

Adult genetic physical assessment Physical examination in an adult client may reveal indicators for further genetic evaluation. For example, in the adult with unusual tallness, scoliosis, and dilatation of the aortic root, Marfan syndrome is suspected. The clinical features of café-au-lait spots or neurofibromas of the skin suggest the presence of neurofibromatosis type I. Past medical history is also assessed. As another example, an adult presenting with renal cell carcinoma, pancreatic cancer, and/or retinal hamartomas should be referred for genetic evaluation for von Hippel-Lindau syndrome. Indications for referring an adult for genetic evaluation are listed in Table 6-8.

Finding Peer Support

Making the diagnosis of a genetic condition in a person and family is just the beginning, as Julie points out in Alec's story. What comes afterward is the process of learning about the condition and living with it. Understanding may take time and require additional follow-up with genetics specialists so that information can be repeated, clarified, and reviewed. In addition to the genetic referral for diagnostic evaluation, parents need to be offered additional help and support with medical care that follows a genetic diagnosis. The importance of support groups in assisting individuals and families in which a genetic condition has been diagnosed has become increasingly clear. Peer support groups offer the family guidance and support along the way—the fellow traveler—in a way that clinical professionals are not able to provide (Jorde et al., 2003). In Alec's story, Julie tells of the invaluable resource she was given during the genetic evaluation—the name of a parent with children who also had Fragile X syndrome.

Individuals and families may feel a sense of isolation when a genetic diagnosis is made, particularly if the condition is rare. Connecting the individual and family with a peer support group or person can help to alleviate this. For many individuals, like Julie and her family, a parent-to-parent or person-to-person contact is the most meaningful. For others, participating in ongoing research with support groups is helpful. There are now many more resources available on the Web and Internet where families can go for good information about their particular condition. These resources provide families and providers with a way to keep up with the latest testing and treatments. For professionals, the resources offer ongoing education and management guidelines for children and adults living with genetic conditions.

Over the past several decades, a partnership between genetics professionals and persons with genetic conditions has evolved. Organizations such as the Genetic Alliance and the National Organization for Rare Disorders *(http://www.nord.org)* have provided a much-needed service and promoted the establishment of databases and research studies. These resources are of great help to individuals with genetic conditions and also for families. In some families, for example, it may be difficult for parents to explain a genetic condition to other family members. Printed materials, parent support groups, and individual parent-to-parent contact are all available (see Table 6-7) and are helpful resources. Referral to a support group and the provision of its written information are now routine components of care and management of individuals with genetic conditions. Nurses can assess, in follow-up to the genetics referral, which support resources would be the most helpful.

Summary

Accurately recognizing and identifying a genetic concern allows nurses to provide individuals with basic genetic information, referral for genetic evaluation, education, and counseling, and support as individuals incorporate new genetic diagnostic and treatment information. With each encounter, the nurse considers:

- Is the individual–family voicing a genetic concern?
- What is the genetic component to the present condition?
- Is the genetic component–condition inherited or familial?
- Whether genetic testing and/or pharmacogenomic interventions are available?
- Has the individual been offered and referred for genetic counseling and evaluation to help with diagnosis and related questions and concerns?
- What can be done to help the individual understand and incorporate new genetic information?

Knowledge of how genetic heritage and variation are passed on in families or through ethnicity and how the combination of genetic and environmental influences affect the expression of a particular condition will aid the nurse in effective evaluation of the individual and family. Taking medical and family histories, conducting physical exams, and utilizing information from developmental assessments provides relevant information to consider when assessing whether a genetic referral is indicated. Knowing what testing and interventions may be available enables the nurse to make that information more accessible to individuals. Making appropriate and timely referrals for genetic evaluation and counseling allows nurses the opportunity to work together with individuals and families and collaborate in health management (Lea & Smith, 2003).

Chapter Activities

1. As described in this chapter, clinical genetics services are provided by medical geneticists, genetic counselors, and advanced practice nurses with a credential in genetics. These services are often provided in a pediatrics genetics division at academic medical centers or community hospitals and through state-sponsored genetics services programs. Go to the GeneClinics Web site *http://www.geneclinics.org/* and click on the Clinic Directory. Locate the nearest pediatric genetics services nearest you. Check to see if there is an advanced practice nurse in genetics on staff.
2. Jane is a 32-year-old pregnant woman who comes to the prenatal clinic for her first visit. When taking the family history, Jane tells the nurse practitioner of the following family history. Her husband has a history of learning difficulties. He has a sister who was born with a cleft palate. His sister has a daughter who was born with a heart defect and who has learning issues. Her husband's father has a history of depression. Her father-in-law had a sister who was born with a cleft palate and a heart defect and died in early infancy. Jane is concerned about the chance for her baby to be born with a heart defect. Construct Jane's husband's family tree. What could you tell Jane about the constellation of birth defects and learning issues in her husband's family? What referrals would be appropriate to discuss with Jane?

References

American College of Medical Genetics. *www.faseb.org/genetics/acmg*

American College of Medical Genetics Foundation Clinical Guidelines Project (1999). Evaluation of the newborn with single or multiple congenital anomalies: A clinical guideline. Published by the American College of Medical Genetics under the auspices of a grant from the New York State Department of Health to the American College of Medical Genetics Foundation.

American College of Obstetrics and Gynecology & American College of Medical Genetics. (October 2001). *Preconception and prenatal carrier screening for cystic fibrosis.* Washington, DC: Author.

Bennett, R. L. (1999). *The practical guide to the genetic family history.* New York: Wiley-Liss.

Cohen, L. H., Fine, B. A., & Pergament, E. (1998). An assessment of ethnocultural beliefs regarding the causes of birth defects and genetic disorders. *Journal of Genetic Counseling 7,* 15–29.

Fisher, N. L. (Ed.) (1996). *Cultural and ethnic diversity: A guide for genetics professionals.* Baltimore, MD: The Johns Hopkins University Press.

International Society of Nurses in Genetics, Inc (2003). Credentialling Commission. *http://www.isong.org*

Jorde, L. B., Carey, J. C., Bamshad, M. J., & White, R. L. (2003). *Medical genetics* (3rd ed.). St. Louis, MO: Mosby, Inc.

Kingsley, P. (1987). Welcome to Holland. Accessed September 15, 2004, from *http://www.angelfire.com/ky/touristinfo/holland.html*

Lea, D. (2002). AAN News & Opinion: Position statement: Integrating genetics competencies into baccalaureate and advanced nursing education. *Nursing Outlook, 50*(4), 167–168.

Lea, D. H., & Smith, R. (2003). *The genetics resource guide: A handy reference for public health nurses.* Scarborough, ME: Foundation for Blood Research. Written under a grant from the State of Maine Department of Human Services Genetics Program. Grant #BH-01-166B and C.

National Society of Genetic Counselors. *http://www.nsgc.org*

Nussbaum, R. L., McInnes, R. R., & Willard, H. F. (2001). *Thompson & Thompson. Genetics in medicine.* (6th ed.). Philadelphia: W. B. Saunders Company.

Peters, J. A., Djurdjinovic, L., & Baker, D. (1999). The genetic self: The human genome project, genetic counseling, family therapy. *Families, Systems & Health, 17*(1): 5–25.

Williams, J. K., & Lea, D. H. (2003). *Genetic issues for perinatal nurses* (2nd ed.). Wieczorek, R. R. (Ed.). White Plains, NY: March of Dimes.

Chapter 6 Interdisciplinary Commentary

"Breaking the News" Talking with Parents about Their Child's Birth Defect or Genetic Condition

Alan E. Guttmacher, MD
Physician

For most health care providers, informing parents that a child has a birth defect or genetic syndrome is an infrequent role. If it is infrequent and the provider also does not have wide exposure to the specific condition involved, this experience tends to provoke anxiety for the provider. When the discussion is about a condition with major impact for the child, it is also highly anxiety provoking for the parents. Nonetheless, this discussion is important. Largely anecdotal evidence suggests that how parents first hear of their child's diagnosis and prognosis has a real impact on how they deal with the condition. This is an important episode in a family's life, and an important part of the effective provider's contribution to patient and family well-being.

Many medical geneticists and genetic counselors actually engage in this discussion fairly frequently. Indeed, this is a key part of their clinical "art." If done well, this is an opportunity to make a valuable and unique contribution to the life of an individual and family. It can become a source of pride and satisfaction, rather than anxiety.

Each conversation that "breaks the news" is truly unique. Variables making it so include the child's age and specific medical condition, the parents' medical sophistication, their life experiences, their financial situation, their personality styles and world views, the location of the discussion, local supports, etc. Nevertheless, certain generalities can serve as a helpful guide in planning and conducting this conversation. Following are suggestions that come from experience with many parents. These are meant simply as points to consider and to react to. Ignore or contradict any, as best fits your judgment and personal style. Many are designed with the example of a newborn in mind, but most are applicable to children of any age. Many are true even if the patient is an adult.

- It is impossible to "break the news" perfectly. Expecting to do so perfectly will, at best, make you feel that you have failed; at worst, it will prevent you from breaking the news at all. It is possible, important, and rewarding to break the news well. It also is a valuable learning experience—like almost everything in health care, you learn something each time you do it and gradually get better at it.
- Do it. Often, this discussion is put off. However, once an important diagnosis is considered seriously around the hospital or the office, parents almost always hear about it. Honestly discussing a probable, even if unconfirmed diagnosis, is always better than parents learning of it by accident. The longer you wait, the more likely such inadvertent knowledge will become public. Of course, balance this demand for timeliness against the benefits of having both parents present and enough time available to do the discussion well.
- You are starting a process, not completing a one-time task. This may be an acute situation, but it is not an "acute visit"—you are almost always dealing with a chronic issue that cannot be "fixed" in one visit.
- You are modeling an informed reaction to this event for the parents. In fact, this is probably the most important function you will fulfill. Long after this discus-

sion, the parents will have forgotten much of its intellectual content, even if they understood it at the time. They will carry with them, however, memories of how you regarded their baby. To the parents, even if not to yourself, you represent someone who they trust and who is an expert about this condition. Your reaction is extremely important.

- If at all possible, have the child and both parents present. It is difficult for both parents if only one is present when the news is broken. If the parents desire, a *few* other family members or close friends may be helpful additions to the conversation.
- Touch the child. Treat the child as desirable, as an individual, not as a diagnosis—refer to the child by name. Remember, you are modeling informed reaction. If the very person who the parents regard as an expert distances the child, this may leave an indelible impression that those "in the know" view a child with this condition as an "untouchable."
- Know about the child's current condition. This is part of both treating the child as an individual and maintaining the parents' trust in your ability to stay on top of what will seem to them a difficult situation. However, try to separate formally the discussion of acute medical problems from the discussion of the "big picture" of the diagnosis and its implications.
- You are there to help, not to be remembered fondly. Luckily, these are not mutually exclusive; in fact, being "nice" can be a real advantage here. Just remember what is paramount—the well-being of the patient and family, not how nice you seem.
- Don't let the parents feel that this is a test. There is no "right" way to react, but parents often try to act as they think you want. Give them permission to react however works best for them.
- Be multidisciplinary, yet minimize the number of health care providers present. If schedules permit, it may be helpful to have other staff members, or others who will work with the child, involved in the discussion. At the same time, realize that this is an intimate, emotionally charged discussion for the parents, so minimize both interruptions and unnecessary individuals.
- Allow parents control, dignity, and respect. Almost all of us want to feel in control of our lives and our family's lives. However, this is a time at which parents feel that external forces have wrested away control of key parts of their lives. If you conduct the discussion so that the parents maximize feelings of control and of being respected by someone they trust, you will help them.
- Each individual and family is different. This is not a "one size fits all" conversation. Both because of this and the last point, let the parents help define the conversation. Different individuals and families will have different questions, different areas of particular concern, and different ways of processing what you tell them.

- Make this a dialogue, not a lecture. Parents get more from a discussion in which they engage actively than from a soliloquy to which they are a passive audience. Such interchange will also allow you to learn more.
- Break the news in person, in a private and comfortable setting, and at an appropriate time. Both the literature and anecdotal experience are clear that this cannot be done successfully by phone, in a rush, or when called out of the room frequently.
- If feasible, the parents should already know and trust the "breaker." If the parents relate well to a primary care physician or other health professional, try to include that person in the discussion.
- Allow sufficient and uninterrupted time. This is one of life's most intense experiences for families, and it isn't easy for professionals. Its length often cannot be scheduled exactly. Both you and the parents may need time afterward to regroup.
- Be knowledgeable. Your having correct information will be important to the parents and make you feel better. It does not have to be exhaustive, but see if you can gather some recent information about the condition before you speak with the parents.
- Inform knowledgeably, but do not overwhelm with facts. While facts are important, this discussion is also about perceptions, emotions, and fears. As you feel it appropriate, explore these. Discourage overly intellectualized reaction but don't remove intellectual support—intellectualizing can be an appropriate short-term mechanism in dealing with an overwhelming situation. It may, for instance, help one feel in greater control of the situation.
- Early in the conversation, ask the parents what they already know about the condition. Knowing where you are starting from improves almost any conversation. Remember that correcting misinformation may be even more helpful than providing new information.
- Consider being nondirective. Different providers have different styles; some are more comfortable providing their own advice along with the facts. Some families, of course, are more eager for such advice than are others. In general, I have found it best in these situations to inform families and allow them to make decisions based on the information.
- Be honest but optimistic. Tell the truth. However, parents usually see the downside of their child having a birth defect or genetic condition without your belaboring it. On the other hand, they may not see the upside of the situation unless you point it out. You may want to bring up the many ways in which their child's life will be "normal" or happy. Make sure to discuss not only what the child will not do, but also what he or she will do.
- Every child is an individual, not a diagnosis. Stress that with time this will become more apparent to the family—it always does. Especially when a child is a

newborn, and we know nothing about them as an individual, it is easy to forget that they are one. After all, we don't yet know whether they prefer vanilla or chocolate ice cream, but we can read volumes about their condition. With time, however, the relative importance of the diagnosis usually recedes, as the relative importance of all those other pieces that make the whole person increases.

- Be appropriately certain, but allow parents to take time to believe. Sometimes one needs to await confirmation of the diagnosis. However, if you have clinical certainty, be honest and share it with the parents. Don't get swept up in their need to deny the reality of the diagnosis. On the other hand, don't feel that you must destroy all vestiges of denial, which can be a healthy temporary coping mechanism. Health professionals sometimes overreact to its temporary use by parents.
- Silences are appropriate. This is not a radio talk show; time to reflect and gather oneself can be quite useful. It is also appropriate to smile and even to laugh. If everyone acts as though this is an unmitigated disaster, it may become one.
- Create permission for parents to have any emotions and to ask *any* questions. If the family feels they can share all reactions and concerns, you will be able to help them deal with the situation more fully.
- Consider family dynamics. Recognize that each individual in the couple will have different reactions, thoughts, etc. Try to help each as an individual, and the dyad as a unit, to deal effectively with the situation.
- Point out that, for those with "differences," in many ways the present is better than the past, and the future *may* be even better. Whether in terms of medical prognosis or functioning in society, we often base our expectations for those with disabilities on past experience. Realize that, by definition, our knowledge of how adults with any condition do is based on those born 20, 40, or even 60 years ago. For most conditions, it is reasonable to believe that both medical and "social" prognosis is decidedly better for those born today. This is important for both health professionals and families to keep in mind.
- Discuss others' reactions and how one might deal with them. This can help prepare parents to deal with grandparents, neighbors, coworkers, etc. Remind parents that misinformation exists "out there," and that the parents will educate others about the child's condition. This reminder is useful in giving parents an active role that they can take on immediately to help their child have a better life. Many parents latch onto this opportunity to become proactive rather than purely reactive—it helps them be in control of the family's life.
- Explain, as you can, causation. Many health care professionals are loath to ask parents what they feel they did to cause their child's problems. We shouldn't be. Almost all parents of children, even of newborns, with birth defects or genetic conditions have started to construct theories about what they did wrong to cause the child's problems. If professionals ignore this dynamic, they simply further its

festering. Explore and expunge parental guilt—it may be one of the most helpful things you do for the well-being of the child and family.

- Express confidence in parents' abilities. Especially from a trusted professional, honest reassurance can be a needed boost. Your belief that the parents can manage the situation and provide a nurturing environment will count for a lot to them. Go ahead and express it.
- Provide options. Especially at times of stress, it helps to feel that options are available and that one is making a choice. This furthers feeling that one is in control of the situation, rather than vice versa. For instance, it is helpful for families of newborns with Down syndrome to know that, although times have changed, they still have a choice about their child's future. While society used to offer the choice of rearing the child in the biological home or in an institution, it now offers the choice of rearing the child in the biological home or in an adoptive home. For some families this is important information, since they decide that the best interests of their child and family are met through such adoption. For other families, it is helpful in simply making them aware that the decision to raise their child at home is an active choice, rather than a default position (and learning that there are families that want to adopt only children with Down syndrome provides an important message about the desirability of such children).
- Make clear that the parents are not alone. Even families that like to be self-reliant and private may benefit from knowing that supports are available at times of stress. Offer other parents, medical specialists, clergy, literature, and organizations as resources.
- Before you leave, summarize and restate key information. However, do not expect any factual recall. Especially in the newborn period, this may be the prototype of the poor learning situation—a couple who expect perhaps the greatest celebratory moment of their lives and instead find themselves sitting through a technical discussion of scientific terms and concepts that they have not recently, or ever, heard. If one adds to their natural distress about the child's condition, sleep deprivation, and often postoperative pain, drugs, and hormonal flux, don't expect this to be an easy time for the parents to absorb new facts.
- Come back or provide further access. Let the parents know how to reach you for questions and plan your next meeting with them—whether it is in the hospital, your office, or their home (this may be one of the most appropriate situations for a house call that you will ever face).
- Right after the session provide the family time alone to be a family. It is often important to allow the family time and space to sit together as a family after the professionals leave. They may have private thoughts to contemplate, the need to discuss or cry together as a family, or want to regroup before facing other people. Provide the opportunity to do so.

For further reading:

Buckman, R. (1992). *How to break bad news: A guide for health professionals.* Baltimore, Johns Hopkins University Press.

Farrell, M., Ryan, S., & Langrick, B. (2001). 'Breaking bad news' within a paediatric setting: An evaluation report of a collaborative education workshop to support health professionals. *Journal of Advanced Nursing, 36,* 765–775.

Lynch, E. C., & Staloch, N. H. (1988). Parental perceptions of physicians' communication in the informing process. *Mental Retardation, 26,* 77–81.

Pueschel, S. M. (1985). Changes of counseling practices at the birth of a child with Down syndrome. *Applied Research in Mental Retardation, 6,* 99–108.

Quine, L., & Pahl, J. (1987). First diagnosis of severe handicap: A study of parental reactions. *Developmental Medicine and Child Neurology, 29,* 232–242.

Sharp, M. C., Strauss, R. P., & Lorch, S. C. (1992). Communicating medical bad news: Parents' experiences and preferences. *Journal of Pediatrics, 4* (121), 539–546.

Strauss, R. P., Sharp, M. C., Lorch, S. C., & Kachalia, B. (1995). Physicians and the communication of "bad news": Parent experiences of being informed of their child's cleft lip and/or palate. *Pediatrics, 96,* 82–89.

Alan Guttmacher, MD
Deputy Director, National Human Genome Research Institute, NIH
Bldg. 31 Rm 4B09
Bethesda, MD 20892

CHAPTER 7
Finding What I Need

Having an understanding of available genetic resources and professionals and their diverse responsibilities enables the nurse to better assist clients seeking genetic information or services. Knowledge of the components of the genetic counseling process and the indications for making a referral to a genetics specialist helps the nurse to appropriately identify those individuals who could benefit from genetic consultation and prepare them for what to expect. Recognizing the nursing role in referral of clients for genetic services, or for provision, follow-up, and quality review of genetics services, contributes to nursing collaboration with other health care team members. This recognition includes an awareness of the importance of providing genetic education and counseling fairly, accurately, and without coercion or personal bias. To do this, nurses must also appreciate the importance of sensitivity in tailoring information and services to clients' culture, knowledge, and language levels.

Introduction

During the twentieth century, understanding of genetic conditions, variability, mechanisms of inheritance, and contributions to common diseases has grown tremendously. Emerging laboratory technologies have significantly advanced knowledge of the role of genetics in health and disease, allowing scientists to see how genes control every process the human body performs. The American Society of Human Genetics (ASHG), a professional organization whose membership includes genetic scientists, researchers, and clinicians, was founded in 1949 to create a forum for members to share their laboratory and clinical research and to further genetic advances. Since its establishment, services, resources, and education in genetics have evolved into a genetics medical specialty—largely in response to the need for the health care system to use this information for diagnosis, treatment, and prevention of genetic disorders. The genetic counseling profession, for example, was started in the early 1970s in order to meet the need for more professionals to provide genetic services. Not long afterward, the National Society of Genetic Counselors (NSGC) was formed (1979) to promote a network of communications within the genetic counseling profession (Baker, Schuette, & Uhlmann, 1998; Heimler, 1979). Nurses working in genetics formed their own society in 1987, the International Society of Nurses in Genetics, Inc. (ISONG), in recognition of the need for all nurses to incorporate a genetic focus into nursing practice. Both ISONG and NSGC have created mechanisms to certify genetics nurse specialists and genetic counselors (Baker, Schuette, & Uhlmann, 1998; ISONG, 2003). In 1990, ASHG recommended the establishment of the American College of Medical Genetics (ACMG), an organization that now provides certification boards for geneticists with PhD and MD degrees. ASHG has since become the primary professional organization for human geneticists in North America, consisting of more than 6,000 members. Each of these societies—ISONG, NSGC, ASHG, and ACMG—holds annual meetings at which both laboratory and clinical advances in genetics are presented (**Table 7-1**).

Genetic services, provided by medical geneticists, genetic counselors, and genetics advanced practice nurses, were initially housed in academic medical settings with a research and service focus (Andrews et al., 1994). In 1978, federal funding was made available for genetic services to ensure access to comprehensive genetic health care for all populations (Forsman, 1998). Statewide systems were created with the funding to establish genetic services. In many states, for example, satellite genetics clinics were established in primary and secondary sites such as community hospitals and health departments to increase accessibility. Genetic services continue to be supported by state and federal funding and have expanded as newer genetic technologies and testing has evolved. Such services are now provided by many more public and private entities, including commercial genetic testing companies, freestanding maternal and fetal medicine clinics, and via the Internet (Gollust, Wilfond, & Hull, 2003; Guttmacher, Jenkins, & Uhlman, 2001; Khoury, 2003).

Genetics, genetic education, counseling, and treatments are now recognized as the common thread of knowledge that all health care providers must have among their skills when offering health care services (Lea, Jenkins, & Francomano, 1998). This knowledge base and these genetic services were not always so accessible to individuals and families. As Vicki's story clearly illustrates in the following section, in earlier times, families were given very little information or support when a new diagnosis of a genetic condition was made. Genetic services were fairly new in the 1970s, and primary care providers had not had the education and training to recognize when to make a genetic referral and how to connect individuals and families to support resources that would help them cope with a new and life-changing genetic diagnosis.

In Vicki's story we learn that she and her family did not get the support and assistance they needed for ongoing management of the genetic disorder, sickle cell

TABLE 7-1 Genetics Professional and Resource Organizations

ACMG—American College of Medical Genetics

Clinical arm of the medical genetics community. Oversees Board Certification of PhD and MD geneticists. Annual meetings, journal, position, and policy statements.
http://www.faseb.org/genetics/ACMG

ASHG—American Society of Human Genetics

Research arm of the genetics community and the largest, oldest genetic organization nationally. Annual meeting, journal and other publications; position statements.
http://www.faseb.org/genetics/ASHG

Genetic Alliance

A consortium of support groups for patients and others with a focus on genetic conditions; publishes consumer/advocacy information and a Directory of Organizations.
http://www.geneticalliance.org

ISONG—International Society of Nurses in Genetics

Professional society for nurses working in genetics; annual meeting; position statements, newsletters; membership available to all nurses.
http://www.isong.org

NSGC—National Society of Genetic Counselors

Genetic counselors' professional organization; annual meetings, journal, newsletter; position statements, and other publications.
http://www.nsgc.org

Office of Genetics and Disease Prevention at the Centers for Disease Control

Division of the CDC providing access to current information and global resources on the impact of human genetic research on epidemiology, public health, disease prevention, and health promotion.
http://www.cdc.gov/genetics

disease (SCD), to effectively cope with a new diagnosis. Vicki and her family had to seek out such information and support, and learn to advocate on their own. Families receiving a new diagnosis of a genetic condition or birth defect are often taken by surprise and shocked, becoming anxious with the unfamiliar and fearful for their own and their children's future (Baker, Schuette, & Uhlmann, 1998). Vicki speaks of receiving the diagnosis of SCD over the phone and of discovering the unavailability of services to help her and her family manage and care for her daughter's genetic condition. They were not offered genetic counseling, or support with ongoing management of the condition in their daughters, until after her second daughter was diagnosed. As Vicki poignantly explains, she and her family continue to deal with the "emotional, social, physical, and psychological challenges related to its effect on their lives." They had to deal with rejection of life insurance coverage for her girls. Vicki expresses concern about the potential for abuse of genetic information, fearing that unauthorized disclosure of her daughters' diagnoses will cause them to experience job and further insurance discrimination. Along the way, they have had to learn to be self-advocates, but would have liked and greatly benefited from support from the very beginning.

Sandy's story, in contrast, illustrates the vitally important role that genetic specialists now play in so many patients' lives. Sandy's diagnosis of a rare form of cancer came in the 1990s at a time when genetic counseling and evaluation services had become better established. From Sandy, we learn about the role of a genetics advanced practice nurse in the genetic counseling process and the important link this professional can provide among the patient, primary care, genetics services, and genetics research. Sandy, in contrast to Vicki, finds ongoing support and guidance from her genetics advanced practice nurse and genetic counselors, including the opportunity to share her story with the public through a *Genetics Quarterly Newsletter* (Southern Maine Genetics Services, 2000).

Vicki's Story: Finding What I Need

We begin with Vicki's story when times were different, and learn of Vicki's struggle to live with SCD in the absence of genetics professionals, support groups, and genetics counseling.

The genetic condition in my family is one type of SCD, a disorder called sickle cell anemia (**Box 7-1**). We learned about it several months after the birth of our second child, who is our first daughter with SCD. She went for a well-child visit at the health department and was found to be anemic. The nurse asked questions about family history of sickle cell trait or disease. Then they did some more blood work on our daughter to look into this further.

We received the report that our daughter had SCD from the nurse over the phone. There were no specialty services or clinics for SCD available to us at that time. We went to our pediatrician to find out more. He did not have any information about SCD, but told us to do our best to care for our daughter when she gets sick. We did not know, nor were we told of any other families we could talk to. We got most of our information from the college library.

We were not offered genetic counseling at any time. We moved to another state when I was about 6 months pregnant with our third child. I requested that she be

Box 7-1 Sickle Cell Disease

General Information

Hemoglobinopathies, or inherited disorders of hemoglobin, are easily the most common genetic disorders in the world population. All normal hemoglobin (Hb) molecules have four globin chains: 2 α globin and 2 non-α (β δ γ) globin chains. In normal adults, 97% of Hb is HbA (α2 β2), and the rest is HbA2 (α2 β2).

Sickle Cell Disease (SCD) is the most common hemoglobin disorder. SCD is caused by Hemoglobin S (Hb S). Hb S was the first hemoglobin variant discovered, and is caused by a gene mutation in the β-globin chain. Under very low oxygenation conditions, this alteration leads to the "sickle" deformity in the red cells. Vaso-occlusion shortens survival of red cells, and chronic anemia results.

Clinical Symptoms

A person who carries one copy of Hb S is said to have *sickle cell trait* (HbAS). Individuals with *sickle cell trait* (only one functional gene) are usually in good health. Individuals who have two copies of Hb S-Hb SS, in contrast, have a constant anemia. Although the phenotype is variable, individuals with SCD (no functional genes) have a lifelong hemolytic anemia with acute exacerbations ("crises," which are rapid-onset episodes of pain in the limbs, back abdomen, and chest), increased susceptibility to infection, and the detrimental effects of repeated vaso-occlusive events.

Prognosis

SCD is a chronic, lifelong condition with acute exacerbation. Children with SCD have increased susceptibility to potentially life-threatening bacterial infections and meningitis. In women with SCD, preterm labor and growth retardation in the developing baby are common.

Inheritance Pattern and Recurrence Risks

SCD is inherited in an autosomal recessive pattern in families. In autosomal recessive inheritance, parents who each carry a copy of Hb S (have sickle cell trait) have a 1 in 4 or a 25% chance with each pregnancy to have a baby born with SCD. Carrier parents also have a 50% chance with each pregnancy to have a child who is a carrier and has sickle cell trait, and a 25% chance to have a child who inherits the normally functioning hemoglobin gene from each parent and is neither affected with SCD nor a carrier.

Prevalence

SCD is found primarily in persons of African descent, but it is also present in those of Mediterranean, Caribbean, Latin American, or Middle Eastern descent. Approximately 8% of Americans carry the Hb S gene.

Box 7-1 continued

Relevant Testing

The best carrier testing for SCD and other hemoglobinopathies is a Complete Blood Count (CBC) in combination with hemoglobin electrophoresis, which will screen for both the thalassemias and qualitative hemoglobin variants. This testing helps to ensure accurate hemoglobin identification, which is needed for genetic counseling. Solubility testing (such as sickle prep or sickledex) is not an adequate screen for SCD, except in an emergency setting. Prenatal diagnosis using DNA analysis is available for parents identified to be carriers who have a 1 in 4 risk of recurrence. Newborn screening for SCD is available in many states.

Treatment

Treatment is symptomatic and may include the use of antisickling agents. Infections should be treated promptly. Daily prophylactic penicillin has been shown to be effective in reducing the incidence of serious pneumococcal infections. Only bone marrow transplantation, at this time, is considered to be truly curative, although gene therapy holds promise for a curative approach in the future.

Prevention

Genetic counseling is available to carrier couples (couples in which both have sickle *cell trait*) for discussion of recurrence risks (25%), and available reproductive options such as prenatal diagnosis.

Patient Resources

Sickle Cell Disease Association of America, Inc. (SCDAA)
200 Corporate Pointe, #495
Culver City, CA 90230-7633
Information for affected individuals and families
Peer support; referral to local chapter/group; crisis intervention
Telephone helpline
Phone: 800-421-8453 or 310-215-3722
scdaa@sicklecelldisease.org
http://www.sicklecelldisease.org

Box 7-1 continued

American Sickle Cell Anemia Association
10300 Carnegie Avenue Cleveland Clinic
Cleveland, OH 44106
Information for affected individuals/families; peer support; crisis intervention
Phone: 216-229-8600
Fax: 216-229-4500
Irabragg@ascaa.org
http://www.ascaa.org/

Reference: Adapted with permission: Lea, D. H. and Smith R. (2003). *The genetics resource guide: A handy reference for public health nurses.* Scarborough, ME: Foundation for Blood Research. Written under a grant from the State of Maine Department of Human Services Genetics Program. Grant #BH-01-166B and C.

tested at about 6 months of age. She also has SCD—our second daughter with SCD. Again, there was no specialty clinic available, but this time the pediatrician knew about penicillin prophylaxis. We were finally connected to a sickle cell clinic (with a hematologist) about 70 miles away from us after our second daughter had a mild stroke.

We became more knowledgeable about SCD after we went to the teaching hospital with a sickle cell clinic. We had to learn about and avoid all of the predisposing and precipitating factors that cause multiple sickle cell episodes. We had to learn and prepare for the unpredictable nature of the episodes and respond accordingly. We have learned to be self-advocates, obtaining both primary and specialty care for the girls. We would have liked to have this kind of information and support given to us in the beginning when we first learned that our daughter had SCD.

We continue to deal with the emotional, social, physical, and psychological challenges related to its effect on the quality of our lives. We have dealt with rejection for life insurance coverage for the girls and approval for our son who has sickle cell trait. We have faced challenges in setting limitations and justifying the rational for every action between the siblings, having to explain why the girls have SCD, and their brother carries the trait. Our biggest concern is the potential for abuse of genetic information. It includes unauthorized disclosure, discrimination (sports, job, insurance, and other entitlements), and potential for eugenics.

In the future of genetic medicine, we would like to see the acceptance of genetic variation as a universal characteristic of the human species. We would like to see prevention of misuse and abuse of genetic information. There should be more funding for genetic research and services so that it can be included in the general health care panel. A cure—for us, for SCD—would be the ultimate goal of genetic medicine. Where no cure is found, we would at least expect that genetic discoveries would lead to ways to lower the disability and early death associated with so many genetic conditions.

Vicky

How Genetics Professionals Became a Part of My Life

Vicki's story illustrates the challenges and difficulties that arise for families who are trying to understand and cope with a new genetic diagnosis. There were no specialty clinics for SCD available to her or her family at the time of diagnosis with both of her daughters, nor were any resources made available to them. Sandy's story, on the other hand offers an alternative view of the diagnostic process and how genetic professional resources available to her as she learned about her diagnosis of a rare form of cancer, Cowden disease, became an integral part of her life.

Sandy's Story: A Life Touched by Genetics

I hadn't spoken with Grace, my genetic counselor at the Regional Genetics Program, in about 7 years, and I suddenly had the impulse to call her. Not only did she remember me, but she also remembered "Cowden syndrome"—the rare genetic disease I have. I can't tell you how good that made me feel.

The Regional Genetics Program had been recommended to me by my dermatologist. I had been living with the diagnosis of "Accro-keratosis verruciforms of Hopf" for about 10 years. However, when the growths began to appear internally, the diagnosis was changed to Cowden syndrome. Cowden syndrome was something new to me, and the only information I had about it was from an awful-looking picture in a big, green book. So, when I first walked into Grace's office, I was scared, confused, and sure that in time I would look like the picture in the big, green book.

Some people feel initial impressions are very important. Grace had a picture of her daughter on a horse in her office; this was an immediate groundbreaker for me, as I had also ridden in my youth. I had no idea what a genetic counselor was. Grace explained that she was both a genetics advanced practice nurse specialist, and a genetic counselor, and was there to help answer my questions and to help me find what I needed. My dermatologist recommended I consider a genetic evaluation, so I made an appointment. Grace was to start me on a journey that is ongoing to this day. After our initial conversation, Grace explained some of my situation to the medical geneticist. He made some telephone calls to a major cancer center two hours away and my journey had begun. A day or so later, I spoke with Dr. N., who was then a fellow at the cancer center. I have been her patient ever since. Dr. N. was at that time the only doctor in the United States who was researching Cowden syndrome.

My relationship with genetic counselors did not stop with Grace. She opened the door for my long-term relationships with genetic counselors. After the Regional Genetics Center there was the genetic counselor, Karen, who was working with Dr. N. at the cancer center. (On a lighter note—beware; genetic counselors involved in research love completing forms and taking notes, particularly about family histories!) Karen filled out many forms and sat in on the first couple of visits with Dr. N. A few years later, as a patient, I participated in a Cowden research study at the National Institutes of Health. Again, I encountered lots of forms, kind words, and helping hands on my journey.

Dr. N. is now at a large university where she is chief of a clinical cancer program. I saw her there in March. I met with her genetic counselor, Helen, before Dr. N. joined us. Helen sat in on the entire visit, taking copious notes. I also had the opportunity to meet with two other genetic counselors working with Dr. N. and Helen. All of these professionals bring with them important knowledge about Cowden syndrome, so their patients will have a guiding presence as they begin—what has the potential for anyone having a rare genetic disease—a bumpy ride.

I imagine every person's experience with a genetic counselor will be different. However, I have found each individual I have met along the way to be a very caring and willing fellow journeyer. Over the years, I have been to many hospitals, seen physicians and researchers, and attended numerous medical meetings. I recite my family history in my dreams, I give blood at every corner, and I have been poked and prodded many times. Amidst all of this, I felt special knowing that after 7 years I could, in essence, "phone home" back to where my journey began.

Sandy (Adapted with permission, Southern Maine Genetics Services, 2000).

Genetic Counseling and Evaluation: What Is It and Who Are the Providers?

Genetic counseling services, not available to Vicki and her family for either their first or second child with SCD, are now widely available in a variety of prenatal, pediatric, and adult clinical settings. Genetic counseling, as defined by the ASHG in 1975, is a communication process that deals with the human problems associated with the occurrence, or the risk of recurrence, of a genetic disorder in a family (ASHG, 1975). Genetic counseling is based on a number of principles—the guiding principle being that the decision to use or participate in genetic services should be entirely voluntary. Other principles include equal access to genetics services, patient education, complete disclosure of information, and nondirective counseling (Baker, Schuette, & Uhlmann, 1998). **Table 7-2** provides a summary of genetic counseling principles.

In the past, genetic evaluation and counseling were provided to families of children with birth defects and genetic disorders. These genetic services are now

TABLE 7-2 Guiding Principles in Genetic Counseling

- **Voluntary Utilization of Services.** The decision to seek and use genetic counseling should be entirely voluntary. Patients and families have the right to make health decisions, especially regarding genetic testing and reproduction, without coercion or suggestion that a particular course of action is socially irresponsible.
- **Equal Access—Genetic Services.** Counseling, diagnosis, and treatment should be equally accessible to all who need and choose to use them.
- **Client Education.** A core feature of genetic counseling. Patient education regarding a particular genetic condition generally includes providing the patient with information about the features and natural history, the range of variability of the condition, its genetic or nongenetic basis, how it can be diagnosed and managed, the chance for the condition to recur in a family, the economic, social, and psychological impacts—positive as well as negative—it may have, resources available to help families cope with the challenges presented by the condition, and the various strategies for treatment or prevention that the family may wish to consider.
- **Complete Disclosure of Information.** The counselor discloses and provides any information relevant to decision making in ways that the patient can understand and act upon.
- **Nondirective Counseling.** Using a nondirective (nonprescriptive) approach to providing genetic information and counseling is a defining feature of genetic counseling.
- **Attention to Psychosocial and Affective Dimensions in Counseling.** To help patients cope with a genetic condition or risk, or to make difficult health care decisions, the genetic counselor encourages patients to see themselves as competent and helps them anticipate how various events or actions could affect them and their family. This is done with attention to the patient's social, cultural, educational, economic, emotional, and experiential circumstances.
- **Confidentiality and Protection of Privacy.** Genetic information—family history, carrier status, diagnosis of a genetic condition—has the potential for stigmatization and discrimination. Therefore, it is critical that a person's genetic information be kept private and confidential. A special circumstance arises when knowing a person's genotype can, in some situations, provide important information about other family members' risk. When the risk is substantial and when treatment options to prevent harm exist, then the counselor may have an ethical duty to warn relatives.

(ASHG, 1975; Baker, Schuette, & Uhlmann, 1998).

becoming a part of health care for individuals throughout the life span. Genetic counseling is provided in a variety of clinical settings—primary care as well as specialty health care centers—for reproductive planning and diagnostic purposes. Individuals with advanced education in genetics—medical geneticists, genetic counselors, genetics advanced practice nurses—provide genetic counseling that includes risk assessment and interpretation (Williams & Schutte, 1999). Genetic specialists usually practice in large tertiary-care or community hospitals where they are involved in providing genetic evaluation and counseling for a specialty clinic, such as cystic fibrosis, cancer clinics, or regional genetics public health clinics. A growing number of genetic specialists are providing services in state health departments, health maintenance organizations, and commercial genetic companies.

Components of the Genetic Counseling Process

Genetic specialists providing clinical genetics evaluation and counseling obtain and interpret comprehensive family history information, establish or verify a genetic diagnosis, and provide detailed explanations of genetic concepts and testing. During the counseling process they provide information regarding the process to ensure protection of patient confidentiality and privacy. When genetic testing is involved, genetic specialists interpret complicated genetic test results. Many genetic specialists are also involved in coordinating services for families with genetic conditions and collaborate in managing care of individuals with genetic conditions (Lea, Jenkins, & Francomano, 1998).

Genetic counseling Genetic counseling is a communication process designed to help individuals and families understand the nature of the genetic disorder in themselves and their family, and the chance for the disorder to occur in other family members including children. The first step of genetic counseling is learning from the individual and family what they understand about the condition or concern and what they expect from the genetic counseling process. Although families often come to genetic counselors seeking information, they may experience strong emotional reactions to the genetic information that they receive. Exploring with patients their experiences, emotional responses, cultural and religious beliefs, family and interpersonal dynamics, coping style, and financial resources are all integral parts of the genetic counseling process (Baker, Schuette, & Uhlmann, 1998; Williams & Schutte, 1999).

Family history assessment Family history assessment is a primary component of the genetic counseling process. The history includes information about the family's concerns, and the assessment may occur during the prenatal period, labor, and delivery, or when an infant or child is being evaluated. Past medical history is collected from adults anytime, but may occur when undergoing a genetic evaluation. The history and documentation of the family relationships is collected by constructing a family pedigree or family tree. (See Chapter 2 for more details on family history evaluation.) Family history is recorded in the form of a pedigree to clarify relationships and clinical features that may be relevant to the diagnosis. The family history information is often used to obtain an accurate diagnosis of the genetic condition (Baker, Schuette, & Uhlmann, 1998). For example, a family history of early-onset breast and ovarian cancer could indicate that a gene for hereditary breast–ovarian cancer is present in the family.

Family history information is also used to guide the estimation of recurrence risks by helping to determine whether a particular genetic condition is being transmitted in a family or has occurred as a new mutation in an individual (Jorde et al., 2003). Other family history of significance includes the individual's ethnic background, the presence of consanguinity, infertility, birth defects, late-onset diseases, and mental disability. For example, today it is routine to inquire about a person's ethnic background so that appropriate carrier testing for conditions such as SCD, other hemoglobinopathies, Tay-Sachs disease, and cystic fibrosis can be offered to couples preconceptionally or prenatally. In Vicki's situation, this was never offered, nor was newborn screening offered to her for her daughters to be tested for SCD, which is now a routine component of many newborn screening programs.

Establishing the genetic diagnosis Establishing and verifying the diagnosis of the genetic condition in the family is another important element in genetic counseling. This can be achieved, in some families, by reviewing medical records. In prenatal settings, it may require a prenatal diagnostic procedure, such as amniocentesis, to obtain genetic information about the developing baby. Physical examination may be required by a clinical geneticist in infants, children, or adults to confirm or rule out a genetic diagnosis. For Sandy, physical examination by the genetic specialist was crucial in confirming the diagnosis of Cowden syndrome. Confirmation of a clinically suspected diagnosis may also require additional genetic testing or procedures such as imaging or examination of other family members (Jorde et al., 2003; Baker, Schuette, & Uhlmann, 1998).

Risk assessment Once a diagnosis is made, risk assessment for other family members can be determined. In many instances, the individual is concerned about implications for future reproductive decisions or personal risk for a genetic condition. Analysis of the family pedigree can sometimes provide this information by taking into account the inheritance pattern and the individual's relationship to the family member with the genetic condition. For complex diseases such as heart disease, dementia, and many cancers, the risk assessment and counseling may have to be clarified by looking to the scientific literature regarding outcomes in families with similar conditions. Evaluation of data collected from the family pedigree, medical histories, and risk assessments is carried out by the genetic counselor, genetics advanced practice nurse, and/or the geneticist (Baker, Schuette, & Uhlmann, 1998; Lea, Jenkins, & Francomano, 1998).

Providing genetic information Providing information about the specific genetic condition (communication of genetic information) is a central component of genetic counseling. Information about the implications of the genetic condition for the family member with the condition and for other family members is given, including the natural history of the condition, its variability, and various treatment options. At this point, there is detailed discussion about the availability of medical, financial, and social support, as well as reproductive options (when appropriate). Sandy describes the detailed information gathered by the genetic counselor and geneticist to confirm the diagnosis of Cowden syndrome and the in-depth evaluation and discussion that followed confirmation of the diagnosis. Discussion of the manifestations of Cowden syndrome and recommendations of appropriate screening and management options, as well as ongoing support, were provided to Sandy.

Supportive counseling Receiving a diagnosis of a genetic condition in oneself or a family member can have a powerful impact on the family (Schutte, 1999). Counseling centered around preparation for such responses and assistance with coping, often over a period of months and years, should be given. In Vicki's situation, she and her family did not receive such support until they finally found an SCD center. In contrast, Sandy was given support and guidance by her genetic counselors all along the way. In addition to knowing about resources that can help individuals and families adjust to a new diagnosis or risk status, genetic specialists are attuned to any adverse psychological reactions that would warrant a referral for more comprehensive psychosocial support (Baker, Schuette, & Uhlmann, 1998).

Medical genetic evaluation For many genetic conditions, a comprehensive medical genetic evaluation performed by a trained clinical geneticist is required so that accurate genetic information can be given to patients, families, and other health care providers. The purpose of the medical genetic evaluation is to help establish or confirm a genetic diagnosis for an individual or several individuals in a family (Baker, Schuette, & Uhlmann, 1998). When a genetic diagnosis is verified, questions such as "Why did it happen?" "Who else in the family may be at risk?" "What are the chances the condition will happen again?" and "How can the condition be managed?" can be addressed (Baker, Schuette, & Uhlmann, 1998). Sandy's genetic evaluation did just that for her, helping answer her questions and connecting her with lifelong management and support services. Genetic evaluation, like genetic counseling, involves the review of past and present family medical history. In Vicki's family, for example, the family history evaluation could have focused on the fact that she had a child with SCD and, given this information, the chance that she might have other children with SCD. Prenatal and newborn screening options available for SCD after the birth of her first daughter and before the birth of her second daughter could have been discussed and offered.

During a genetic evaluation, the medical geneticist carries out a detailed physical examination looking for variations (from normal) of clinical features and other findings that together may represent a genetic condition or syndrome in an individual. Additional family members may need evaluation to investigate and confirm the presence or absence of a genetic disorder. Recording of specific physical measurements and photographs are standard components of a genetic evaluation. The medical geneticist reviews the findings with the individual and family and makes recommendations for further genetic or other testing. In some instances, referral to another specialist is warranted, for example, referral to a cardiologist when Marfan syndrome is suspected.

Routinely, the medical geneticist sends the referring provider and the family a letter that summarizes the evaluation, including the diagnosis, natural history of the genetic condition, and reproductive risk information. When testing has been completed, a follow-up appointment is held to discuss test findings. If and when a genetic diagnosis is made, referral to support organizations is made and available clinical research is given (Jorde et al., 2003). Sandy's story illustrates the support that she has received and how participation in ongoing research for Cowden syndrome became part of her care.

Cultural Considerations

Genetic counseling takes into account individual and family culture, traditions, and values. Cultural influences are encountered when exploring patients' understanding of health, illness, and methods of treatment and prevention. In some cultures, for example, illness is interpreted as a punishment for having violated religious morals or codes. In other cultures, such as the Hispanic-American communities, good health is believed to be to a gift from God or good luck, and is not taken for granted. Keeping a healthy lifestyle and balance is important. Illness is thought to be caused by an imbalance in the body or punishment. Therefore, prevention activities center on prayer and wearing of religious medals and protective charms. Folk healers, *curandero,* are often used exclusively or in combination with Western medical care (Aguilar et al., 2001). Genetic specialists provide genetic counseling in the context of each patient's culture and belief system to gain an understanding of culturally learned coping practices and support networks. Genetic specialists working with culturally different families need to consider alternative family structures, hierarchies, values, and beliefs. Working with traditional healers and religious and spiritual leaders in a given community builds trust and extends support for the family. Genetic specialists working in these communities use translators with appropriate professional background as well as cultural knowledge about the family's background to aid with genetic counseling and evaluation (Baker, Schuette, & Uhlmann, 1998).

Genetic Specialists: Who They Are and What They Do?

The components of genetic counseling—risk assessment, education about genetic conditions and reproductive options, offering of resources and psychological support related to these issues—are provided by a variety of genetics professionals. In many centers, genetic counseling and evaluation services are provided by a team of specialists who have distinct disciplinary backgrounds, roles, and areas of expertise. These include clinical geneticists, genetic counselors, genetics advanced practice nurses, and other genetic specialists.

Clinical geneticists are physicians who have completed an accredited fellowship program in genetics. Once fellowship is completed, physicians are eligible for certification in clinical genetics by the American Board of Medical Genetics (ABMG) or the Canadian College of Medical Genetics. To become board certified in clinical genetics, physicians must have knowledge and experience with diagnosing and treating genetic conditions and birth defects, as well as an in-depth knowledge of human genetics and underlying principles. Clinical geneticists usually have a specific focus in their practice, such as prenatal diagnosis, dysmorphology, metabolic disorders, or cancer genetics, but they are trained to provide expertise on diagnosis and management of a wide spectrum of genetic disorders (Baker, Schuette, & Uhlmann, 1998).

Genetic counselors have a master's degree in genetic counseling from one of more than 25 programs in the United States, Canada, England, and South Africa. Most genetic counselors provide direct clinical services in prenatal, pediatric, and adult genetics clinical settings. A smaller number are involved in teaching, research, clinic coordination, program administration, public health and policy, marketing, and client services in commercial laboratory settings. The American Board of Genetic Counseling was incorporated in 1992 to certify genetic counselors and oversee accreditation of genetic counseling programs. Candidates for the genetic counseling board examination must have completed their degree at an accredited genetic counseling program and have provided genetic counseling in 50 diverse cases (Baker, Schuette, & Uhlmann, 1998). As of 2003, there were 1500 to 2000 genetic counselors in practice in the United States *(http://www.nsgc.org)*.

Genetics advanced practice nurses often work in genetics clinics and programs. Nurses practicing at the advanced level work in collaboration with other genetics specialists (medical geneticists and genetic counselors) to provide genetic evaluation and counseling for individuals and families who have genetic concerns and who are considering genetic testing and therapeutics (Greco & Mahon, 2003). The American Nurses Association (ANA), recognizing the important role that nurses have in providing genetic services, has established genetics as an official specialty of nursing, and approved the International Society of Nurses in Genetics' document, *The Scope and Standards of Genetics Clinical Nursing Practice* (ISONG, 1998). This document outlines basic and advanced levels of genetics nursing practice and contains both Standards of Care relating to patient management issues and Standards of Professional Performance relating to how nurses represent themselves to the public, the profession of nursing, and other health professionals.

Building on this document, ISONG and its subsidiary, the Genetic Nursing Credentialing Commission *(http://www.geneticnurse.org)* created a mechanism that recognizes nurses who have a strong knowledge base in genetics and genomics—a credential for the Advanced Practice Nurse in Genetics. Further work with the ANA toward American Nurses Credentialing Center certification of the advanced practice nurse in genetics is ongoing and will serve to validate nursing specialization in genetics (Lea & Monsen, 2003).

Several other specialists also participate in the provision of genetic services. These include PhD scientists who work in the areas of cytogenetics, molecular genetics, and biochemical genetics. Many of these professionals direct genetic diagnostic laboratories. Medical geneticists are those genetic specialists who are skilled in the quantitative aspects of genetic analysis and gene mapping. The ABMG certifies PhDs as well as MDs in these areas. In some cases, geneticists who are certified in laboratory subspecialties counsel and treat the patients with the diseases that they diagnose. Clinical geneticists who specialize in diagnosing and treating individuals with metabolic or chromosomal disorders may have additional certification through the ABMG in these subspecialties (Baker, Schuette, & Uhlmann, 1998).

Subspecialty Genetics Resources

In response to the growing knowledge of genetics as a component of all health and disease, subspecialty organizations have begun to create genetics resources for health professionals. As an example, the National Cancer Institute has created as one of its Physician Data Query (PDQ) databases a Cancer Genetics Section *(http://www.cancer.gov/cancerinfo/pdq/cancerdatabase)*. The database contains evidence-based, peer-reviewed cancer information summaries, including genetic counseling. The summaries are made available to health professionals, patients, and the general public as a service of the National Cancer Institute. Each PDQ section has an editorial board responsible for producing and maintaining the information summaries. This Cancer Genetics Editorial Board consists of experts in the fields of epidemiology, primary care, ethics, law, psychology, and the social sciences, as well as medical genetics. The editorial board meets four times a year, and the primary purpose is to develop and maintain PDQ cancer information summaries pertaining to the rapidly growing field of cancer genetics.

The American College of Obstetrics and Gynecology (ACOG) now recommends carrier testing and screening for cystic fibrosis for all couples who are planning a pregnancy or who are already pregnant. To provide support and guidance to health professionals working in preconception and prenatal settings, ACOG, in collaboration with the ACMG, published Clinical and Laboratory Guidelines, which are physician and patient education support materials regarding preconception and prenatal carrier screening for cystic fibrosis (ACOG & ACMG, 2001). These have been made widely available to prenatal care providers.

Genetic texts targeting specific subspecialties are now being written to address the educational and clinical needs of subspecialty physicians struggling with new genome biology. The Cold Spring Harbor Laboratory, for example, has published a series of genetics texts for specialists—cardiologists, dermatologists, ophthalmologists, and orthopedic surgeons etc. Each book provides highly focused and relevant genetic information about a range of diseases within the specialty. The hope is that this series "will fill the gap between individual chapters in large textbooks and general introductory books, and it should help clinicians to become more informed and aware in their fields." (Remedica Genetics Series, Cold Spring Harbor Laboratory, USA, 2004).

Nursing and Genetic Education and Counseling

As illustrated by Sandy's story, genetic counseling is an interactive helping process. The genetic counselors working with Sandy throughout her journey provided information, support, and continuity of care. Sandy's initial encounter with genetic services was with an advanced practice nurse in genetics. Professional nurses and advanced practice nurses who are not genetic specialists can also incorporate aspects of the genetic counseling process into their clinical practice. The nursing focus is on the "identification and prioritization of strategies to promote a desirable outcome for an individual, family or group manifesting or at risk for developing or transmitting a birth defect or genetic condition" (Williams & Schutte, 1999, p. 424). Nursing activities include assisting individuals and families to learn about and cope with genetic information and to manage problems associated with genetic disorders.

Nursing and Genetic Counseling Interventions

In 1992, Genetic Counseling was recognized and included in the Nursing Interventions Classification (McCloskey & Bulechek, 1992). The Genetic Counseling nursing intervention consists of nursing activities that incorporate interactive discussions of genetic information and support of an individual's or family's coping abilities. Key components of the Genetic Counseling nursing interventions classification are presented in **Table 7-3.**

Nursing Outcomes and Genetic Counseling

A core component of the nursing process when caring for individuals with genetic conditions or genetic-related concerns is the identification of desired outcomes of

Table 7-3 Key Components of the Genetic Counseling Nursing Intervention

- **Health Teaching.** Providing genetic information in ways that are appropriate to the patient's educational, social, and cultural background. Nurses assess factors that influence understanding of genetic information and create new ways to present genetic information that overcome barriers that limit patients' abilities to understand and retain Genetic Counseling information.
- **Decision Making.** A core element of Genetic Counseling, when patients consider their reproductive, diagnostic, and management options for genetic conditions. Nurses understand the complex relationships between Genetic Counseling, decision making, and health behavior.
- **Psychological Well-Being.** Psychological adjustment and emotional well-being are of importance when patients and families participate in Genetic Counseling. Assessment of adjustment to and coping with genetic information and diagnoses are nursing activities with regard to the psychosocial aspects of genetic information and Genetic Counseling.

(Williams & Schutte, 1999)

genetic counseling services in order to evaluate their effectiveness. A nursing-specific outcome is "a measurable client or caregiver state, behavior, or perception that is influenced by and sensitive to nursing interventions" (Williams & Schutte, 1999, p. 434). **Tables 7-4** and **7-5** list nursing protocols and outcomes of genetic counseling that apply to individuals and families participating in the genetic counseling process.

Health teaching is integral to the Genetic Counseling intervention. Nurses provide information about genetic conditions to persons in a variety of settings, including diagnostic and disease management programs such as those for hemophilia and SCD. Oncology nurses have become increasingly involved in health teaching regarding the genetic aspects of specific cancers in adults and children. One outcome of genetic counseling related to health teaching is enhanced knowledge. Patient knowledge of the disease process, health resources, a plan for follow-up care, and use of community resources is monitored to determine the extent to which the patient has understood the information and can make a plan for further care and support. Assessing the patient's interpretation and evaluation of the information in light of his or her expectations is another important nursing activity (Williams & Schutte, 1999).

Decision making is a core element of Genetic Counseling when individuals consider diagnostic, reproductive, and management options for genetic conditions. Participation and decision making—personal involvement in selecting and evaluating health care outcomes—are nursing outcomes that relate specifically to the nondi-

Table 7-4 Genetic Counseling and Nursing Intervention Protocols

- **Obtaining Comprehensive Family History.** Nurses use family history assessment to identify individuals in the family who may be at risk to have an inherited condition or have the potential to pass a gene for a condition on to their children. When gathering family history information, the nurse can observe family interactions and assess attitudes, beliefs, and understanding about a genetic condition. A nonjudgmental approach is used to communicate respect for the individual and a desire to gather correct information, as genetic information is personal in nature.
- **Discussing Diagnostic Tests.** Nurses will encounter genetic testing in all areas of clinical practice. Discussions of genetic testing involve ensuring informed consent. Informed consent includes discussion of the potential benefits, risks, and limitations of the testing. Risks include possible discrimination or loss of employment and insurance. A part of the nursing intervention includes talking with individuals and families about strategies that may be used to inform other family members.
- **Assessing Family Support.** Genetic conditions may have a psychological impact on families. The family often has caregiving responsibilities, and is the transmitter of genetic health information among its members. Nursing awareness of family knowledge, culture, attitudes, beliefs, and feelings is important when nurses are involved with Genetic Counseling activities.
- **Providing Referrals.** Nurses assist patients and families to evaluate and access available services for management and support. This includes referring families to genetics specialists for further assessment and counseling, and explaining the purposes of Genetic Counseling. Another important aspect of referral is to facilitate the family to contact others who have the same genetic condition. The Genetic Alliance *(http://www.geneticalliance.org)* and the National Organization for Rare Disorders (NORD—*http://www.nord.org*) are two such organizations.

(Williams & Schutte, 1999)

Table 7-5 Nursing Outcomes Related to Genetic Counseling

- Enhanced Knowledge
- Participation in Health Care Decisions
- Risk Detecion
- Health Seeking Behavior
- Acceptance of Health Status
- Anxiety Control
- Coping
- Hope

(Williams & Schutte, 1999)

rective and participatory nature of the genetic counseling process. These outcomes encompass both the patient's ability to collaborate and negotiate with health care providers in determining and evaluating care options and to choose between alternatives (Williams & Schutte, 1999).

Concern for personal, reproductive, and family risk for inheriting or passing on a genetic condition is a main reason why many individuals seek or are referred for Genetic Counseling. The Risk Detection and Health Seeking Behavior outcomes are nursing outcomes associated with this aspect of genetic counseling. For the Risk Detection outcome, the nurse monitors the patient's ability to identify health risk factors, including family history and ethnic background. The nurse also monitors the patient's actions aimed at promoting health, recovery, and rehabilitation for the Health Seeking Behavior outcome (Williams & Schutte, 1999).

Whenever an individual or family has received genetic counseling, psychological and emotional well-being are considered. Nurses can use several nursing-sensitive outcomes to monitor the effectiveness of Genetic Counseling in meeting a patient's psychosocial needs as related to the genetic aspects of their health. These include Acceptance of Health Status, Anxiety Control, Coping, and Hope. Acceptance of Health Status relates to the patient's ability to adjust to or come to terms with his or her genetic health issues. Indicators of the Anxiety Control outcome can be used to measure a person's anxiety level, and relief from or exacerbation of anxiety following genetic counseling and testing. The nurse also monitors the patient's actions directed at managing stressors for the Coping outcome. Maintaining hope when facing uncertainty or loss is an important outcome of genetic counseling that the nurse can monitor through indicators of the Hope outcome (Williams & Schutte, 1999).

Future Models for Genetic Counseling and Evaluation

Genetics in Primary Care

Human genome discoveries are rapidly being applied to clinical practice, particularly in the area of genetic testing and screening. Genetics professionals have traditionally provided genetic counseling and evaluation services to individuals and families in need of or seeking genetic testing. Genetics professionals, however, are limited in number with uneven distribution throughout the United States, thus making the delivery of genetics services a challenge for health care (Guttmacher, Jenkins, & Uhlman, 2001). Increasingly, physicians and other health care providers including nurses are offering genetic testing and education in primary care settings. Genetic testing with and without the support of genetic counseling is also being widely offered on the Internet (Gollust, Wilfond, & Hull, 2003). In the post-genome era, the increase in available genetic tests and the increasing number of individuals seeking genetic testing and services requires evaluation of current models and development of newer models of delivering genetic services for optimal genetic health care delivery (Khoury, 2003).

Genetics and Health Workforce

In 2000, the Genetics and Health Workforce Research Center was created to address this important public health issue with the aim of developing a framework for monitoring change in practices over time among genetics professionals and primary care clinicians. The Center is conducting a 3-year national study assessing the provision of genetics services in various practice settings across the country. This project is funded by Health Services and Resources Administration and the National Human Genome Research Institutes, Ethical, Legal and Social Implications Program, and is based at the University of Maryland, Baltimore.

One important component of this project has been surveys of genetics professionals—medical geneticists, genetic counselors, and advanced practice nurses in genetics—and primary care physicians to assess their current and emerging genetics-related practices. The overall goal is to describe the various practice roles and genetics services models to inform future workforce needs in genetics. The survey of medical geneticists was completed in 2004. Results showed that medical genetics was a small specialty with a downward trend in new diplomats. Geneticists responding to the survey noted that they have limited time in patient care with almost 50% of time in research education and administration (Cooksey, 2004)

The genetics nursing survey, conducted with members of ISONG, is under development. The approximately 300 ISONG members to be surveyed are clinical, educational, and research genetics nurse specialists with various policy, education, research, and clinical roles. Key informant interviews of ISONG members conducted in preparation for developing the survey revealed that ISONG members are well educated, with most having a masters degree in nursing or another specialty, such as public health. On-the-job training is the most common way that these nurses have learned genetics due to limited opportunity for formal instruction. Most have split duties, like the medical geneticists, filling roles as clinicians, educators, and researchers. Genetics clinical nurse specialists are involved in long-term patient care and advocacy, as opposed to a one- or two-session clinical involvement. As a group, the ISONG key informants are concerned about multiple issues including political, social and ethical concerns in nursing, education in the field of genetics nursing, and a desire for recognition of their efforts (financial and otherwise). The ISONG survey once completed (projected 2005) will provide a more detailed account of ISONG nurses' roles and responsibilities in genetics nursing practice.

Genetic testing and information is also becoming a routine component of everyday health care and disease prevention. Genetic testing is often just one part of genetic counseling and evaluation; however, not every genetic test requires genetic counseling (Guttmacher, Jenkins, & Uhlman, 2001). The traditional provision of genetic counseling and evaluation services as described in this chapter will always be appropriate for a small proportion of individuals and families that need ongoing support services for the management and understanding of specific disorders. Delivery of genetics services for "genetic disorders," however, is undergoing a transformation with an extension of the genetics services continuum to include "communication of genetic information" as a routine component of primary care practice (Khoury, 2003).

Primary care physicians have a tradition that emphasizes working with patients to improve health behaviors and relating to patients based on family and social contexts over time. These traditions lend themselves to genetic health risk assessment and the provision of genetic information and testing. Use of the family history to identify individuals at risk for common, chronic conditions with a genetic component (e.g., heart disease, cancer, diabetes) is one tool that is being re-introduced to primary care physicians as a means of personalizing health risks and assisting individuals to improve their health based on family history of chronic disease. Most individuals in the general population will have a moderate family history (as opposed to high or average risk). For these individuals, primary care physicians can use a holistic approach to provide family-centered interventions that complement but do not replace existing prevention strategies. Dr. Muin Khoury, Director of the Office of Genomics and Disease Prevention, Centers for Disease Control has proposed that, "Family history can begin to build a genomic bridge between a public health approach to prevention (one size fits all) and a clinical genetic approach (one at a time)" (Khoury, 2003, p. 266).

Nurses have a long history of involvement in genetics beginning with case finding and managing care of infants and children with genetic disorders (Forsman, 1998). Nurses participate in a variety of roles in primary and specialized care, and in a broad range of health care settings. Nurses skilled in genetics "may be uniquely qualified to integrate genetics into virtually all corners of the nation's healthcare." (Guttmacher, Jenkins, & Uhlman, 2001, p. 220). As genetic testing becomes more widespread, nurses have become more involved in genetic counseling and risk assessment activities. The Oncology Nursing Society (ONS), for example, has created core skills, knowledge, and required competencies in genetics for oncology nurses (Calzone, Jenkins, & Masny, 2002). In a position statement, ONS holds that "because oncology nurses possess the skills and are well-suited to assume expanded roles in cancer genetics and genetic counseling, oncology nurses at both the general and advanced levels must have a foundation in genetics" (ONS, 2000). As another example, the Association of Women's Health, Obstetrics and Neonatal Nurses has created a clinical position statement on the role of the registered nurse as related to genetic testing *(http://www.awhonn.org)*. The position statement outlines the essential components related to genetics and genetic testing and the role of the nurse in supporting clients considering genetic testing.

Practicing nurses across the profession will become involved in genetic services as discoveries of complex disease are applied to health care and begin to change the clinical management of patients with complex disease. As genetics becomes more fully a component of all health and health care, nurses will be transitioning from a traditional approach of managing patients with complex disease that often uses universal treatment and prevention protocols to a tailored approach that will treat disease with interventions specifically tailored to the molecular processes of genetic expression that cause that disease. Nurses will take part in disease prevention through early assessment of susceptibility and appropriate environmental interventions that include behavioral–lifestyle factors, modifications of surroundings, and reduction or elimination of exposure to specific toxins or agents (Frazier et al., in press).

Nurses who already have an understanding of counseling and support and who have genetic knowledge can become midlevel case coordinators, expanding their role in supporting patients and families to manage their own health. This new role—genomics nurse case coordinator—allows nurses to participate as interdisciplinary team members to help meet the growing demands of genomic health care, particularly care of individuals with complex diseases. The genomics nurse case coordinator can carry out a number of activities that support genetic health care in primary and specialty practice. These activities are listed in **Box 7-2.** The midlevel genomics case coordinator can function in primary settings for early identification of genetic risk and referrals for genetic consultation, and in specialized care settings such as prenatal, cancer, or cardiac clinics where referrals are received and health status assessments and interviews are conducted (Lea & Monsen, 2003).

Genetic Services and the Internet

Access to information about genetics is needed by the public to prepare them to understand and make use of genetic services and testing. Implementing new genetic medicine requires information, education, and new tools to educate providers and the public and to make their interactions more productive (Billings, 2000). The Internet is becoming an important source of this information. A number of state and national Internet sites have been developed to provide the public with information about genetic testing and services. The Centers for Disease Control, Office of Genetics and Disease Prevention, provides genetic resources for professionals and the public *(http://www.cdc.org)*. The National Center for Biotechnical Information

Box 7-2 Genomics Nurse Case Coordinator

Nurses who already have an understanding of counseling and support and have genetic knowledge can become midlevel case coordinators, expanding their central role in supporting patients and their families to manage their own genetic health. This new role—genomics nurse case coordinator—allows for nurses to participate as interdisciplinary team members to help meet the growing demands of genomic health care. The genomics nurse case coordinator carries out for the following activities:

1. Conduct family history risk assessment including constructing and reviewing a three-generation family tree
2. Participate in the informed consent process by offering information about the risks, benefits, and limitations of genetic testing, screening, and therapeutic interventions
3. Prepare clients for genetic counseling and evaluation
4. Coordinate care for patients undergoing gene-based diagnostics and therapies
5. Provide ongoing support and case management with attention to the implications and impact of genetic conditions, testing, and treatment on the individual, family, and community.

Clinical settings where the midlevel genomics nurse case coordinator would function include primary care settings for early identification of risk and referrals for genetic consultation, and in specialized care settings (e.g., cancer, prenatal) where referrals are received and health status assessments and interviews are conducted.

The nurse, functioning as the genomics nurse case coordinator, can apply genetic knowledge, combined with traditional expertise in disease and treatment, to guide the patient and family. The focus is on educating the patient and family about the disease process and treatment, explaining the interaction of genetic factors and environmental factors in the disease process, answering questions, and discussing issues of genetic testing including informed decision making and privacy and confidentiality of genetic information gained from testing. This role of the genomic nurse case coordinator differs from that of the certified genetic counselor and goes beyond the traditional role of nurse as patient educator by introducing a genetic component to health education that requires additional education for the nurses. In response to the demand for recognition and credentialing of the genetics clinical nurse practicing at this level, the Genetic Nursing Credentialing Commission (GNCC) has created a second certification for the baccalaureate-prepared nurse that recognizes nursing competency at this level in genomic-based health care.

(Bankhead, Emery, Qureshi, Campbell, Austoker, & Watson, 2001; Greco & Mahon, 2003; Lea & Monsen, 2003).

at the National Library of Medicine and National Institutes of Health provides genetics informational and educational Internet sites to the public (Ouellette, 1999). The information site for the public called Genes and Disease *(http://www.ncbi.nlm.nih.gov/disease)* also includes an information site for the public. These information-rich databases, in addition to enhancing genetic counseling and educational consultative services, are helping large numbers of primary care providers to integrate medical genetics into their practice, and informing the public of the role of genetics in all health and disease (Billings, 2000).

Genetics testing and services are also offered on the Internet. Their presence has raised concerns for the safety of the public accessing these resources. In a recent study of genetics services sold directly to consumers on the Internet (Gollust, Wilfond, & Hull, 2003), 105 Internet sites were identified that offered a genetic service to a consumer. The methods by which consumers would receive their test results (specified only in 60 sites) were mail (47%), e-mail (13%), online (11%), and by telephone (6%). Only 14 of these sites offered health-related genetic testing, and only seven sites described genetic counseling services available to consumers. Other uses of the Internet for genetic testing included parentage testing, forensics, geneology, child support, newborn identity, and "Gift for the person who has everything" (Gollust, Wilfond, & Hull, 2003, p. 334). The authors raise three important concerns about pursuing genetic testing via the Internet: insufficient information for pretest decision making, the potential for misunderstanding results, and the availability of genetic tests without clinical value. Genetics and health professionals need to be aware of the growing use of the Internet and the potential for inaccurate and harmful information that patients may receive.

Public Health and Genetic Services

Public health professionals are taking on more and varied roles in genomics health care. One major role in public health and genetics is ensuring safe and effective genetic testing. Public health professionals have the ability to construct outcome studies that validate the clinical utility of genetic tests and to conduct the types of epidemiological studies needed to understand the clinical manifestations of gene-environment interactions (Guttmacher, Jenkins, & Uhlman, 2001). The Centers for Disease Control, Office of Genomics and Disease Prevention, is currently funding a project—the ACCE Project—to evaluate the clinical validity, specificity, and utility of a number of genetic tests, including genetic testing for hereditary breast–ovarian cancer, hereditary colorectal cancer, and the Factor V Leiden *(http://www.cdc.org)*.

The Genetics Services Branch of Maternal and Child Health Bureau promotes federal and state partnerships to support access to genetic services for pregnant women, mothers, and children. National projects and the sponsorship of state projects by the Genetics Services Branch have strengthened the public health infrastructure and access to services including specialty treatment centers. As an example, the Genetics Services Branch developed a national newborn Screening and Genetic Resources Center *(http://genes-r-us.uthscsa.edu/)* in 1999 to provide professionals and the public with genetics resources and educational information and to promote genetic literacy of the general public (Khoury, Burke, & Thomson, 2000).

Federal and state agencies play an important role in assessing the population's genetic and environmental risk factors. The practice of genomics is emphasized as related public health issues increasingly center on information resulting from variations at one or multiple genetic loci and interactions with environmental factors such as diet, drugs, infectious agents, chemical and physical agents, and behavioral factors. A more general approach to assessment of genetic information than that used in traditional genetics services is therefore needed for promoting health and for diagnosing, treating, predicting, and preventing diseases to accomplish this. Information gained from family history can be used for quantifying risk associated with family history. The Centers for Disease Control and Prevention (CDC) is examining how family history can be used more widely as it reflects the presence of shared genes, shared environments, and complex gene-environment interactions (Khoury, 2003). The CDC has also developed competencies in genetics for all public health professionals that will enhance appropriate use of genetic technology and testing and better education of the public *(http://www.cdc.org)*.

Summary

Prior to the 1970s, there were very few genetic services available to individuals learning about and coping with genetic disorders. The establishment of genetic services as a medical specialty and the professions of clinical geneticist, genetic counselor, and genetics advanced practice nurse has created a broad range of diagnostic, counseling, and support services for individuals and families who have or who are at risk for a genetic condition. Genetic services are now available in a variety of clinical settings, both public and private, and via the Internet. Culturally appropriate genetic resources and services now exist where until recently there were none. Nurses, both at the advanced practice and basic levels, are incorporating genetics into daily practice, including participation in the genetic counseling process. Genetic counselors, advanced practice nurses in genetics, and clinical geneticists are taking leading roles in providing guidance and education to other health care professionals as they incorporate a genetic focus into their clinical practice.

Chapter Activities

1. As Vicki's story illustrates, families receiving a new diagnosis of a genetic condition or birth defect are often taken by surprise and shocked—becoming anxious with the unfamiliar, and fearful for their own and their children's future. Vicki speaks of receiving the diagnosis of SCD over the phone and of the unavailability of services to help her and her family manage and care for her daughter's genetic condition. Today, there are many more genetic conditions that are included in the newborn screening panel. Go to the National Newborn Screening and Genetics Resource Center Web Site *http://genes-r-us.uthscsa.edu/*. Click on the Listing of Screening Conditions by State. At this time, are all states screening for SCD? Go to the Genetic Alliance Web site *http://www.geneticalliance.org*, and identify resources for families who have a child with SCD.
2. Sandy talks about the value of genetics health professionals. She tells about how all genetics professionals bring with them important knowledge about Cowden syndrome, "so their patients will have a guiding presence as they begin—what has the potential for anyone having a rare genetic disease—a bumpy ride." Using the GeneClinics Web site *http://www.geneclinics.org*, locate a genetics clinic in your state where you would refer a patient who is in need of further evaluation for Cowden syndrome. Use the GeneClinics Laboratory Directory to find out whether gene testing is available for confirmation of the diagnosis of Cowden syndrome.

References

Aguilar, M., Visio, P., Kolb, S., Livingstone, J., Aguirre, C., & Kay, C. (2001). Cultural and linguistic considerations in development of genetic educational materials in a predominately Mexian-American population in south Texas. Presented at the National Institutes of Health's A Decade of ELSI, Bethesda, MD.

American College of Obstetrics and Gynecology & American College of Medical Genetics (2001). Preconception and prenatal carrier screening for Cystic Fibrosis: Clinical and laboratory guidelines. Washington, D.C.: American College of Obstetrics and Gynecology.

American Society of Human Genetics Ad Hoc Committee on Genetic Counseling. (1975). Genetic counseling. *American Journal of Human Genetics, 27,* 240–242.

Andrews, L., Fullarton, J., Holtzman, N., & Motulsky, A. (1994). *Assessing genetic risks: Implications for health and social policy.* Washington, DC: National Academy Press.

Baker, D. L., Schuette, J. L., & Uhlmann, W. R. (1998). *A guide to genetic counseling.* New York: Wiley-Liss.

Bankhead, C., Emery, J., Qureshi, N., & Campbell, H. et al. (2001). New developments in genetics-knowledge, attitudes and information needs of practice nurses. *Family Practice 18*(5), 475–486.

Billings, P.R. (2000). Applying advances in genetic medicine: Where do we go from here? *Healthplan, 41(6),* 32–35.

Calzone, K., Jenkins, J., & Masny, A. (2002). Core competencies in cancer genetics for the advanced practice oncology nurse in practice. *Oncology Nursing Forum, 29*(9), 1327–1333.

Cooksey, J. A. (2004) Genetics workforce concern: Limited supply of medical geneticists. 2004 Annual Research Meeting, Academy of Health. San Diego, *http://www.academyhealth.org/2004/list.htm.*

Forsman, I. (1998). Education of nurses in genetics. *American Journal of Human Genetics, 43,* 552–558.

Frazier, et al. (in press) article on cardiovascular genomics.

Gollust, S. E., Wilfond, B. S., & Hull, S. C. (2003). Direct-to-consumer sales of genetic services on the internet. *Genetics in Medicine, 5*(4), 332–337.

Greco, K. E., & Mahon, S. M. (2003). What is a genetic nurse? *MedSurg Nursing, 12*(2), 124; 110.

Guttmacher, A. E., Jenkins, J., & Uhlman, W. R. (2001). Genomic medicine: Who will practice it? A call to arms. *American Journal of Medical Genetics (Semin. Med. Gene.), 106,* 216–222.

Heimler, A. (1979). From whence we've come: A message from the president. *Perspectives in Genetic Counseling* 1, 2.

International Society of Nurses in Genetics, Inc. (1998). *Scope and standards of genetics clinical nursing practice.* Washington, DC: American Nurses Association.

International Society of Nurses in Genetics, Inc. (2003) ISONG news. *MEDSURG Nursing, 12*(2), 124.

Jorde, L. B., Carey, J. C., Bamshad, M. J., & White, R. L. (2003). *Medical genetics* (3rd ed.). St. Louis, MO: Mosby, Inc.

Khoury, M. J. (2003). Genetics and genomics in practice: The continuum from genetic disease to genetic information in health and disease. *Genetics in Medicine, 5*(4), 261–268.

Khoury, M. J., Burke, W., & Thomson, E. J. (2000). Chapter 1: Genetics in public health: A framework for the integration of human genetics into public health. In *Genetics and public health in the 21st century: Using genetic information to improve health and prevent disease.* Khoury, M. J., Burke, W., & Thomson, E. J. (Eds.). New York, NY: Oxford University Press.

Lea, D. H., Cooksey, J. A., Forte, G., & Salsberg, E. (2003) Nurses in genetics: ISONG survey: The first of a series of nurse surveys. Presented September 29, 2003, to the Genetics in Nursing Advisory Board, Baltimore, MD.

Lea, D. H., Jenkins, J., & Francomano, C. (1998). *Genetics in clinical practice: New directions for nursing and health care.* Sudbury, MA: Jones & Bartlett Publishers.

Lea, D. H., & Monsen, R. B. (2003). Preparing nurses for the 21st century role in genomics-based health care. *NLN Perspectives in Healthcare, 24*(2), 75–80.

McCloskey, J., & Bulechek, G. (1992). *Nursing interventions classification (NIC).* St. Louis: Mosby-Year Book.

Oncology Nursing Society (2000). The role of the oncology nurse in cancer genetic counseling. *ONF. 27*(9), 1.

Ouellette, F. (1999). Internet resources for the clinical geneticist. *Clinical Genetics 56*(3), 179–185.

Remedica Genetics Series. (2004). Cold Spring Harbor Laboratory, USA.

Schutte, D. L. (1999). Identification, referral and support of elders with genetic conditions. Iowa City: (IA): University of Iowa Gerontological Nursing Interventions Research Center, Research Dissemination Core, *http://www.guideline.gov/summary/summary.aspx?doc_id+1968&nbr=1194.*

Southern Maine Genetics Services, Foundation for Blood Research. (Summer 2000). Sandy's story. *Genetics Quarterly Newsletter, 1, Patient Issue,* 1–2.

Williams, J. K. & Schutte, D. L. (1999). Chapter 26: Genetic counseling. In *Nursing interventions: Effective nursing treatments* (3rd ed.). Bulechek, G. M. & McCloskey, J. C. (Eds.). Philadelphia, PA: W.B. Saunders Company.

Chapter 7 Interdisciplinary Commentary

Without a Bigger Team, Our "Genomics" Cup Will Runneth Over

Don Hadley, MS, CGC
Genetic Counselor

Working within the field of genetics for over two decades has provided me with an incredible opportunity to observe the growth and utilization of genetics within the field of medicine. During that time, medical genetics and genetic counseling has grown from providing information based on observation and natural history to now including the option of genetic testing for a multitude of genetic and inherited disorders and diseases, thereby clarifying risks. Remarkably, there are a growing number of disorders and diseases for which we can now provide recommendations and preventative strategies to reduce risk or prevent medical complications. The list of diseases for which preventive strategies or treatments are available will continue to increase, bringing genetics further to the forefront of medicine. More recently the term genomics has emerged and will increasingly be utilized within the field. The term genomics refers to the entire collection of genetic information within each of the cells of our bodies. The development and use of this term suggests that consideration and technology is moving beyond the capability of identifying and evaluating the action of single genes to a much broader attempt to understand complex inheritance patterns involving multiple genes and environmental influences. This represents an exciting leap forward and provides hope for understanding and potentially treating or preventing a much broader group of diseases from a genetics perspective.

My work with families throughout that time has also allowed me the privilege of witnessing the amazing resiliency, determination, devotion, and resolve of thousands of people in families struggling to make sense of and make the best of genetic disorders that challenge them on a daily basis. Their desire and efforts to overcome these incredibly difficult situations provide cherished examples of how dedication, selflessness, and hope can combine to set standards that all humankind should strive to reach.

Unfortunately, there are seemingly equal numbers of individuals within these families that face similar challenges but do not possess the good fortune, resources, or facilities to effectively cope alone. At times, their trials and tribulation with genetic disease may overtake their hope for the future and the available resources, leaving them desperate for help and feeling all too alone. It is likely that, for most, there is a continuously shifting balance between the extremes of progress and despair from one life event to the next with varying needs for information, resources, support, and direction.

Medical Genetics and Genetic Counseling have been and continue to be relatively small medical and counseling subspecialties with limited resources to educate, counsel, and follow families at a level that is warranted. It is within this context that we seek to promote a significantly larger cadre of health care professionals who understand the needs of families with or at risk for genetic disease and are able to provide information as well as support as their patients' lives unfold. The envisioned role seems consistent with the roles and responsibilities that nurses have aptly filled throughout time within a broad array of clinical specialties. Although there is a growing number of nurses who specialize or work in collaboration with clinical or research genetics projects, there is a more urgent need to educate nurses in specialty, primary, and public health arenas about the growing role of genetics in health, medical, and preventive care as genetics moves farther into the clinical arena. The infusion of greater numbers of nurses knowledgeable about genetics and genetic counseling practices will provide additional resources for families as they experience needs or challenges.

People's needs for information, counseling, and support vary broadly. For some, the need for information dominates and provides a focus for them to make sense of the disease that threatens them or their family. For others, making sense out of what has happened or is threatening them is best dealt with on an emotional level first, and only then might factual information become relevant. In both cases, opportunities exist to meld the provision of information with the psychological processing of the event in an effort to facilitate their adaptation to the disease.

The provision of information and associated counseling is time-consuming, and in the case of hereditary conditions may recur as others in the family encounters the disease or disorder. One visit or even yearly visits to the specialists, such as the Medical Genetics team, cannot adequately address the array of medical and social issues that present to these families as they seek to cope with the issues before them. The best outcomes are anticipated and most likely result from information and support coming from multiple health care providers who interact with the families on an ongoing basis. This requires increased knowledge about genetics and genetic disorders on the part of all health care providers, as well as interaction between specialists and primary care providers working with the family at the local level.

Many nurses interested in incorporating genetics within their practices will ask how they can begin. A simple and effective way to begin is by collecting family medical histories and sketching out the pedigree. Resources are available to help in learning how to collect a three-generation family history and transform it into a pedigree. Certainly, consultation from genetics specialist will be helpful in interpreting the information and assessing risks; however, there are some guidelines that can help in identifying "red flags" within the pedigree and knowing when to refer on for more specialized assessment.

It is anticipated that the collection of a three-generation family medical history (FMH) will become a routine part of primary health care as an increasing quantity of genetic information unfolds through the efforts of the Human Genome Project and affects a much larger portion of the population. The list of common diseases that have genetics testing and preventive strategies as an option will likely include diabetes, various cancers, heart disease, mental illness, vascular disease, etc. Computerized collection or patient questionnaires can alleviate the time burden for the busy clinician. However, the FMH will serve to screen for individuals who might benefit from counseling and education pertaining to the growing number of conditions (rare and common) for which genetic information is becoming available. Certainly, the provision of information and counseling is paramount in helping individuals and families to understand the genetic aspects of a condition, cope with the psychological effects, and make decisions when relevant. However, we hope that our knowledge and options for treatment and prevention rapidly expand to include a greater repertoire of medications, actions, behaviors, or lifestyles that will reduce or prevent health and medical risks.

Educating people about genetic risks has typically been approached on a case-by-case basis within the clinical setting. As our knowledge of genetics expands to include common diseases, with our efforts to educate and counsel people about genetics, genetic technologies, and the inherent risks and benefits of genetic testing, we will need to consider new approaches to effectively inform larger portions of the population in a comprehensive fashion. Mass communication strategies and more traditional public health approaches will undoubtedly be considered. Furthermore, greater efforts are needed to update practicing health care providers about genetics and the increasing role of genetics within medicine. These efforts are critical because local health care providers play a significant role in encouraging patient compliance with medical recommendations, lifestyle, and dietary modifications, while assisting them in making decisions regarding medical testing and supporting them through the processes. Nurses can and will play a significant role in moving these educational efforts forward.

Over the next few decades, the knowledge of our genome will likely advance to the level of being able to specifically identify the health and medical risks for which each of us has the greatest risks. This kind of knowledge may have profound effects on the individual, their families, and society as a whole. As these technologies find their ways to the public, nurses will be on the front line to receive the questions and inquiries about the utility, potential benefits, limitations, and harms of genetic testing; recommendations for screening, and preventive strategies to reduce one's risk for the threatening diseases. Initially, the identification of persons with moderately increased risk to develop disease may only provide insight into the fact that a person is at increased risk without providing useful ways to prevent the disease from occurring or even reducing the chances that it will. Effective preventive or curative strategies may eventually follow. However, there is likely to be a period when limited options are available, and families will need the support and resources of their local health care providers to cope with the risks identified.

It is exceedingly important that health care professionals not fall into the trap of thinking that information or action is the only approach that helps families deal with the threat of genetic disease. All too often, one of the greatest resources we can provide comes from the art of listening and providing a safe place for the patients and their families to air their concerns, frustrations, or problems related to dealing with genetic disease. Take the time to listen; that "action" may allow the person to relieve emotions or frustrations that inhibit them from taking other positive steps.

I was recently reminded of the critical role that active listening may play in helping families with the emotionally challenging roads they travel. Our Medical Genetics team was asked to act as consultants for a family that was participating in a research protocol for a rare autosomal recessive disorder. The patient was a 30-year-old woman who came to the research center with her mother. As part of their stay at our facility, they had requested an opportunity to discuss the risk for recurrence in her children. When we met with the couple, the mother pointedly asked if she and her husband could have caused her daughter's health problems. Her question was seemingly loaded with angst. Quite reasonably, we presumed that she was wrestling with feelings of guilt that parents who carry a single, silent copy of a mutation sometimes experience. Because our time was limited with the family, we first elected to focus on the explanation of autosomal recessive inheritance and address the potential for recurrence in other family members. We had anticipated that the explanation would make the situation clear that both she and her husband had equally contributed to the etiology of her daughter's illness. We emphasized that the all people carry a small number of recessive genes; thus, each of us has a small chance of having children with autosomal recessive disorders. We stated that prior to the occurrence of an affected child, couples typically do not have any way of knowing the risk beforehand. However, we stressed that it is not unusual for parents to experience feelings of guilt because of their genetic contributions to their children. The mother did not seem upset by the information, as we had anticipated she might. At that point, the patient was called to another evaluation. I promised to return the following day to address any remaining issues.

Upon my return the patient was busy completing her final evaluations and, therefore, I took the time to talk with her mother. During our conversation, we revisited our discussion about autosomal recessive inheritance from the day before.

Recalling her initial question about the role she and her husband played in her daughter's illness, I asked her how she felt after learning about the genetics of her daughter's illness. She, without hesitation, stated that she assumed the "gene stuff" played some role, but that it didn't really bother her. Somewhat surprised, I asked what she worried about regarding their role in "causing" her daughter's disease. She stated that when her daughter was about 13 years old, they were having work done on the roof of their home. One evening during the construction process, a torrential rainfall caused the ceiling to give way, allowing drenched ceiling insulation to fall onto their daughter. When they found her, she was soaked and covered with insulation. For the last 20-some years, the mother had believed that unfortunate accident had contributed to making the disease worse. She blamed herself and her husband for deciding to undertake that project without moving her daughter out of harm's way. The daughter was not injured by the event nor had she developed any illness afterward, yet that one event seemed to capture the responsibility of blame. Notably, that event came shortly after the initial diagnosis had been made in their daughter, and their reaction to the drenching episode was somehow tied to their feelings about the diagnosis. As we began to address the nature of the disease and the unlikely role that this single event played, it was apparent that this mother was beginning to feel some relief from a burden she had carried for nearly two decades. She stated that she was excited to go home and share the information with her husband, as she was sure that he would feel relieved as well.

Certainly, the variety of roles played by the genetic specialist (medical and family history collection, risk assessment, educational roles, resource identification, testing and screening options, etc.) pressures the limited amount of time available to spend with families. However, taking the time to carefully listen to the family's story can provide insights to facilitate the education, counseling, and supportive roles that health care providers, and in particular nurses, can play in this increasingly prominent role of genetics in medicine. We welcome your interest, skills, and enthusiasm to the fields of genetics and genomics.

Don Hadley MS, CGC
Genetic Counselor, Office of the Clinical Director, NHGRI
Associate Investigator, Social and Behavioral Research Branch, NHGRI
Bldg. 10, Rm. 10C103, Bethesda, MD 20892

References

Baker, D. L., Schuette, J. L., & Uhlmann, W. R. (Eds). (1998). *A guide to genetic counseling.* New York: Wiley-Liss.

Bennett, R. L., Steinhaus, K. A., Uhrich, S. B., O'Sullivan, C. K., Resta, R. G., Lochner, D. D., Markel, D. S., Vincent, V., & Hamanishi, J. (1995). Recommendations for standardized human pedigree nomenclature. *Journal of Genetic Counseling, 4,* 267–279.

Croyle, R. T. (Ed). (1995). *Psychosocial effects of screening for disease prevention and detection.* Oxford: Oxford University Press.

Khoury, M. J., Burke, W., & Thomson, E. (Eds). (2000). *Genetics and public health in the 21st century.* New York: Oxford University Press.

Marteau, T., & Richards, M. (Eds). (1996). *The troubled helix: Social and psychological implications of the new human genetics.* Cambridge, England: Cambridge University Press.

Nussbaum, R. L., McInnes, R. R., & Willard, H. F. (2001). *Thompson & Thompson: Genetics in medicine.* Philadelphia: W. B. Saunders Company.

Weil, J. (2000). *Psychosocial genetic counseling.* New York: Oxford.

URLs for Web Resources

Gene Clinics *(http://www.geneclinics.org/)*

A publicly funded medical genetics information resource developed for physicians, other health care providers, and researchers; available at no cost to all interested persons.

Genetics Education Center, University of Kansas Medical Center *(http://www.kumc.edu/gec/geneinfo.html)*

Information for genetic professionals, University of Kansas Medical Center, with clinical, research, and educational resources for genetic counselors, clinical geneticists, and medical geneticists.

National Cancer Institute *(http://www.cancer.gov)*

Provides information to consumers about cancer, diagnosis, treatment, genetic testing, clinical trials, and cancer genetics specialists throughout the United States.

National Society of Genetic Counselors *(http://www.nsgc.org)*

Provides information to consumers about genetic counseling as well as identifies genetic counselors throughout the United States and Canada.

National Human Genome Research Institute *(http://www.genome.gov)*

Provides comprehensive information regarding the HGP and the Ethical, Legal and Social Implications of the HGP efforts. Also contains a listing of federal and state legislation pertaining to genetic discrimination from insurers and employers.

National Coalition for Health Professional Education in Genetics *(http://www.nchpeg.org)*

Promoting health professional education and access to information about advances in human genetics.

The Genetic Alliance *(http://www.geneticalliance.org/)*

The Genetic Alliance is an international coalition of more than 600 organizations representing families, consumer advocates, health professionals, government entities, and the private sector. Founded in 1986, the Alliance works to improve the health of all by fostering the integration of genetic advances into quality health care, public awareness, and consumer informed public policies.

Clinical Studies at the Warren G. Magnuson Clinical Center, National Institutes of Health *(http://clinicalstudies.info.nih.gov/)*

A database of clinical studies being conducted by the National Institutes of Health at the Clinical Center in Bethesda, Maryland.

CHAPTER 8

Transitions and New Understandings: How Genetics Has Transformed My Life

Understanding the impact of genetic information and the potential physical conditions, psychosocial benefits, risks, and limitations of such information for an individual and his or her family helps the nurse in providing the most appropriate care, support, and interventions.

Introduction: The Impact of Genetic Information

Genetic information is defined as any information that can be linked to a person's biological identity (Scanlon & Fibison, 1995). Genetic information can be found in genetic test results such as DNA testing for hereditary cancer susceptibility, a family tree, and medical records including pathology reports or X-ray readings. Genetic information is distinguished from other medical information because it is considered to be private and linked to a person's identity and sense of self. It links us to other biological family members and to communities in ways that other medical information does not. For instance, a person's blood sugar level has meaning for that individual at a particular time if it is elevated or low and it can be altered through diet or medication. For genetic conditions such as Huntington Disease (HD), neurofibromatosis, or Marfan syndrome, on the other hand, diagnosis in a person is permanent. Currently, there is no cure for these conditions. A diagnosis of one of these or other inherited conditions not only has implications for the person's health, but also for his or her reproduction in terms of the chance to pass on the altered gene to a future generation. The diagnosis has relevance to other family members who may be at risk of having inherited the same condition, and to the broader social context in which those individuals live their lives (Sorenson & Botkin, 2003).

Genetic information can now be collected throughout the life span from before birth until after death because of new DNA technologies. Human genome research has led to an increasing number of genetic tests utilized in clinical application for diagnosis, including prenatal diagnosis, presymptomatic diagnosis, susceptibility, and carrier testing. Genetic testing provides an increasing resource of genetic information about individuals and families.

Genetic tests include the specific laboratory analyses of chromosomes, genes, or gene products (e.g., enzymes or protein) to learn whether a genetic alteration for a particular disease or condition is present in an individual. Genetic testing can be DNA-based, chromosomal, or biochemical. At present, genetic testing is clinically available in the United States for more than 300 diseases or conditions, and this testing is taking place in more than 200 laboratories. Development of tests for additional diseases and for targeted pharmacologic interventions is underway (SACGHS, 2004). For some genetic conditions, the genetic test is used to confirm a diagnosis of a genetic condition, such as Fragile X syndrome or Down syndrome, allowing for more specific treatments and interventions. For other conditions, such as cystic fibrosis (CF), a person's DNA test result may show that he or she is a carrier of a gene for CF. The test result in this instance does not have significance for that person's own health. Rather, the result has implications for that person's reproduction, if he or she has children with another CF carrier. The test result has meaning for other family members who may also be CF carriers. Chapter 4 provides more in-depth information about genetic testing indications and uses.

In this chapter, we focus on the impact of genetic information and the associated physical and psychosocial benefits, risks, and limitations for the individual and the family. As genetic testing becomes more widely available for common and rare adult-onset conditions, individuals and families are faced not only with living with the condition in the family, but also with the implications of the decision to have or not to have a predictive genetic test. Chris tells of her family history of HD and how this shaped the way in which she lived her life. Her story illustrates the impact of the discovery of a genetic test for an incurable condition, HD, and the ways an individual and other family members make decisions about whether or not to have the test. Through her story, we learn about how testing for HD transformed Chris's life on the one hand; yet testing also transitioned her to new roles and responsibilities within her family. As Chris tells her story, we follow the history of HD, from when this genetic condition was poorly understood and for which no genetic testing was available to the present day when there is presymptomatic testing for this still incurable disorder. We learn from Chris about some of the ethical and social issues linked to the difficult decision that family members must make—to test or not to test and who to tell or even whether to tell when a presymptomatic test result is positive.

Jerry's story of HD, on the other hand, offers another side to HD testing—when it is used to confirm a diagnosis, and the impact of this information on the family. Through Jerry's story, we see a family approach to sharing of test results and decision making about presymptomatic testing.

Genetic information may be empowering or disabling for an individual and family (Peters, Djurdjinovic, & Baker, 1999). Further research on genetic testing and the family is needed in this important area for a more complete understanding of the implications for the family and to generate more informed clinical and public policy on genetic testing. A better understanding of the role of the family in genetic testing and diagnosis, sharing of genetic information, and the new understandings of individuals and families living with a genetic condition is critical to creating quality genetics-based clinical care (Sorenson & Botkin, 2003).

Nurses have a unique holistic and family-centered approach to caring for individuals along the health–illness continuum. This includes attention to the individual's physical, psychological, social, and spiritual makeup, as well as consideration of their family and social relationships. With this background, nurses have much to offer individuals and families, especially in the area of support for decision making about genetic testing, understanding the implications of a genetic condition for the individual and family, and transitions to new understandings of themselves and family. Nurses need to become knowledgeable about the ways in which genetic information can impact an individual in the context of their family, especially as more genetic testing becomes available. Nurses must be familiar with psychosocial issues that may arise during the genetic testing and diagnostic processes so that they can provide appropriate and holistic care for all individuals and families, as well as make appropriate referrals when needed. Chris's story offers a poignant starting point for developing this understanding.

Chris's Story

Seven years ago, I was tested for Huntington Disease (HD), a hereditary dementia-like disorder. HD is an autosomal dominant disease and the gene is located on chromosome 4, meaning that it affects males and females equally. You only need to receive the genetic defect from your affected parent to get the disease. Your chance of having the disease if one parent is affected is 50%. HD usually begins to manifest in one's 30s or 40s with piano-like finger movements, facial grimaces, clumsiness, and slurring of speech. It gradually progresses over 15 to 20 years with uncontrollable tremor-like movements, dramatic weight loss, and memory loss. Eventually, rigidity of movement and loss of communication develop. An individual's higher cognitive functioning remains intact for a long time, so you remain painfully aware of your decline in physical abilities and memory.

I was 9 years old when my maternal grandmother was in the middle stages of HD. She was 42 at the time, and I remember her being angry and depressed at her situation, yet desperate for understanding and compassion. When she became too ill to manage at home a year or so later, her family placed her in a state mental hospital. At that time, this was the only placement available for a relatively young woman who needed around-the-clock care. She stayed there until her death at age 54. In my family no one really talked about HD, but we had a good library at school, and I knew a lot about HD by the time I was 12 years old. As long as I can remember, I have lived with the possibility that I could suffer the same fate, since my mother had a 50% chance of carrying the genetic defect and passing it on to me. It would be an understatement to say that this knowledge colored my perceptions of who I was and what I could or would be. I determined at an early age that I didn't want to have children of my own, since I would have no way of knowing whether or not I had HD until I was an adult. I considered not marrying—I had seen what HD had done to my grandparents, emotionally and financially. How could I ask someone to take on such a burden? I also did not disclose my family history of HD to many people. My physicians, as a group, did not know much about HD, and none of them counseled me about the disease when I would share this history with them. During my 20s, my mother's younger brother began showing signs of the disease, and we, as a family, watched his HD progress from his late 20s and early 30s until his death in his mid 40s.

I was determined to try to accomplish as much as possible in the unknown time that I had. I graduated from college early and went straight to graduate school, receiving my doctorate at 26. Along the way I did marry, but later divorced, partly over the issue of whether or not to have children. I hoped that the HD gene had died out with my grandmother, but I lived my life with continued caution and anxious time-centeredness, just in case. I followed the progress of genetic testing for HD by reviewing journal articles.

In 1990, when I was 30 years old, it became clear that my mother had HD. She had not been formally diagnosed and was in extreme denial that anything was wrong with her. She did visit a neurologist, with much encouragement. However, he "looked at a horse and saw a zebra." Even with Mom giving him her family history, he did not believe that she had HD, and instead tested her for Wilson's disease—a rarer genetic disorder you can test for with a basic blood test. When this test came back negative, it only strengthened my mother's denial about HD, since, to use her words, "the expert says there's nothing wrong with me." Genetic testing for HD at that time involved extensive family testing of affected and nonaffected family members. My grandmother was dead, and the neurologist never mentioned this option to my mother. Because of these events, it seemed that there would be no way for me to be tested either.

In early 1994, I learned that the HD gene had been located, and a test for HD was developed that would not require extensive family assessment and testing. I began calling testing centers. Each had different pretest assessment procedures and different timelines for how long it would take to be tested. I finally decided to be tested at a large university hospital nearby. Testing involved two visits, one month apart. The first visit was for the assessment—meetings with a genetic counselor, a neurologist, a psychiatrist, and a psychologist. At this first visit, blood would be drawn for the test. One month later, the results would be given, and follow-up counseling and referral would be available. In addition, this program strongly recommended that a nonfamily member accompany the individual to both visits. The individual being tested was to see a psychotherapist at least 6 weeks prior to genetic testing and continue visits with the psychotherapist for some time following testing and receipt of the results.

Once I found out the test existed, I never questioned whether or not to be tested. The 34 years spent not knowing was infinitely worse than the final, definite knowledge of whether or not I had HD. IF I had HD, I would stay in my not-so-great job, since one of the really great benefits it provided was guaranteed long-term nursing home care. I would quit saving for retirement, because there would be none. I was less clear about what I would do if I was HD-free, but that seemed less important to me at the time. In my heart, I suspected that I carried the genetic defect, and I just needed to know if my suspicions were true.

In preparing to be tested, though, there were several things I had to consider. I made sure that the hospital would not bill my insurance company. I did not tell anyone at my workplace, my primary care physician, or most of my family members that I was being tested. I began psychotherapy, but only after making sure that the therapist would not disclose any information about my testing to my HMO. I did not want the news of my status to be available to anyone without my permission. I feared that I would lose my job and my health insurance if I was HD positive and this information was disclosed. I was also not sure how my family would feel about my decision to be tested, and whether they would be supportive of me in the process. I made sure that I had long-term disability insurance before being tested as well; I wouldn't be eligible for this if I tested positive.

As a psychologist, the assessment phase was not as intimidating for me as it might be for others. The meeting with the genetic counselor was most helpful and made the greatest impact on me. She made sure I understood HD and the testing process. She also made sure I had considered both possibilities—that I was free of HD or that I was HD positive—in terms of insurance, implications, emotional reactions, and the testing's effect on my future. She had counseled many people who had not considered these issues and ultimately chose not to be tested after discussing them with her. The neurologist screened me for subtle beginning signs of HD that would make formal testing moot. I underwent psychological and psychiatric evaluations to rule out major depression, anxiety, problems with substance abuse, or other psychiatric problems that would affect my decision making about being tested. And then my blood was drawn and sent away to be tested.

The hardest part, for me, was the month of waiting—of not knowing. I did meet regularly with a psychologist, and this was helpful both before the testing and during the waiting period. She allowed me to try rationalizing not worrying about the results, she let me be scared sick about the results, and she let me fall apart during the waiting time, knowing that all of these reactions were normal ones. When I returned to the testing center one month later, I received the results: I did not carry the HD defect. I did not have Huntington Disease.

It took about a year for the reality of these results to actually hit home. I asked the genetic counselor if I could have a hard copy of the test results, which she agreed to immediately. I needed to read the report with my own eyes. I did not have HD, but for the last 34 years I had lived in the shadow of the disease, and it took a long time to sink in that I was "normal." I had the same chance as anyone else to live to be an alert old person. I also grieved the time I had spent not knowing and the choices I had made based on "preparing for the worst," assuming that I had HD. I began to see the ways in which living with the threat of HD had limited my life choices and had narrowed my view of available options. More importantly, I felt that now I could have a life. The implications of this fact and my options were staggering. For a time, I was overwhelmed by the fact that I didn't have HD and could live my life any way I wanted, just like anyone else. Small issues and big issues arose. On a more humorous note, I became somewhat anxious about flying—after all, you could be killed in those airplanes! Living to an old age was so important to me now. More importantly, I felt I could now *have* a life. I could now really consider the question of whether I wanted to have children, when before I had told myself I didn't want children because of HD. And, as you might guess, I left my not-so-great job pretty quickly.

I kept in touch with my genetic counselor by mail for a few years, letting her know how I was doing with my life. She responded, and even gave me her forwarding address when she left the center. I can't think of anything that she did that was not helpful—she was respectful of me and my reactions, and this has been the thing that continued to impress me through the years as I reflect on my experiences with her.

There is no end to this story. I continue to adjust and be surprised by the effect HD had on my life. I don't tell just anyone about this part of my life—it's still quite personal and private. Recently, I have chosen to speak in public forums, so it's becoming easier to talk about. Every year I celebrate my "re-birthday"—the day I received my results—by spending several days alone reflecting on my past, present, and future.

My family's experience with HD continues. We put my mother in an assisted-living home 2 months ago. Mom is 59 years old. She had been unsafe at home for many years, but continued to insist that she didn't need any help and wasn't going into any home. A little over a week ago, she became malnourished and delusional and was admitted to a psychiatric unit for short-term stabilization. This has tapped into strong feelings for some of my siblings, who remember my grandmother in the state hospital and who swore Mom would never go into a psych hospital for any reason. They are afraid that she will end up staying at the hospital; that is their memory of our grandmother. I have a sister who is 38 years old and showing the beginning signs of the disease—she has three children, one who is 6 years old and two who are old enough to know what HD is and what this means for them. I have a brother who was tested and does not have HD, and two other siblings, each with children, who have made conscious decisions not to be tested.

The effects of HD continue to swirl around me and will have an impact on me for as long as the disease affects the people I love. I will be watching my mother's and my sister's disease progress and worrying about my nephews and my siblings who do not know their status. I do not have HD, but I'm realizing that I'll never be HD-free.

I believe that education of the public, as well as the general community, about genetic medicine's opportunities and potential areas of ethical and medical conflict is and will continue to be crucial. At-risk individuals need to know what their options are regarding testing and treatment, and they also need to make informed choices about the implications of testing and treatment.

I think that closer integration with allied health professionals will be a key factor in addressing the long-term ramifications of genetic testing in individuals and families. Collaboration with and education of psychologists, social workers, family counselors, and other health professionals is going to become increasingly necessary to address the concerns of the many who will, in the future, undergo genetic testing.

I might have liked to have had the chance to meet again with my genetics counselor at regular intervals after testing, to discuss the implications of testing on my family relations. Perhaps genetic medicine will evolve a life span developmental approach to assessment, diagnosis, and treatment, with families having an identified geneticist and/or genetic counselor, much like we have a general family physician. The role of this provider could be to meet the ongoing genetics needs of a family throughout a lifetime, answering questions and providing medical guidance in the same way a primary care provider does with other medical issues.

Chris

Jerry's Story

I was diagnosed with Huntington's about 3 years ago. I had been seeing Dr. M., a psychiatrist, for diagnosed clinical depression and he suspected a neurological problem. He referred me to Dr. S. She examined me and after the first exam had me come back to get results of blood tests she had ordered. The results were negative for what she originally suspected so she wanted me to get another blood test. I was unaware that the test was for Huntington Disease. When I returned to her office for the third visit, she appeared to be very surprised that I had tested positive for Huntington Disease and offered me a cane "to help to keep me from falling." Well, of course, my wife and I were both in shock and, rather than explaining to us about the disease, all she did was tell us to immediately get in touch with Steve at the research center. Thankfully, Steve was in his office and, although we were not able to see him immediately, he talked to us on the phone for about an hour, giving us a lot of information that we were unaware of. In many ways having the information, albeit from someone we had never met, was comforting.

His voice throughout the phone conversation was calm and reassuring. We met with him at the appointed time and date, I was given another exam, and several other people talked with us about the disease as well as Steve. But he was our Guardian Angel through all those days of learning. He did a lot of hand-holding in the beginning and was especially helpful when we knew we would have to tell our five adult children about the disease. He helped us map out a strategy and gave us packets of information to give to each child so they would be getting the correct information and not just something off the Internet. In the meantime I continued to go to Dr. M., and my wife and I both went to Margaret, a psychologist. That was also a big help in grounding us and keeping us focused on what we needed to do in telling the kids. They were part of our support system and, in looking back on those days, asking for their help was also a very good decision.

I was shocked that I could have something that could be that debilitating because I am in very good physical condition. I work out at the Health Fitness Club three times a week and walk about 20 miles a week. I have continued with this routine even now.

Our concerns with telling the children were many but among one of the most immediate was the insurance question. They are all married and three of them have children. We are grandparents to seven little ones. My condition was now a matter of insurance public record because we did not take precautions to prevent it. We were now faced with the question of who in the world would even sell us any long-term care insurance at a price we could afford to pay. The answer was no one. So when we talked with our kids, we did caution them very strongly about this. None of them have been tested, and at this point none of them have indicated a desire to be tested, opting instead to see if symptoms develop.

With the help of my wife I was able to tell our children (we have three girls and two boys) and give them the information packet. I was able to use the strategies that Steve had helped us develop to inform our children. We had originally planned to tell all of them at the same time because we would all be together at Thanksgiving, but Sue apparently suspected something and asked me directly what was wrong. So, I first shared the diagnosis with my youngest daughter, Sue, who lives at home with us. She was very supportive. She had been diagnosed with a seizure disorder and has lived with that since she was 2 years old. She said, "Dad, we all have something." I told my oldest daughter next. She lives a few blocks away from us and was pregnant at the time with her third child, Tom. She was not as surprised as I expected her to be and I suspect, as all our children are very close, that the grapevine had already been put into place. I invited Mitch and Dave and their spouses to the house and told them that I had tested positive for HD. I was then speechless and could only give them the packets from Steve. It was at that point that both boys just gave me a big hug and said, "We will get through this." I then called our second daughter and told her I was sending her a packet of information for her to read. Our oldest daughter had already talked with her by phone. The grapevine is very strong, even after we told each of them that we wanted to let each of them know ourselves. So much for obeying your parents!

We insisted that they make an appointment with Steve at the research center and talk to him and get any questions answered honestly. Thank God he was willing to talk with all of them. I am sure that was not an easy session for him. But I was very thankful that they followed my advice and made the appointment with Steve. Our second daughter was the one experiencing the most difficulty, and she is the one who lives the farthest away from us. She called and was very upset after the meeting and expressed feelings of suicide. I talked with her for over an hour, using the technique Steve used on us. She sounded better after we talked. Fortunately, within a couple of hours a friend of ours was able to put our daughter in touch with someone that was closer to her and able to meet with her that day. She has been fine since then, just knowing that support is near at hand.

I called both my brother and sister and told them about the test diagnosis being positive for HD. I referred them to a Center for Excellence. That is where Steve suggested we go so they could get a baseline and have regular monitoring. My brother and sister both got tested last fall (2002). To my surprise, they both tested positive for HD. We have all been trying to search out which side of the family this gene came from with absolutely no information to go on. We even went to England and Ireland to try to track down some distant relatives to gather information but to no avail. As far as we can tell, no one on either side of the family has shown symptoms of the disease with the exception of an aunt on my father's side who was diagnosed with Parkinson's when she was 80 years old.

We continue to work with the research center and the medical center and are thankful to have such professional and caring individuals working with us. Now we just continue to work toward a cure or something that will slow down the disease.

Jerry

Transitions and New Understandings: Genetics and My Personal Life

As nurses, we know that illness—acute and chronic—impacts individuals in physical, psychological, social, and spiritual realms. Nursing assessments, diagnoses, planning, and interventions are all designed to restore the individual to a state of health and equilibrium. In the case of chronic illness such as diabetes, heart disease, or stroke, the nurse works with the individual to help them adapt to the particular condition and maintain health in the best way possible. Genetic and inherited conditions, like HD, CF, or Fragile X syndrome are also chronic conditions. Nurses need to become knowledgeable and incorporate knowledge of genetic conditions into their scope of practice as the genetic component to all health and disease is realized.

Chris's and Jerry's stories illustrate some of the psychological issues confronting both the individual and families with HD. Families with HD are faced with chronic neuropsychiatric illness and ultimately deaths of family members from generation to generation. Every child of an affected parent—male or female—has a 50% chance of becoming seriously ill, and yet no one knows exactly when or how the disease will present itself in an individual member of the family (**Box 8-1**).

Before Testing: How I Coped

Emotional well-being and psychological adjustment are of concern with individuals and families adjusting to, coping, and living with a genetic condition. Chris's feelings and descriptions of coping with her grandmother's decline from HD and the uncertainty of HD developing in herself, like others who have an HD history, is similar to coping with an acute illness. Coping strategies in both instances include focusing on the immediate rather than the future, and an orientation toward generating options rather than making specific plans (Mishel, 1992; Williams et al., 1999). For families with a child diagnosed with mental retardation or a genetic disorder, a cyclic process of coping characterized as chronic sorrow, has been described (Clubb, 1991; Olshansky, 1962). The process of coping with a genetic condition has also been analyzed in other age groups and for other types of genetic conditions such as hemophilia, CF, and hereditary cancers. (Admi, 1995; Spitzer, 1992; PDQ reference). These studies contribute to better understanding of the emotional responses to genetic information and provide guidance to nurses for implementation of emotional and social support (Williams & Schutte, 1999). **Box 8-2** presents some of the psychosocial issues related to genetic information and hereditary breast–ovarian cancer.

To Test or Not to Test

Making decisions about having a genetic test done is a new option for individuals and families for which limited social or family guidelines currently exist (Williams et al., 1999). The ability to identify disease-causing gene mutations before developing symptoms presents new health care options for which individuals and families may not be prepared (Shiloh, 1996). The clinical availability of gene testing for many more both rare and common diseases means that individuals and families will

Box 8-1 Huntington Disease: What It Is and How It Is Inherited in Families

General Information

Huntington Disease (HD), a genetic condition that appears in midlife affecting men and women equally, is an autosomal dominant trait associated with severe neurological and mental illness. It is a progressive neurodegenerative disorder involving reductions in cerebral enzymes and neurotransmitters with eventual death of the nerve cells in the brain. HD is one of the very first genetic conditions for which molecular genetic methods led to the discovery of a DNA marker closely linked to a disease-causing gene, thus helping to verify the technology.

Clinical Symptoms

The age of onset is variable, but typically occurs in postreproductive years, often between 30 and 50 years of age. Onset has been known to occur in children as young as age 2 and as old as age 80. Death usually follows within 15 years of the onset of symptoms. Primary signs and symptoms are involuntary "choreic" movements and cognitive and psychological impairment that includes severe depression, sometimes leading to violent temper outbursts and suicide. Because HD has a midlife onset and has only been recently diagnosed with any certainty, people with the gene will probably have had children with each one having a 50% chance of inheriting the gene. Many will have been old enough to witness at close range the course of the illness for a parent and sometimes for a grandparent as well.

Pathology

The HD gene is located on the short arm of chromosome 4p16.3. The gene segment for HD contains a three nucleotide, CAG, repeat. The protein product of the HD gene has been named huntingtin, and scientists are working to determine its structure and function. A normal gene for the huntingtin protein typically has 10 to 29 CAG repeats. There are typically between 35 and 86 copies of the CAG repeat in affected individuals. The longer the CAG repeat sequence, the less stable the huntingtin protein. Repeats of 30 to 34 are considered intermediate and are subject to expansion into 35 or more repeats. Age of onset and severity of condition seem to correlate with the number of repeats present in an individual.

New mutations in the HD gene, if they exist at all, are extremely rare, and have never been documented. Equally, the CAG repeat has not been found in other neuropsychiatric conditions, which can be confused with HD.

Prognosis

To date there is no cure for HD, and few palliative measures are available. HD is considered a terminal disease with death occurring within 10 to 20 years of onset.

Box 8-1 continued

Inheritance

Inheritance in families is autosomal dominant, meaning that a person who inherits the HD gene will develop clinical symptoms. Virtually all individuals with HD can be shown to have an affected parent. A positive history is one of the diagnostic criteria. A person who has the HD gene has, with each pregnancy, a 50% chance to pass on the HD gene to a child who will be affected and a 50% chance to pass on the normal version of the HD gene to a child who will not be affected, nor will that child pass on the HD gene to his or her future generations.

There is a "parent of origin" effect, meaning that the onset of symptoms tends to be earlier when the gene is inherited from the father.

Prevalence

One in 10,000 individuals is affected. As of 1994, 30,000 individuals in the United States were affected with HD and approximately 15,000 others were at risk according to the Huntington Disease Society of America (*http://www.hdsa.org*).

Diagnosis

A positive family history is one of the diagnostic criteria. A thorough neuropsychiatric evaluation is necessary to make the diagnosis of HD. Such an evaluation is recommended for individuals who have not developed the condition but who know they have the CAG expansion. This is because other neuropsychiatric disorders can also be present, even in those with HD. Prenatal and pre-implantation diagnoses are possible with DNA techniques, but such diagnoses are limited to special centers and require extensive counseling because of the ethical issues surrounding these tests.

Relevant Testing

Due to the many social, ethical, and legal issues surrounding testing for HD, it is currently recommended that asymptomatic individuals interested in testing be referred to the closest Huntington Disease Testing Center for evaluation, diagnosis, and counseling. Direct DNA analysis of the CAG repeats is available for presymptomatic and prenatal testing. It is useful if the gene expansion (number of repeats) is characterized in at least one affected member of the family.

(Adapted with permission, Foundation for Blood Research. The Genetics Resource Guide (2003), pp. 142–146; Sobel & Cowan, 2000)

Box 8-2 Psychosocial Issues, Genetic Information, and Hereditary Breast–Ovarian Cancer

A substantial body of literature now exists regarding decision making around predictive testing for hereditary breast–ovarian cancer. Like HD, breast and ovarian cancers are of later onset. In theory, predictive testing is intended to help women make informed decisions about health care and to influence their self-care. In actuality, however, some aspects of predictive testing for hereditary breast–ovarian cancer make the potential benefits uncertain and the psychosocial aspects especially challenging.

Breast cancer is the most common cancer among women, excluding skin cancers. Ovarian cancer is a relatively uncommon malignancy; however, most cancers of this type present at an advanced stage of disease and are therefore associated with a higher mortality rate. Overall, American women have approximately a 12% (1 in 8) chance of developing breast cancer and approximately a 2% chance of developing ovarian cancer during the course of a lifetime. Only a small proportion of all breast and ovarian cancers, about 5 to 10%, are hereditary.

Hereditary breast and ovarian cancers are caused by *inherited* (or germline) *mutations* that can be passed from either parent to a male or female offspring (through the egg or sperm cell). In contrast, *sporadic cancers* are associated with mutations that are acquired during a person's lifetime (also called somatic mutations). Somatic mutations only occur in the cells of the affected–malignant tissue, and are not passed from parent to offspring.

Mutations in two recently isolated genes, *BRCA1* and *BRCA2,* have been shown to be a cause of inherited breast and ovarian cancer. Women who inherit a single mutated copy of either *BRCA1* and *BRCA2* have a sharply increased risk of breast and ovarian cancer. Both *BRCA1* and *BRCA2* are tumor suppressor genes. Loss of function in *BRCA1* and *BRCA2* confers an increased risk for breast and ovarian cancer.

How Predictive Testing for Hereditary Breast–Ovarian Cancer Differs from Presymptomatic Testing for Huntington Disease (HD)

One way that predictive testing for hereditary breast–ovarian cancer differs from presymptomatic testing for HD is that not all people with inherited genetic mutations associated with cancer will develop the disease. With HD, all individuals who inherit the HD mutation will develop symptoms and die from complications of the disorder. Furthermore, cancer can develop in those who do not have inherited genetic mutations.

How a negative test is interpreted depends on whether a known mutation has been identified in a family. When a person with a diagnosis of cancer and a family history of the disease has been tested and found to have a gene mutation in either the BRCA1 or BRCA2 gene, the family is said to have a known mutation. If, on the

Box 8-2 continued

other hand, such information is not found in a tested family member, then they are considered to be at population risk for developing breast and ovarian cancer, and not at increased risk to pass on a known mutation to their children. In families where no previous BRCA1 or BRCA2 gene mutation has been identified, a negative test result is not informative. A negative result may thus be a true negative or a false negative, since a woman may have a mutation in a gene other than BRCA1 or BRCA2. In some families, the BRCA1 or BRCA2 result shows a change in the gene of unknown significance. Women with this type of result who are expecting a clearly positive or negative result are left with ongoing uncertainty and psychological distress, the nature and extent of which are just beginning to be studied and understood.

Like HD, the number of women who actually undergo testing for BRCA1 and BRCA2 mutations is substantially lower than reported levels of interest (Lerman et al., 1995; Biescker et al., 2000). Several factors have been identified as predictive of whether women of families with hereditary breast and ovarian cancer will actually pursue predictive genetic testing. These include:

- The presence of a known familial mutation
- A woman's age (those under age 50 are reported to have a higher testing rate for BRCA1 and BRCA2 than those women older than 50)
- Being a woman or parent, or of Ashkenazi Jewish ancestry
- Psychological status (women with a high baseline level of stress)
- A lower level of optimism
- Desiring information for the benefit of other family members—clarification of risk for siblings and children
- Concerns about insurance or employment discrimination.
- Greater family cohesiveness, which may motivate members to help other family members by undergoing testing

(Kash, K.M. et al., 1992; Lerman, C. et al., 1993; Rosenthal, G. et al., 1995)

Psychological Issues

Like individuals with a family history of HD, women at risk for hereditary cancer often have an exaggerated and elevated perception of their risk (Lerman, 1994). This may be due to their anticipatory fear and psychological distress resulting from their experience of living with relatives ill with cancer. Subjective assessment of risk may lead a woman to choose the drastic measure of prophylactic surgery because of disproportional estimation of risk. For other women, longstanding depression and anxiety may actually interfere with compliance with health care recommendations such as self-breast examination and mammography (Lerman, 1994). Psychological and genetic counseling and patient education can help address fears and distress among these patients (Baum et al., 1997).

Box 8-2 continued

Some women undergoing predictive genetic testing for hereditary breast–ovarian cancer experience severe psychological distress (Bredart et al., 2001). Women generally have higher rates of depression and anxiety than men, and compounded anxiety about cancer risk can cause further distress. In addition, many women perceive their breasts as intrinsic to their femininity, to their self-esteem, and to their sexuality; and the risk of losing a breast can produce anxiety. Women carrying inherited gene mutations may experience psychological distress, manifested as persistent worry, depression, anxiety, confusion, and sleep disturbances (Lynch, 1993). Some of the highest levels of stress have occurred in those women with a BRCA mutation who had not had cancer nor cancer-related surgery (Croyle & Lerman, 1993). For these women, the anticipation of developing breast cancer seems more distressing than actually having it. This is in contrast to people who already have a diagnosis of breast cancer and have the disease actualized. Sometimes there is more distress for those who continue to live with the risk of developing it, which may or may not occur (Pascreta, Jacobs, & Cataldo, 2002).

Researchers are beginning to identify possible risk factors in people who may experience sustained psychological distress. Identified risk factors include age—younger women are faced with a longer period between receiving positive test results and a possible diagnosis of breast cancer, meaning a longer period of uncertainty and anticipation; history of cancer in the patient's mother; a family history of breast and ovarian cancer and the unfavorable prognosis associated with such a history; and a diagnosis of cancer in many relatives combined with a history of the disease in a mother (http://www.cancer.gov/cancertopics/pdq/genetics/breast-and-ovarian). Women who test negative may also need psychosocial services, as they may still worry about getting cancer and feel guilty if they test negative (Lerman, 1995).

be increasingly faced with decisions regarding presymptomatic gene testing, and prenatal, carrier, and susceptibility testing as well. Any individual may be the first one in his or her family to participate in the decision-making process involved in genetic testing.

The consideration of genetic testing to learn about one's future health status is eloquently explained by Chris as she moves through the process of deciding whether or not to pursue gene testing for HD. The identification of the HD gene in 1983 made this disorder one of the first autosomal dominant inherited disorders for which predictive testing was possible (Gusella et al., 1983). Recommendations for presymptomatic testing for HD have been developed and are outlined in **Box 8-3.**

The desire to reduce uncertainty is often given by individuals, such as Chris, as the reason why they are undergoing genetic testing for HD. People seeking

Box 8-3 Presymptomatic Testing for Huntington Disease: Recommendations and Considerations

Current Guidelines

The Huntington Disease Society of America (*http://www.hdsa.org*), together with the world Federation of Neurology Research Group on Huntington Disease and the International Huntington Disease Association, has developed guidelines for presymptomatic testing for HD.

Designated HD Testing Center

It is recommended that individuals considering presymptomatic HD testing be referred to a designated HD testing center. Testing centers can be located through the Huntington Disease Society of America *(http://www.hdsa.org)*.

Team Approach

A team approach to presymptomatic HD testing is used. Team members include medical geneticist, neurologist, psychiatrist, genetic counselor, advanced practice nurse in genetics, psychologist, and social worker. Primary care providers are advised to refer individuals interested in testing to the nearest designated HD testing center where trained personnel, who are best equipped to provide counseling and administer any recommended testing, are available.

Components of Recommended HD Testing Programs

It is recommended that the referring provider look for the following components in any HD testing program under consideration:

- Initial contact is by telephone and includes a prescreening interview with the individual who is exploring the idea of having HD presymptomatic testing for self or as part of a family.
- Three, in-person pretesting sessions are held at the testing center where genetic counseling, and neurological and psychological examinations will be offered and are accompanied by the provision of additional information sources, such as reading materials, to further ensure informed consent.
- A fourth session is held at the testing center to discuss the test results with tested individual(s) as a routine part of the testing program.
- Post receipt of the test results, follow-up counseling sessions are held over a 2-year period.

Box 8-3 continued

Prior to any participation in a testing program, individuals are encouraged to identify a close friend or partner who can serve as a companion throughout the process. Also recommended, and especially important if the individual lives at a distance from the testing program, is identification of a local counselor (e.g., psychologist, social worker, psychiatrist, or other mental health specialist).

Pretest Counseling

Pretest counseling is one of the most important aspects of testing. Each person considering testing needs to be informed of certain facts.

- Clinical, psychological, and genetic implications of HD
- Their chance of having inherited the HD gene
- The possibility for them to transmit/have transmitted the HD gene to their children
- The available testing options
- The limitations of current testing technology
- The accuracy of the test results
- The possibility of ambiguous test results
- The negative implications of being tested and receiving results
- The need to carefully consider all risks, benefits, and limitations of the currently available diagnostic testing technology
- The need to include in those considerations any possible implications after being tested and receiving actual test results, whether positive or negative, for the future (i.e., ramifications of a social, legal, ethical, or personal nature).

Testing of Minors

Testing of minors (children under 18 years of age) is not currently recommended unless there is a medically compelling reason, such as the appearance of symptoms. Otherwise, testing is not considered to be in the child's best interest. Differences of opinion exist, and testing centers and programs are left to develop their own policies with regard to testing of children.

Patient Resources

Huntington Disease Society of America
(HDSA) 140 West 22nd Street, 6th Floor
New York, NY 110011-2420
292-242-1968 or 800-345-4372
http://www.hdsa.org

(Potter, Spector, & Prior, 2004)

presymptomatic testing for HD expect to gain relief from uncertainty (Tibben, Timman, Bannink, & Duivenvoorden, 1997). Most individuals choosing presymptomatic testing, like Chris, anticipate that the test will confirm that they have the HD gene, but want relief from not knowing. Many people monitor themselves and watch for indications that the disease is beginning. Many seek testing to make decisions about their own health care and to state their preference early for a nursing home at the later stages of their disease. They expect that knowing will help them to plan for the future. Chris, for example, took out life insurance before undergoing testing, and had decided to remain in a job that she didn't really like, if she tested positive, because they offered long-term disability insurance. Individuals wanting to plan for the future also desire information to better predict their children's chances of developing HD.

As we see in Chris's story, the anticipated benefits of presymptomatic HD testing are identified within a context of acknowledging the potential for negative outcomes. Expectations of adults who have made the decision to seek HD presymptomatic testing often include concern about the potential for loss of genetic privacy. Like Chris, these individuals did not share their concerns or decision to pursue testing with their health care provider. They express fears of loss of health insurance and employment. Individuals deciding to pursue HD presymptomatic

TABLE 8-1 Presymptomatic DNA Testing for Huntington Disease: Benefits and Risks

Positive Effects of Positive and Negative DNA Test Results

- Relief from uncertainty of not knowing
- Increased sense of control
- Greater reliance and faith in spiritual and religious beliefs.

Negative Effects of a Positive DNA Test Result

- Guilt at having possibly transmitted HD gene to children
- Depression
- Risk for decline in psychological well-being and coping over time.

Negative Effects of a Negative DNA Test Result

- Psychological distress due to having built life around being affected with HD
- Feelings of guilt for being spared HD
- Increased concerns regarding caregiving for other affected family members.

(Decruyenaere et al., 1996; Taylor & Myers, 1997; Williams & Schutte, 1999; Williams et al., 2000)

testing have also described venturing into an area where other family members had no experience or interest. Some anticipate withdrawal of family support based on family reactions to their decision to pursue testing, while others feel that the decision to disclose test results will change relationships that had been formed around the common bond of sharing a risk to have the disease (Williams et al., 1999). **Table 8-1** outlines identified risks and benefits of presymptomatic testing.

Since the discovery of the HD gene, researchers have investigated the impact of predictive genetic testing for HD in individuals who have been identified to have the HD gene mutation. The initial concern was the potential for suicide after confirmation of the presence of a gene mutation for a neurodegenerative disease for which there is no effective treatment. These fears have not been confirmed, however, and the rates of adverse outcomes including suicide are reported to be similar to that of family members living at risk for HD who have not had testing (Almqvist & Bloch, 1996; Lawson et al., 1996). One reason suggested in the literature may be that family members who request testing, such as Chris, believe that they can cope with the results, while others avoid testing because they are afraid that they cannot cope with the information (Codori, Hanson, & Brandt, 1994).

How My Test Result Affected Me

The experience of waiting to learn about one's genetic makeup, as Chris points out, is intense and is only the beginning of a soul-searching process (Williams et al., 2000). For Chris and others, learning that one has a normal HD gene is a life-changing event. It causes people to question their view of their past history, consider the meaning of the present, and begin to revise their future life plan. These strategies have been compared to those typically used by individuals approaching old age (Williams, 2000).

A "Redefinition" process begins (Williams, 2000). Throughout this process, those individuals testing negative for the HD mutation search for meaning and a desire to resolve guilt over one's good fortune when compared to the potential fate of other family members. For some individuals, this new relationship with family involves revising expectations. This is the result in some families of preselection—the singling out in advance of an asymptomatic family member who it is believed will eventually develop the disease (Kessler, 1988). The individual receiving normal results may experience survivor guilt as he or she receives information that contradicts personal and family expectations. The concept of survivor guilt has been related to trauma and raises the question of the traumatic aspect of experiencing predictive genetics testing and learning that one is not "at risk" (Williams et al., 2000).

Those who receive normal results, like Chris, do not always experience uncomplicated relief. As Chris notes:

It took about a year for the reality of these results to actually hit home. I asked the genetic counselor if I could have a hard copy of the test results, which she agreed to immediately. I needed to read the report with my own eyes. I did not have HD, but for the last 34 years I had lived in the shadow of the disease, and it took a long time to sink in that I was "normal." I had the same chance as anyone else to live to be an alert old person. I also grieved the time I had spent not knowing and the choices I had made based on "preparing for the worst," assuming that I had HD. I began to see the ways in which living with the threat of HD had limited my life choices and had narrowed my view of available options. More importantly, I felt that now I could have a life. The implications of this fact and my options were staggering. For a time, I was overwhelmed by the fact that I didn't have Huntington Disease and could live my life any way I wanted just like anyone else. Small issues and big issues arose.

Some who have received normal results express feelings of guilt and feel they have betrayed their families. Other difficulties reported include disappointment that the test result did not resolve prior life problems, such as marital or employment difficulties or failed attempts to change poor eating and exercise habits. Qualitative studies by Williams et al. (1999) support what we learned in Chris's story—that individuals testing negative for the HD mutation were not prepared for the implications of a "good news" outcome. The receipt of normal results constitutes a loss of one's former self-definition and provides life-changing information in terms of career, marriage, and reproduction. The hope for relief from uncertainty is replaced with a new set of concerns that reflect the loss of the old self and a "very personal search for understanding one's changed identity. The redefinition process encompasse[s] one's view of self and the view of one's place within the family and society" (Williams et al., 1999, p. 266). Chris, for example, took several days alone to celebrate and reflect upon her "re-birthday." Knowing that she does not carry the HD gene has meant a change in employment and a newly discovered option for her life—living a long life. Chris has also transitioned into a role of caretaker, watching her mother and sister become symptomatic. Although she is free of the HD gene, she expresses that she will never be "HD-free."

Although presymptomatic testing resolves some uncertainties, we see through Chris's story that it raises others. For some, receipt of negative results for degenerative neurologic disorders causes them to experience new uncertainties regarding how to live their lives as persons with the potential for a healthy future. The negative test results may challenge lifelong-held beliefs. Having the opportunity to learn about one's future risk for disease or freedom from an inherited disease is new and expanding in rare and common diseases, posing new challenges and questions regarding traditional views of life span psychological processes (Williams et al., 1999).

Learning that one has inherited the HD gene and will develop symptoms of HD has other consequences as Jerry's story illustrates. Long-term follow-up studies in programs offering HD presymptomatic testing report that both carriers and noncarriers experience emotional reactions after disclosure of their test results, but no increase in long-term, adverse emotional consequences has been documented (van't Spijker & ten Kroode, 1997). Adverse reactions—suicide plans or attempts, psychiatric hospitalizations, depression, or breakdown of an important relationship—have occurred in nearly 15% of a group of people receiving a test result indicating increased risk for HD as well as those receiving results indicating a decreased risk for HD (Lawson et al., 1996).

Individuals who receive results indicating that they have inherited a mutation within the HD range report occurrences of feelings of emotional numbness, sadness, depression, and anger (Codori, Hanson, & Brandt, 1994; Decruyenaere et al., 1996). Some individuals with HD gene mutations, on the other hand, have expressed a low stress response, which may reflect a denial type of coping strategy, the long-term consequences of which are not known (Dudokdewit et al., 1998).

The process involved for those individuals who choose not to participate in presymptomatic HD testing is also complex. The decision-making process in this situation is influenced by risk perception and by perceived costs and benefits of testing. Personality characteristics such as ego, strength, and style of coping with threatening information are other factors that play a role in decision making about a presymptomatic HD test. Intrapersonal aspects, including concern about a partner and children, and emotional and/or unconscious mechanisms such as family dynamics and defense mechanisms, may also contribute to nonparticipation. The expected burden of pretest counseling for HD is another influence (Decruyenaere et al., 1997). For Jerry's children, concerns about long-term health benefits have played a significant role in their choice not to pursue presymptomatic testing.

More Decisions: Sharing My Genetic Information

As Chris tells us, sharing personal genetic information with others, both within and outside of the family, is of major concern. For Chris and others, where a genetic test identifies a presymptomatic individual or susceptibility to a condition, there are concerns about potential stigmatization and employer and insurance discrimination. Chris never shared her family history of HD for fear of discrimination. She made sure that she had applied for and received life insurance before undergoing genetic testing due to concerns about discrimination. Even after testing negative for the HD gene, Chris has maintained privacy.

Sharing genetic information with family presents additional challenges. Communication of genetic test results to relatives is not universal in families, and seems to be influenced, at least in part, by the genetics of the particular disease

(Sorenson, Jennings-Grant, & Newman, 2003). In addition to anticipating the effect that test results will have on their own personal lives, individuals with a family history of HD consider with whom they will share their test results. Participants in HD testing programs describe a range of expected reactions of family members, including anticipation of withdrawal of family support. These individuals have also expressed feelings that a negative or positive test result would make them feel different from other family members or change relationships that had been formed around the common bond of sharing a risk to have the disease. Those choosing to have HD testing have described strategies they plan in deciding to whom they will disclose their test results, reflecting their assessments of potential consequences of the news for family members. Some have indicated a desire not to reveal test results to family members who already have the disease or to their caregivers (Williams et al., 1999). The decision to have presymptomatic HD testing has upset family equilibrium, and the receipt of normal results has set people apart from the family.

In contrast to Chris's and other family situations, Jerry's family situation has been more supportive. With help from his genetic counselor and psychologist, he and his wife were able to prepare for sharing information with their children and siblings. Jerry had a support plan in place—referral to Steve—so that his children would receive additional information from a knowledgeable specialist.

New Roles and Relationships: Genetics and My Family

Disclosure of genetic test results and genetic information has the potential to alter family relationships in both positive and negative ways. The impact may vary depending upon the mode of inheritance. With autosomal dominant conditions such as HD, each child has a 50% chance to inherit the gene for the disorder, while for X-linked or autosomal recessive inherited conditions, children may inherit a gene that either confers carrier status or causes the condition.

Genetic conditions are a concern for families, not only from the perspective of inheritance and risk, but also from a psychological perspective in terms of the impact on the family itself. Families share caregiving responsibilities for members with genetic conditions, as Chris describes in her story. The family is also the transmitter of genetic health information among its members, as we see in Jerry's story. Families are social systems, and within these systems information, including genetic information, is selectively communicated—influenced by the values, beliefs, and life experiences of the individual family members. Poor understanding of genetic inheritance, family myths, and attitudes may limit the ability of family members to share information with each other. Decisions to have presymptomatic testing for HD, as related by Chris in her story and others, may be made without discussions or support from other family members (Williams & Schutte, 1997). Knowledge, attitudes, feelings, and communication patterns about genetic health information among family members are all-important factors that affect the family as a unit.

The nature of the genetic test also influences the discussion of genetic information, including test results, within families and has psychological effects. Individuals having genetic testing for cancer susceptibility, for example, may share their test results, in part so that other family members can have the option to undergo appropriate risk management strategies and screening. The desire to help family members is related to the desire to have genetic testing for cancer, such as colon cancer (Vernon et al., 1999). With regard to presymptomatic testing for HD, on the other hand, the anticipated impact that genetic test results could have on family members is an important factor for some individuals' decision making (Binedell & Soldan, 1997).

The impact of DNA testing for HD on the family is profound (Sobel & Cowan, 2003). Chris's story gives us a glimpse into the impact of genetic information on the family—their beliefs about HD, patterns of communication, and how the family members live their lives with and around an incurable genetic condition. Research studying the impact of genetic information and testing has usually focused on the individual and how individuals seek, select, and respond to genetic testing, while recognizing that "genetic diseases are family diseases" (Sorenson & Botkin, 2003, p.1). This initial research paradigm is now expanding to study the family and the broader social context in which people live their lives as an additional important means for understanding the impact of genetic testing. Newer studies look at the role of the family in awareness of genetic testing, in decision making with regard to genetic testing, in adjustment to the results of genetic testing, and in communication with relatives about genetic testing. The ethical and legal responsibilities of investigators, providers, and family members regarding genetic testing are also being explored (Sorenson & Botkin, 2003).

Newer research studies of HD are addressing the gap between offering the predictive test and understanding its impact on the wider scale of family functioning (Feetham, 1999; Peters, Djurdjinovic, & Baker, 1999). Defining the family as the unit of analysis within the framework of the Systems Theory links interactions of individuals, families, and the social environment (Sobel & Cowan, 2003). The Family Systems Theory emphasizes that family members' relationships influence each other in such a way that a change in one member influences a change in others and in the system as a whole that, in turn, affects the initial member. Measures used to describe family systems include family structure, communication patterns, cohesion, beliefs, life cycle norms, and multigenerational patterns (Sobel & Cowan, 2000).

For families with HD, the existence of a genetic test exerts influence on the family system. Areas of family functioning that are affected include family membership, family patterns of communication, changes in their current relationships in response to genetic test results, and future caregiving concerns (Sobel & Cowan,

2000). Some of those tested, for example, particularly those with a normal result, feel that they have lost their membership in their families. In other families, differing test results among siblings causes a rupture in the relationship. From the family's perspective, factors that influence the impact of testing on their membership, communication, and caregiving concerns are the test results of other siblings and the testing process itself, for example, whether the family approached the testing process as a unit or whether individuals approached it alone. Families that approach the possibility of genetic testing as a unit seem to experience less disruption in membership, maintain more open communication, and are less traumatized by caregiving issues. In contrast, families in which the individual's choice to be tested was a solitary matter experienced more disruption (Sobel & Cowan, 2000).

Families with chronic and terminal illness such as HD and cancer are often pervaded with a sense of loss. In families with HD, the sense of loss is magnified by a history of premature loss and also by the knowledge of genetic information, now available through DNA testing, of future loss (Sobel & Cowen, 2003). The anxiety in a family has been described as being both vertical and horizontal (Carter, 1978). The vertical flow in a family system includes patterns of relating and functioning that are transmitted down through generations of a family. In families with HD, this flow includes how the family deals with the history of HD: Is it one of secrecy, shame, guilt or openness? Horizontal flow and stressors include the developmental tasks of a family: life cycle transitions, unpredictable events (e.g., untimely death and chronic illness), or accidents. Families are often stressed at transition points in the family developmental process. Issues and conflicts arise when there is an interruption in family developmental life cycle (Brouwer-Dudokdewit et al., 2002). Families with a history of difficulties in communication, for example, keeping secrets, taking action without consulting others, competition, or feuding as a means to avoid difficult issues, can carry these patterns of behavior into the HD illness spectrum and the HD genetic testing process. These observations of HD and the impact on family are likely to be applicable to individuals at risk for other autosomal dominant, mind-life onset genetic conditions (Dudokdewit et al., 1998).

Nursing Implications

The Individual: Nursing Interventions

The increasing capacity to predict genetically increased risk for adult-onset, chronic, and life-impacting diseases has tremendous implications for nurses in all practice settings. Genetic information, once associated with birth defects and children with rare diseases, is now pertinent across the life span. Chris's story illustrates that genetic information, both the family history of and genetic testing for HD, makes deep inroads into the individual and family life over generations. Through

Chris's story, we learn that genetic information has the potential to change an individual's entire life perspective. The challenges of changing lifelong-held beliefs and coping with the meaning of a normal predictive test result for HD are unique in a health-related testing situation (Williams et al., 2000). Chris emphasizes this point when she says "I might have liked to have had the chance to meet again with my genetics counselor at regular intervals after testing to discuss the implications of testing on my family relations. Perhaps genetic medicine will evolve a life span developmental approach to assessment, diagnosis, and treatment, with families having an identified geneticist and/or genetic counselor, much like we have a general family physician." Nurses can help facilitate this process by ensuring that individuals and families have access to ongoing support throughout the life span for coping and living with genetic information. This begins with nursing assessments and referrals for genetic counseling, psychological counseling, and support services. **Table 8-2** presents topics for nursing assessment.

Genetic counseling is based on the identification and analysis of potential risk factors in an individual, family, or group. Components of genetic counseling include health teaching, decision making, assessing, and ensuring emotional and psychological adjustment. Nurses participate in the genetic counseling process when they obtain a comprehensive health history to identify individuals in the family who may be at risk of having an inherited condition or of passing on a gene for the condition to their children. Taking a family history allows the nurse to observe family interactions and patterns of communication, and to assess attitudes and understanding about a genetic condition (Williams & Schutte, 1999).

Genetic counseling includes discussing advantages, risks, and costs of genetic testing as well as minimizing any coercive actions that could force an individual to choose testing or to feel guilty because they choose not to have testing. Genetic testing will be encountered by nurses in all practice settings (see Chapter 4 on Genetic Testing). Assessing individuals who may benefit from genetic testing, discussing genetic testing and screening, and helping individuals become more psychologically prepared for either positive or negative testing outcomes is an important nursing role in the testing process. From Chris, and through other studies of individuals with HD, we learn that the decision to seek presymptomatic testing is often made without discussion with health care providers. This suggests that nurses may not be asked for information by those considering genetic testing, leading to the loss of opportunity to assess the individual's knowledge and expectations. Nurses could use a pretest assessment that would help at-risk individuals identify areas in which their expectations may not be met and refer those individuals to genetic counseling or other psychosocial counseling services for further information and preparation for testing. Discussing potential benefits and risks of genetic testing and strategies to share genetic test results and information with family members is part of the nurse's intervention (Williams et al., 2000; Williams & Schutte, 1999).

TABLE 8-2 Topics for Nursing Assessment with Individuals and Families Undergoing Testing for a Genetic Condition

- *Family values, beliefs, and rituals.* Inquire about favorite stories in the family, and important celebrations and rituals that family members attended; who is in the extended family; what adjectives describe the family (e.g., close, warm, distant, flexible, inflexible).
- *The Family Story about the genetic condition and how its legacy has influenced family relationships and coping.* Inquire about what relatives were affected before the disorder was able to be diagnosed, the family's response, and how the disorder was explained to others; have family members made subsequent life decisions based on the diagnosis (e.g., genetic testing, marriage, children, occupation).
- *Family involvement in the test participants decision to have the genetic test.* Inquire about what made the participant decide to have the test; whether other family members have considered testing; which family members have been involved in discussions about the testing and which have not; how did you (or do you wish to) receive results, with or without family; with whom in the family did you (or do you plan to) share genetic test results.
- *Impact of test results on relationships with sisters–brothers.* Ask individual to describe his or her relationship with brothers and sisters; inquire about what role the individual has played as youngest, oldest, middle; ask whether the individual's role in the family has changed as a result of new genetic information and what was their reaction.
- *Impact of test results on other family members.* Inquire whether the individual has told a parent or child and what were their reactions; whether the extended family has been informed and, if so, their reaction; inquire about changes in family membership and whether these occurred before or after testing; whether family rituals and celebrations have changed after testing; have family friends been told of the testing and, if so, their reactions.
- *General inquiries for individuals–families having had genetic testing.* Inquire in what other ways genetic testing has made a difference in the family; would the individual encourage other family members to be tested.

(Adapted from a questionnaire used by Sobel & Cowan in assessing impact of genetic testing for HD on the family system, 2003)

Chris recommends that closer integration with allied health professionals and collaboration with psychologists, social workers, and family counselors will be an essential factor in addressing long-term ramifications of genetic testing. Jerry talks about his ongoing relationship with genetic counselors and medical professionals attuned to caring for individuals with HD. Providing referrals for specific counseling, testing, management, and supportive services is a nursing intervention that meets this need. Referrals for appropriate consultation help individuals clarify any misinformation, identify expectations that may not be met by genetic testing, and create a support network (Williams et al., 1999). Many genetic conditions, like HD, are relatively rare, and locating providers with necessary expertise may be challenging. Nurses with knowledge about what services are available and how to access them can help individuals in considering and evaluating desired services. When nurses refer individuals to services, such as genetics specialists for further assessment and counseling, they can explain the purpose. This helps individuals understand why they are being referred and what they can expect.

Identification of desired outcomes of genetic counseling services in order to evaluate their effectiveness is a critical component of the nursing process when caring for individuals with genetic-related concerns. A nursing-sensitive outcome of a measurable client or caregiver state, behavior, or perception is influenced by and sensitive to nursing interventions (Johnson & Maas, 1997). **Table 8-3** lists nursing outcomes of genetic counseling that apply to individuals and families who are participating in the genetic counseling process.

TABLE 8-3 Nursing Outcomes for Individuals and Families Undergoing Genetic Counseling and Testing

A nursing-sensitive outcome is one that is a measurable client or caregiver state, behavior, or perception that is influenced by and sensitive to nursing interventions. Nurses can link outcomes and monitor care of individuals and families pursuing and receiving genetic information through the Genetic Counseling intervention. Nursing outcomes include:

- **Enhanced Knowledge**
  - *Knowledge: Disease Process.* The extent to which the individual–family understands information conveyed about the genetic condition, its pattern of inheritance, and the health implications for family members
  - *Knowledge: Health Resources.* Knowledge of health resources, including plan for follow-up care and use of community resources
- **Participation**
  - *Health Care Decisions.* The ability of the individual–family to collaborate and negotiate with health care providers in selecting and evaluating health care options
  - *Decision Making.* The client's ability to choose between alternatives
- **Risk Detection.** The individual–family's ability to identify health threats, including family history and genetic background
- **Health Seeking Behavior.** The individual–family's actions aimed at promoting wellness, recovery, and rehabilitation
- **Acceptance**
  - *Health Status.* The individual–family's ability to adjust to or reconcile his or her health circumstances
  - *Anxiety Control.* Relief from anxiety from genetic testing can be measured with indicators of the Anxiety Control outcome
  - *Coping.* The individual–family's ability to manage stressors
  - *Hope.* The individual–family's ability to maintain optimism and hope in the face of unexpected loss

(Johnson & Maas, 1997; Williams & Schutte, 1997; Williams & Schutte, 1999)

The Family: Nursing Interventions

Genetic information, such as the family history of a genetic condition or positive or negative genetic test results, has been shown to interfere with family communication pathways among members. With genetic testing, for example, it is often the case that what persons learn about themselves from the genetic test can have significant implications for other family members. When one family member undergoes testing, other family members, in a sense, are also tested. A common assumption in many clinical genetics services is that family members do, or will if requested, communicate their genetic test results with relevant family members. This may or may not be the case, and it may be affected by concerns such as stigmatization within the family and the emotional coping ability of those family members with whom they choose to share the information (see Table 8-3) (Sorenson, Jennings-Grant, & Newman, 2003).

Nurses can assist individuals who are undergoing genetic testing with communicating that information to at-risk family members by including such education and counseling in their discussions about genetic testing. As a part of this discussion, nurses can suggest that families be involved in the process of a member's decision to have the predictive test. Nurses can initiate and participate in research and translate their knowledge into ways of helping people communicate the results of their testing to family members. This can be accomplished with the aim of developing more effective counseling and educational approaches. This type of research would recognize the familial context of genetic testing whether that be carrier, diagnostic, or predictive genetic testing (Ormond et al., 2003; Sorenson et al., 2003).

Greater family involvement in the genetic testing process must be balanced by respect for an individual's right not to know of his or her status or the status of other family members. A pretest interview for family members wishing to be involved in the testing process will help address differences in coping among family members and support the interest of maintaining family unity (Sobel & Cowan, 2003). Potential topics covered in the interview are listed in Table 8-2.

With the advent of predictive testing, it has been observed that it is not so much the test result itself that may disrupt a family; rather it is the changed expectations and possibilities for the future. Families are forced to cope with disruptive events and untimely deaths of family members. While some families can make certain transitions, others get stuck at particular transition points. In addition to understanding the personal nature of genetic information, Chris's and Jerry's stories help us see that genetic information is not only of concern to the individual, but it is also information that affects others in the family. The process of Redefinition experienced by Chris and others is also a process that applies to the family framework. Nurses can have a central role in helping families find support for this process, in identifying outcomes of new bonds and roles within the family resulting from gene identification in family members, and in investigating strategies that promote family well-being (Williams et al., 2000). Helping families find psychosocial support to learn what it is that hinders them from making the transition into the next life stage will also help them resolve these issues so they can move on is an important nursing intervention.

Nurses may be the first health care professionals to hear individual and family stories about the impact of illness and disease such as HD or cancer. For many families it may be helpful, and even transforming, to stimulate a family to tell a more coherent story about its experiences with a genetic condition. Individuals participating in psychological studies often report that the interview helped them to put things in perspective again, to make a more or less coherent story of their experiences with HD. Many individuals and families may be still working on creating their coherent story and need to come and talk, while others may have their coherent story and wish for further support (Brouwer-Dudokdewit et al., 2002). Nurses can support these families by making a referral to other health professionals such as psychologists, family therapists, and psychiatrists.

Summary

Nurses will become increasingly involved in managing genetic information, particularly with patients who are making the decision to have or who have already had genetic testing. It is important for nurses to have an understanding about the psychological and social consequences of genetic information gained from genetic testing to be able to fully support both individuals and families throughout the process. This includes recognition of situations where an individual may feel coerced into testing or knowing, and supporting the individual's right not to know. Nurses, with their family-centered approach to care, can participate in future research that will provide insights and consideration to everyone involved in the process of testing and to the issues that individuals and families have identified.

Chapter Activities

1. Chris describes her family history of HD and how the diagnosis of HD not only has implications for her own health, but also for her reproductive decisions in terms of the chance to pass on the altered gene to a future generation. The diagnosis of HD also has relevance for her other family members who may be at risk to have inherited the same condition.

   Read the following case history:

Susan, age 30, reports to you a family history of colon cancer in her mother at age 45. Susan's brother, age 38, has recently been diagnosed with colon cancer. Susan tells you that in talking with other relatives she has learned that her mother's brother (her uncle) died from colon cancer at age 50, and her mother's sister has a history of endometrial cancer at age 29. Susan expresses her concern to you about her risk for cancer. "I don't want to have children if I have the gene for this cancer." She tells you that she has been pressing her brother to have genetic testing so that she can find out more about her risk, but he has refused to have testing.

Construct Susan's family pedigree. How would you respond to Susan's concerns? What resources and referrals would you offer Susan?

2. Jerry's story of HD describes HD testing when it is used to confirm a diagnosis and the ramifications of this information to the family. Through Jerry's story, we see how a family approaches sharing of test results and decision making about presymptomatic testing. Discuss how Jerry's diagnosis of HD may be empowering or disabling for his children and family.

References

Admi, H. (1995). Nothing to hide and nothing to advertise: Managing disease-related information. *Western Journal of Nursing Research, 17*(5), 484–501.

Almqvist, E., & Bloch, M. (1996, October–November). World-wide survey on catastrophic events following predictive testing for Huntington Disease. Poster session presented at the annual meeting of the American Society of Human Genetics meeting. San Francisco, CA.

Baum, A., Friedman, A. L., & Zakowski, S. G. (1997). Stress and genetic testing for disease risk. *Health Psychology, 16,* 8–19.

Biescker, B. B. et al. (2000). Psychosocial factors predicting BRCA1/BRCA2 testing decisions in members of hereditary breast and ovarian cancer families. *American Journal of Medical Genetics, 93*(4), 257–263.

Bindell, J., & Soldan, J. (1997). Nonparticipation in Huntington's Disease predictive testing: Reasons for caution in interpreting findings. *Journal of Genetic Counseling, 6,* 419–432.

Bredart, A. et al. (2001). Psychological dimensions of BRCA testing: An overshadowed issue. *European Journal of Cancer Care (England), 10*(2), 96–99.

Brouwer-Dudokdewit, A. C., Savenije, A., Zoeteweij, J. W., Maat-Kievit, A., & Tibben, A. (2002). A hereditary disorder in the family and the family life cycle: Huntington Disease as a paradigm. *Family Process, 41*(4), 677–692.

Carter, E. A. (1978). The transgenerational scripts and nuclear family stress: Theory and clinical implications. In R. R. Sager (Ed.). *Georgetown family symposium* (vol.3). Washington, DC: Georgetown University.

Clubb, R. (1991). Chronic sorrow: Adaptation patterns of parents with chronically ill children. *Pediatric Nursing, 17,* 461–466.

Codori, A., Hanson, R., & Brandt, J. (1994). Self-selection in predictive testing for Huntington's Disease. *American Journal of Medical Genetics, 54,* 167–173.

Croyle, R. T., & Lerman, C. (1993). Interest in genetic testing for colon cancer susceptibility: Cognitive and emotional correlates. *Preventive Medicine, 22*(2), 284–292.

Decruyenaere, M., Evers-Kiebooms, G., Boogaerts, A., Cassiman, J., Cloostermans, T., Demyttenaere, K., Dom, R., Fryns, J., & Van den Berfhe, H. (1996). Prediction of psychological functioning one year after the predictive test for Huntington's Disease and impact of the test result on reproductive decision making. *Journal of Medical Genetics, 33,* 737–743.

Decruyenaere, M., Evers-Kiebooms, G., Boogaerts, A., Cloostermans, T., Cassiman, J., Demyttenaere, K., Com, R., Fryns, J., & Berghe, H. (1997). Non-participation in predictive testing for Huntington's Disease: Individual decision-making, personality and avoidant behavior in the family. *European Journal of Human Genetics, 5,* 351–363.

Dudokdewit, A., Tibben, A., Buivenvoorden, H., Niermeyer, M., & Passchier, J. (1998). Predicting adaptation to presymptomatic DNA testing for late onset disorders: Who will experience distress? Rotterdam Leiden Genetics Work Group. *Journal of Medical Genetics, 35,* 745–754.

Feetham, S. L. (1999). Families and the genetic revolution: Implications for primary health care, education and research. *Families, Systems & Health, 17*(1), 27–43.

Gusella, J., Wexler, N., Conneally, P., Naylor, S., Anderson, M., Tanzi, R., Watkins, P., Ottina, K., Wallace, M., Sakaguchi, A., Young, A. Shoulson, I., Bonilla, E., & Martin, J. (1983). A polymorphic DNA marker genetically linked to Huntington Disease. *Nature, 306,* 234–239.

Johnson, M., & Maas, M. (1997). *Nursing outcomes classification* (NOC). St Louis: Mosby-Year Book.

Kash, K. M. et al. (1992). Psychological distress and surveillance behaviors of women with a family history of breast cancer. *Journal of the National Cancer Institute, 84*(1), 24–30.

Kessler, S. (1988). Preselection: A family coping strategy in Huntington Disease. *American Journal of Medical Genetics, 31,* 617–621.

Lawson, K., Wiggins, S., Green, T., Adam, S., Bloch, M., Hayden, M., & the Canadian Collaborative Study of Predictive Testing (1996). Adverse psychological events occurring in the first year after predictive testing for Huntington's Disease. *Journal of Medical Genetics, 33,* 856–862.

Lerman, C., Daly, M., Masny, A., Balshem, A. (1994). Attitudes about genetic testing for breast-ovarian cancer susceptibility. *Journal of Clinical Oncology, 12,* 843–850.

Lerman, C. et al. (1993). Mammography adherence and psychological distress among women at risk for breast cancer. *Journal of the National Cancer Institute, 85*(13), 1074–1080.

Lerman, C. et al. (1995). Interest in genetic testing among first-degree relatives of breast cancer patients. *American Journal of Medical Genetics, 57*(3), 385–392.

Lynch, H. T. et al. (1993). DNA screening for breast/ovarian cancer susceptibility based on linked markers: A family study. *Archives of Internal Medicine, 153*(17), 1979–1987.

Mishel, M. (1992). Commentary. *Western Journal of Nursing Research, 14,* 46–68.

Olshansky, S. (1962). Chronic sorrow: A response to having a mentally defective child. *Social Casework, 43,* 190–193.

Ormond, K. E., Mills, P. L., Lester, L. A., & Ross, L. F. (2003). Effect of family history on disclosure patterns of cystic fibrosis carrier status. *American Journal of Medical Genetics Part C (Semin. Med. Genet.), 119C,* 70–77.

Pascreta, J. V., Jacobs, L., & Cataldo, J. K. (2002). Genetic testing for breast and ovarian cancer risk: The psychosocial issues. *American Journal of Nursing, 102*(12), 40–48.

PDQ® Cancer Information Summaries: Genetics Genetics of Breast and Ovarian Cancer (PDQ®) *http://www.cancer.gov/cancerinfo/pdq/genetics/breast-and-ovarian*

Peters, J. A., Djurdjinovic, L., & Baker, F. (1999). The genetic self: The Human Genome Project, genetic counseling and family therapy. *Families, Systems & Health, 17*(1), 5–25.

Potter, N., Spector, E., & Prior, T. (2004). Technical standards and guidelines for Huntington Disease testing. *Genetics in Medicine, 6*(1), 61–65.

Rosenthal, G., et al. (1995). The Strang National High Risk Registry: A program for delivery of cancer risk information and a resource for research. *Annals of the New York Academy of Sciences, 768,* 317–322.

Scanlon, C., & Fibison, W. (1995). *Managing genetic information.* Washington, DC: American Nurses Association.

Secretary's Advisory Committee on Genetics, Health, and Society. Accessed September 15, 2004 from, *http://www4.od.nih.gov/oba/sacghs.htm.*

Shiloh, S. (1996). Decision-making in the context of genetic risk. In T. Marteau and M. Richards (Eds.), *The troubled helix: Social and psychological implications of the new human genetics,* 82–103. Cambridge, MA: Cambridge University Press.

Sobel, S. K., & Cowan, D. B. (2000). Impact of genetic testing for Huntington Disease on the family system. *American Journal of Medical Genetics, 90,* 49–59.

Sorenson, J. R., & Botkin, J. R. Guest Editors (2003). Genetic testing and the family. *American Journal of Medical Genetics Part C (Semin. Med. Genet.), 119C:* 1–2.

Sorenson, J. R., Jennings-Grant, T., & Newman, J. (2003). Communication about carrier testing within hemophilia A families. *American Journal of Medical Genetics Part C (Semin. Med. Genet.), 119C:* 3–10.

Spitzer, A. (1992). Coping processes of school-age children with hemophilia. *Western Journal of Nursing Research, 14*(2), 157–169.

Taylor, C., & Myers, R. (1997). Long-term impact of Huntington Disease linkage testing. *American Journal of Medical Genetics, 70,* 365–370.

Tibben, A., Timman, R., Bannick, E., & Duivenvoorden, H. (1997). Three-year follow-up after presymptomatic testing for Huntington Disease and partners. *Health Psychology, 16,* 20–35.

van't Spijker, A., & ten Kroode, H. (1997). Psychological aspects of genetic counseling: A review of the experience with Huntington Disease. *Patient Education and Counseling, 32,* 33–40.

Vernon, S. W., Gritz, E. R., Peterson, S. K., Perz, C. A. et al. (1999). Intention to learn results of genetic testing for hereditary colon cancer. *Cancer Epidemiology, Biomarkers & Prevention, 8,* 353–360.

Williams, J., & Schutte, D. (1997). Benefits and burdens of genetic carrier information. *Western Journal of Nursing Research, 19*(1), 71–81.

Williams, J. K., Schutte, D. L., Evers, C., & Forcucci, C. (1999). Adults seeking presymptomatic gene testing for Huntington Disease. *Image: Journal of Nursing Scholarship, 31,* 109–114.

Williams, J. K., & Schutte, D. L. (1999). Chapter 26: Genetic counseling. In *Nursing Interventions: Effective Nursing Treatments* (3rd ed.). Bulechek, G. M. & McClosky, J. C. (Eds.). Philadelphia: W. B. Saunders Company.
Williams, J. K., Schutte, D. L., Evers, C., & Holkup, P. A. (2000). Redefinition: Coping with normal results from predictive gene testing for neurodegenerative disorders. *Research in Nursing and Health, 23*(4), 260–269.

Chapter 8 Interdisciplinary Commentary

The Role of the Psychologist in Genetics Clinical Care

Andrea Farkas Patenaude, PhD
Psychologist

The expansion of genetics and its integration into the clinical care of patients has created new opportunities and new responsibilities for many health professionals (Collins & McKusick, 2001). Psychologists do not provide frontline genetic counseling, but they can serve a number of useful roles in concert with the physicians, nurses, and genetic counselors who do provide genetic services (Patenaude, Guttmacher, & Collins, 2002).

Applied genetics is at the crossroads where science and social science meet. Genetic medicine incorporates the latest scientific findings from a field marked by rapid advances with direct clinical implications and aspects of psychology, ethics, and health policy. Advanced genetic knowledge is not useful for patients unless they understand the implications of genetic information. If prevention or early detection is possible, patients must at least understand the recommended steps to prevent or detect the diseases they are at increased risk to develop. If, as with Huntington's Disease (HD), positive mutation status means certainty that the individual will develop an incurable disease, patients undergoing genetic testing must understand the social and psychological implications of this information. They may still choose to be tested to better plan for their future. Questions about whether or when to undergo genetic testing or counseling involve more than discussions of health and disease; they involve an understanding of the emotional, financial, and quality of life implications of hereditary illness predisposition. People vary enormously in their attitudes toward health information (Miller, Shoda, & Hurley, 1996). Some truly believe that "knowledge is power" and actively seek all available forms of information relevant to their health. Others take a more cautious approach, believing that sometimes there is harm in the information itself or that it is not so useful in predicting what will actually happen to them. It is often important to recognize when it may be advisable to restrain someone who is making decisions without the necessary time to process the feelings that have been aroused by the actions they plan to take. Similarly, with others, it may be important to help them lower their barriers to acquiring the knowledge they need in order to make informed decisions about genetic testing, screening, or surveillance for mutation carriers or prophylactic surgery.

Nurses and psychologists have a long history of working closely together and understanding and respecting the overlap and distinctions in their professional roles. Both groups are concerned about the patient's well-being and are also concerned about family implications of the patient's condition. Genetic nurses have a heavy burden of varied tasks to accomplish with patients concerned about hereditary disease predisposition. This includes informing patients about complex genetic information, guiding them through a decision about genetic testing, overseeing the actual testing (the taking of their blood, the wait for results, and the disclosure of results), giving advice about screening and surveillance and possible risk-reducing measures, helping with informing family members, and making appropriate referrals for related services. While nurses are trained in recognizing signs of psychological distress in patients, they may have too little time in the context of their other roles to explore in depth the emotional concerns that surface in the face of exploration of the past and future implications of hereditary disease predisposition. Psychologists can focus on the immediate emotional concerns, on the arousal of early experiences of illness and loss, on fears about future illness for the patient and for other family members, and on the myriad issues and feelings that arise in the sharing or failure to share genetic information within families.

Not all patients require or want in-depth discussion about their feelings with regard to genetic testing. But some patients clearly ask for and are helped by the opportunity to focus on the range of feelings, which can arise in this context. Other patients who come for genetic consultations raise questions for the staff who see them about the advisability of their undergoing genetic testing at this point. While this concern may center on the question of whether testing would make any difference in the management of the patient, sometimes the concern raised is about the emotional stability of the patient or their internal resources to handle the stress of waiting for and receiving genetic test results. Psychologists can see these patients for more extensive evaluation of their current psychological status, can review their ways of handling prior stressors, can assess their vulnerability or resilience, and can discuss their sources of social support. The distressed individual might be depressed or anxious, have a borderline personality disorder, or be manic-depressive. They might evidence suicidal or psychotic thinking. The psychologist or other mental health provider can determine the extent of the distress or disordered thinking and the likely role that learning one's test result would have on further deterioration of the patient's emotional status.

Concurrent stressors are not unusual in the families of individuals with inherited disease predisposition. Life-event stressors, which may complicate the decision about when or if to undergo testing, include terminal illness or recent diagnosis of a close relative, illness in the patient him or herself, divorce, recent pregnancy, adoption, miscarriage, or job loss, with attendant issues about insurability. In rare cases, there are also questions about cognitive competence, when individuals with low average IQ or other neurological compromise present for testing. Some individuals can handle multiple stressors and might even be aided by the relative certainty testing offers. For others, the added stress of testing might make maintenance of current life functioning difficult. For some, it may be preferable to wait until other events have resolved, such as a custody battle or the death of a parent. In the very rare event that a decision is made on emotional grounds to postpone offering of genetic testing, the psychologist can help present the rationale and the need to opt for safety to that individual and can recommend psychotherapeutic or supportive services for the interim.

In cases where there is no unusual concern on the part of the staff, psychologists can still be of great help to individual patients in their decisions about whether they want to undergo genetic testing. They can also help some patients endure the waiting time for results and help prepare patients for the life-altering test result to come. Chris, the patient undergoing HD testing (see page 211–215), saw a psychologist, for the six weeks preceding her result, who helped her to consider the meaning of the possible answers she might get from being tested. In her case, the psychologist also administered a battery of psychological measures, a common component of research programs evaluating outcomes of genetic testing. (Psychologists often also do research on outcomes of genetic testing in collaboration with other team members.) The psychologist helped Chris put in perspective the wide range of feelings, which the prospect of getting a test result brought to the fore, including some wishes to rationalize and to bury all feelings related to HD and, alternatively, deep sadness about all that had been taken from her and other family members by the disease.

Psychologists can also help patients cope with the results they receive from genetic testing. While sadness is expected with the news that an individual is a carrier of a disease-predisposing mutation, psychologists can help patients bear their fears and sadness and can help them decide when they are able to consider the next step—the planning of a program to reduce their personal risk through screening or prophylactic surgery. Some patients who receive positive results are hit much harder than they anticipated by the news and must reconsider anew the implications for themselves and others in their families. Cancer patients, for example, who were undergoing *BRCA1/2* testing were less likely than others to accurately forecast what knowing their results would mean to them (Dorval et al., 2000). Unexpected or atypical reactions often spur referrals for further evaluation to the psychologist from the nurse or genetic counselor. One of the surprises in this field has been that some patients who test negative experience disappointment or even depression (Huggins et al., 1992). They may feel separated from other family members who have developed the illness or who are mutation carriers, or they may feel excessively guilty for their good fortune in avoiding the inheritance of a deleterious mutation. Learning one is not at increased risk for a disease common in one's family can represent a reversal of life course. In such a circumstance, it would be natural to have lived one's life preparing to be struck by the same illness at around the same age it struck other family members. Such preparation may have restored some measure of control to a painful and difficult situation where one's family seemed singled out for unusually bad fortune. To suddenly have those assumptions undermined by news that one is not under such a threat can leave an individual feeling defenseless, foolish, or regretful about opportunities not undertaken. Working through those emotions may help patients find relief in the news, rather than frustration.

It is not surprising that complex, emotional issues arise for people undergoing genetic testing. The immediate situation is a stressor of high magnitude. It is rare in one's life that a single event can have such far-reaching implications for the expected course of a life and the lives of one's offspring. Genetic counseling involves receipt of complex information, often delivered over a relatively short period of time. The counseling may occur in close proximity to one's own diagnosis or that of a close family member, further heightening the emotional background against which the information is received and decisions made. The consideration of family history may also arouse painful memories of earlier illness, often unresolved grief, and, if the patient has him or herself been affected by the disease, memories of his or her own diagnosis or treatment experience. Gathering family history may also necessitate communication with family members about such events, which, in itself may be painful and difficult. Family relationships are complex, and close biological kinship does not necessarily imply open communication about sensitive topics, like illness. The need to ask relatives about their own or family history, or, in some cases, to ask affected individuals to be tested themselves to identify the deleterious, familial mutation, can elicit anger or disappointment or lead to the discovery of family secrets. Once genetic testing results are known, further issues may arise about who will be told the results. The implications of the information for children may stress marital relationships when parents hold differing views about when or if children should be told a parent's result. Individuals, especially those who test positive for a disease-predisposing mutation, often find solace in talking to others in a similar situation. Psychologists may work with nurses or genetic counselors to develop and run support groups or one-day support meetings for such patients. They may also make referrals for individuals who have, in the course of undergoing testing, uncovered emotional issues which they would like to explore in ongoing psychotherapy. For all of these reasons, it is useful to have a psychologist or other mental health provider available with whom to discuss cases that present with unusual amounts of stress or problematic interpersonal relationships.

Psychologists can work in a variety of models within a genetics clinic. They may be on-site, integrated members of a multidisciplinary genetics team. In such a case, they would likely be well-informed about the work of the clinic and the roles of the other staff and be easily available to answer questions and see patients the staff is concerned about. In some cases, the clinic might actually routinely schedule patients to meet with a psychologist as part of the initial clinic visit. Psychologists might also participate in clinic rounds, making it easy to comment on cases where staff concern exists. Another more likely model, given the current scarcity of mental health clinicians working in a genetics clinic, is a consultative one. Here, a psychologist would not be in the clinic, but would have some ongoing relationship to it and could be called if there were patients to refer or cases the staff wanted to discuss.

To work in the genetic setting, psychologists have much to learn about basic genetics, the syndromes that patients are seen for in the clinic, and the meaning of test results—positive, negative, and indeterminate. They should be aware of the implications for patients and family members, the growing literature on psychosocial outcomes of genetic testing for disease predisposition, and the resources available to patients with such concerns. Like other health professionals, psychologists and mental health professionals may wish to get training from the growing number of workshops run by the American Cancer Society or other professional organizations, from CD-ROMs produced for patients with genetic issues, from Web-based resources like GROW *(http://www.geneticsresources.org)* and the NIH PDQ Cancer Genetics summaries *(http://www.cancer.gov/cancertopics/prevention-genetics-causes/genetics)*, which include reviews of much of the psychosocial literature in cancer genetics. They may be interested to learn about Web-based support groups for those at-risk including FORCE *(http://www.brca.com/)*, the Genetic Alliance *(http://www.geneticalliance.org/)*, and for The Huntington's Disease Society *(http://www.hdsa.org/)*. They also may want to learn more about relevant ethical concerns, especially those that relate to the privacy of genetic information, through conferences on the topic or through reports of research supported by the Ethical, Legal, and Social Implications program of the Human Genome Project.

In summary, there is much to be gained from the close working relationship between nurses and psychologists in the genetics clinic. The fascinating questions, which arise in the hopes of utilizing genetic information to reduce the morbidity and mortality of patients from families with hereditary syndromes, can best be solved through the multidisciplinary cooperation of professionals with interest and training in helping to improve the physical well-being and emotional resilience of those at risk.

References

Collins F. S., & McKusick, V. A. (2001). Implications of the Human Genome Project for medical science. *Journal of the American Medical Association, 285,* 540–544.

Dorval, M., Patenaude, A. F., Schneider, K. A., Kieffer, S. A., DiGianni, L., Kalkbrenner, K. et al. (2000). Anticipated versus actual emotional reactions to disclosure of results of genetic tests for cancer susceptibility: Findings from *p53* and *BRCA1* testing programs. *Journal of Clinical Oncology, 18,* 2135–2142.

Huggins, M., Bloch, M., Wiggins, S., Adam, S., Suchowersky, O., & Trew, M., et al. (1992). Predictive testing for Huntington Disease in Canada: Adverse effects and unexpected results in those receiving a decreased risk. *American Journal of Medical Genetics, 42,* 508–515.

Miller, S. M., Shoda, Y., & Hurley, K. (1996). Applying cognitive-social theory to health-protective behavior: Breast self-examination in cancer screening. *Psychological Bulletin, 119,* 70–94.

Patenaude, A. F., Guttmacher, A. E., & Collins, F. S. (2002). Genetic testing and psychology: New roles, new responsibilities. *American Psychologist, 57,* 271–282.

Andrea Farkas Patenaude PhD
Dana-Farber Cancer Institute
44 Binney Street
Boston, MA 02115

CHAPTER 9
Connecting Genomics to Society

Nursing cannot be viewed in isolation; instead the profession must be viewed within the social, cultural, economic and political context of the health care system and wider society.

Garfield et al., 2003

The completion of the human genome raises expectations and hopes for health benefits for all populations, while at the same time challenging societal, health, and cultural beliefs with the use of genetic information. In order to provide culturally competent care, nurses need to become aware of the influence of ethnicity, culture, and related health beliefs in the clients' ability to use genetic information and services. Knowledge of the history of misuse of human genetic information is a part of that understanding. Recognizing the cultural and ethical perspectives that influence the use of genetic information and services, the issues that undermine clients' rights to informed decision making, and voluntary action supports the ability of nurses to provide culturally sensitive and competent care.

Introduction

In April 2003, the National Human Genome Research Institute (NHGRI) published a landmark scientific report describing the future of the field of genomics. The document outlines the role of NHGRI and the role of the National Institutes of Health and other government agencies in translating the comprehensive sequence of the human genome into health benefits for the individual, family, community, and society. This document, a result of discussions, workshops, and consultations, outlines a vision for the future of genomics research. Genomics to society is one of three main themes for research. The vision for genomics and society includes research on social issues and the extent of understanding more completely "how we define ourselves and each other" (Collins et al., 2003, p.11; *http://www.nature.com*). Exploration and research in this area will involve analysis and understanding of the relationships between genomics, race, and ethnicity, and the consequences of uncovering these relationships. Genomic discoveries are also unraveling the complicated pathways that underlie human traits and behaviors. Having an understanding of the consequences of uncovering the genomic contributions to human traits and behaviors is essential to avoid serious problems of stigmatization and to find a deeper understanding of the contributions and interactions between genes and environment to contribute to various traits and behaviors (Collins et al., 2003; *http://www.nature.com*).

Genomics research contributes to knowledge and understanding of biology, health, and life. Although many aspects of this understanding are anticipated to benefit individuals, families, and communities, some applications are controversial and call for societal dialogues to define appropriate and inappropriate uses of genomics. Nurses need to contribute to that dialogue. Different individuals, cultures, and religions view the ethical boundaries for uses of genomics in different ways. This vision includes research efforts to look at which sets of values determine attitudes toward the appropriateness of applying genomics to reproductive testing, for example (Collins et al., 2003). Use of genomic information and technology is expected to expand beyond biology and health care into areas such as long-term care insurance, the legal system, educational institutions, and adoption agencies. Further research and policy development are recommended for these important areas (Collins et al., 2003).

In this chapter, we explore current considerations in understanding the relationships between genomics, race, and ethnicity as they affect self-identity and group identity, and in forming what role individuals and cultures believe genes or other biological factors have in these concepts. How the scientific community, including nurse researchers, understands and uses concepts of race and ethnicity in designing research and interpreting findings is reviewed. Ethical boundaries in the use of genomics, including genomics in non-health care settings, are discussed. Culturally safe nursing care that includes cultural competency is described. We begin with Arielle's story about her family history of early onset dementia, how it affected her sense of "self," how her family viewed the illness, and her concerns regarding the potential for discrimination (**Box 9-1**).

Genomic Data and Race, Ethnicity, and Culture

Genome discoveries have led to increasing amounts of data on populations from around the world, and are adding to a deeper understanding of the differences among and between populations and individuals. Genome sequencing research has

Box 9-1 Arielle's Story

My Family and Me: Our Roots and Our Concerns

I have lived with knowledge of the particular illness—early onset dementia—in our family for 30 years. Sometimes it feels like it was just yesterday that my mother began losing her "self," as we knew her. It started out with little things—short-term memory, forgetting where she parked the car—then gradually took over her sense of humor and sense of propriety. She would say things that were so socially inappropriate. My family is Black American and we lived in a small rural town in South Carolina, so you can imagine how some of her comments were taken back then in the early 1970s. As my mother's condition worsened, she became more disoriented. My father and her sisters held strong beliefs that the spirit of her mother (my grandmother), who had suffered memory loss and confusion—in my family they called it "the dizzies"—before her death at an early age, caused her memory loss. They believed that since my mother looked like her mother, my grandmother's spirit had entered my mother. When they finally took her to a doctor, they were told after evaluation and testing that she had an early form of Alzheimer's disease. They could not accept the medical meaning of this, and took her home. My mother's sisters and my father cared for her until her death at age 50.

I was in college and pursuing my career in history during the period of my mother's illness and decline. I did well in school and received a scholarship to attend a private college in Virginia. It was a wonderful experience for me—so much to learn. During that time, I made new friends from various parts of the country. One of my required courses was biology. It was here that I began to learn about genetics and inheritance. I could not help but wonder whether my mother had perhaps inherited her condition from her mother. I didn't dare bring these thoughts up at home as it would have been considered heresy in my family.

A few years after I finished college, my mother's brother, owner of a local coal business, began to have some of the same problems as my mother. He became forgetful to the point where he could no longer run his own company. My father and aunts again talked about my grandmother's spirit and how it had entered my uncle's body too. This time I tried to talk with them about inheritance, but they did not want to hear that explanation.

I wanted a medical explanation because I was afraid for myself. I did not believe the spiritual explanation given by my family. I looked into research, but there was none at that time. Over time, and with the increasing information about genetics and disease, I began to have second thoughts about wanting to know.

Today, I am the same age as my mother was when she became bedridden with her illness. I have watched the developments of genetic discoveries with concern and interest. As an historian, I am aware of the misuses of genetics, especially with American Black populations and the Tuskeegee experiments and the sickle cell testing. As much as I would like to believe that research would help unravel the mysteries of my

Box 9-1 continued

family's illness, I am afraid of what that knowledge will mean to my family and to me. I have concerns about discrimination against us if it is found that we have a hereditary dementia in our family. Now that I am older, I wonder how my aging relatives would understand genetic answers to my mother and her brother's illness. I used to think that I would want to have a genetic test to find out if I have a susceptibility to dementia. I do not want that test any more, as it might be used against me. There might be some link between my race and the disease, and how would that be for me in terms of my insurance or if my employers found out. I think that as a society we have a lot to learn about how genetics is becoming so much the focus of explaining health and illness. By focusing so much on genetic explanations, we may be missing important environmental causes like poverty or malnutrition; and we will lose what I have come to see as the richness of cultural beliefs and explanations about how illness occurs in families and what it means.

Arielle

revealed that humans are 99.9% alike in their genetic makeup. Furthermore, genetic variation is seen largely between individuals of the same group or population rather than among populations (Guttmacher & Collins, 2004). Some scientists hold that this finding means that the concept of "race" may have no genetic or scientific basis. However, in the world of genomic research, a growing number of scientists want to use the 0.1% of variation between individuals to learn more about genes that cause susceptibility to genetic disorders in specific groups, such as cystic fibrosis in Northern European Caucasians or sickle cell anemia in African-American individuals. This knowledge of variation can then be utilized to design specific pharmacogenomic treatments. These scientists believe that race and ancestry are critical to interpreting outcomes of biomedical research. Others believe that the idea that there is a genetic component to race is dangerous and could lead people to declare one group superior or inferior to another (*http://www.dnaprint.com,* 10/29/03).

The challenge put forth by the NHGRI vision is for society to understand the relationship among genomics, race, and ethnicity. The terms race, ethnicity, and ancestry are frequently used interchangeably; however, specific distinctions between the three have been made (Drevdahl, Taylor, & Phillips, 2001; Risch, Burchard, Ziv, & Tang, 2002). Further clarification of the relationship between genomics, race, and ethnicity requires coming to an understanding of one's own definitions and beliefs about these terms, dialogue among researchers about their differing opinions, and exploration of societal views and beliefs.

Race

Race has been traditionally understood as an innate biological feature of human beings with foundations in human evolution. Most people think of race as a biological category and as a way to separate and label groups according to a set of common biological traits such as skin color, and eye, nose, and facial shape. However, no biological criteria for dividing races into distinct categories have been discovered. A very small proportion of genes defines visible physical traits and is associated with race, but they do not reliably differentiate between social categories of race. Although variations between groups are being identified, these variations cannot be used to distinguish groups from one another. Recent genome discoveries have shown, in fact, that there is greater genetic variation within a racial group than across racial groups.

Race, although not a biological category, has meaning as a social category. Various cultures classify people into racial groups according to a set of characteristics that are socially significant. In the United States, for example, skin color is the indicator most frequently used for race. Although skin color is a continuous variable, it is used to separate groups; and there are no specific guidelines for determining the point at which the boundary between colors is made (Ferrante & Brown, 1999).

The concept of race is especially controversial when certain social groups are separated, treated as inferior or superior, and given differential access to valuable resources. As an example, there has been a low level of participation of African Americans in biomedical research over the last several decades. The Tuskegee Study is usually cited as the source of the mistrust and skepticism that pervade the African-American community today. The study monitored low-income African-American males with syphilis for 40 years. Even when a proven cure—penicillin—became available in the 1950s, the study continued. The U.S. Department of Health, Education, and Welfare eventually stopped the study after its existence was leaked to the public, and it became a political embarrassment.

The Tuskegee syphilis study is one of the most widely used examples of research in which human subjects were not adequately protected. This study, and other similar studies, provided the impetus for federal regulations that now restrict the treatment of human subjects in research *(http://www.gpc.edu~shale/humanities/composition/assignments/experiment/rivers.html)*.

The experience of African Americans with sickle cell disease is another historic example of genetic discrimination and stigmatization related to a disease state in this population. The events involved in sickle cell screening in the 1970s have left an indelible mark on the minds and lives of many and have increased concerns about the impact of genetic testing and screening on African Americans today. During the 1970s, sickle cell screening programs were developed to identify individuals with the sickle cell trait in an effort to reduce the incidence of sickle cell disease.

Inadequate education and counseling led to confusion about the differences between sickle cell trait and sickle cell disease, often resulting in stigmatization and unfair discrimination from misinterpretation of test results. Problems with the early sickle cell screening programs in the United States included mandatory screening laws directed at African Americans only. The most significant problem was that the purpose of the public health screening programs was not clear. In the 1970s there was no treatment available, and prenatal diagnosis was not yet developed for sickle cell anemia. Without prenatal diagnosis or any potential therapies, why have a screening program? *(http://www.sph.unc.edu/nciph/phgenetics/sicklecell.htm)*

Ethnicity

Ethnicity generally refers to a common heritage shared by a particular group (American Anthropologic Association, 1997). Heritage includes a group's sense of identification surrounding common characteristics such as shared history, language, religion, and/or common ancestry. Race and ethnicity are understood as separate terms, and even though they may overlap (e.g., race as defined as a social category), each has a different social meaning. As an example, Hispanics are considered an ethnicity, not a race. Individuals identifying with different Hispanic ethnic subgroups such as Cubans, Dominicans, Mexicans, Puerto Ricans, and Peruvians include all races *(http://www.mentalhealth.org)*.

Culture

Culture, as defined by the Department of Health and Human Services (DHHS, 2004), is a common heritage or set of beliefs, norms, and values. Culture typically encompasses a collection of nonphysical traits such as values, beliefs, attitudes, and customs shared by a group of people or families. Cultural identity refers to the culture with which a person identifies and to which he or she looks for standards of behavior (Cooper & Denner, 1998). Since there are many ways to define a cultural group, many people consider themselves to have several cultural identities, for example, the culture of clinicians or the culture of a gang. Culture is often a significant factor in how individuals and families seek, receive, and act on genetic health information. Since cultures vary, family interpretations and values associated with genetic information will also vary, affecting a family's perceptions of health risk and illness (Bassetti, 2002).

An important aspect of culture is that it is dynamic. Culture changes continually and is influenced by people's beliefs and the demands of their environment. For example, when immigrants come to the United States, they bring their own culture. They may gradually adapt or "acculturate" through a socialization process by which they gradually learn and adopt selective elements of the dominant culture. At the same time, the dominant culture is transformed by its interaction with minority groups. In some minority groups, members grow away from their own culture and create a culture that is distinct from both the country of origin and the dominant culture, for example, Chinatowns in major cities (Lopez & Guarnaccia, 2000).

The dominant culture for most of the United States has centered until recently on beliefs and norms of the white Americans of Judeo-Christian origin. Today, the United States has a much more multicultural character. In spite of this evolution, the cultural legacy has left its imprint on how health professionals respond to patients in all facets of care, beginning with the first encounter and continuing through diagnosis and treatment *(http://www.mentalhealth.org)*.

Nurses are accustomed to viewing individuals in a holistic way that takes into account each person's physical, mental, spiritual, social, and cultural attributes. New genomic discoveries and technologies used to classify or categorize race and ethnicity for the purpose of understanding human variation and to tailor diagnostics and therapeutics have the potential to redirect individual, family, and community identity from the social domain into the physical aspects of the body. The prominent focus on the genetic aspects may overshadow other factors such as poverty, access to health services, and how each of these affect outcomes of health and disease. Nurses and other health care providers will need to seek ways to utilize and interpret all aspects of health information in caring for individuals, families, and communities.

Genomics and the Concept of "Self"

Genomic research and technologies have led to increased use of genetic information, including race and ethnicity, to assess risk for genetic conditions and target individualized treatments. Genetic information can also be viewed as an important factor in defining self, family, and community. Unique to genetic information gained from genetic tests based on a person's race, for example, sickle cell anemia testing in black populations or hereditary breast cancer testing in Ashkenazi-Jewish populations, is its ability to generate information with implications for a person's sense of self. For some individuals and communities, genes are personal and linked to a person's identity. On the other hand, gene testing may reveal results that may be inconsistent with commonly held beliefs about a person's biological and cultural sense of self (Peters, Djurdjinovic, & Baker, 1999). As an example, genetic marker testing is now available via the Internet to identify specific genetic markers associated with race. In a recent article, an African-American professor decided to pursue such testing. He was greatly surprised when the test results revealed that his racial origins were primarily Caucasian (Wooten, 2004). These genetic test results challenged his previously held beliefs about his race, his ethnicity, and culture. As another example, genetic testing can now identify individuals who carry predisposing genes to hereditary breast, ovarian, and colorectal cancer. In some families, it is believed that genetic traits "skip a generation," or are present in those family members who resemble an affected family member. Genetic test results identifying "at risk" family members who did not expect these results can be disabling to some, as they do not fit with the family culture and beliefs (Peters, Djurdinovic, & Baker, 1999).

Genomic advances, in the context of the cultural concept of a person's self, raise ethical, social, and legal concerns regarding eugenic misuse, possible discrimination, stigmatization, and disruption of families and societies (Durant, Hansen, & Bauer, 1997). Healthy carriers of genetic alterations, which make them susceptible to developing certain diseases in the future such as breast or colon cancer, have created a new class of "at risk" individuals. The individual "at risk" is not ill at present, but may not remain as well as individuals who do not have the same alteration. In addition, the individual may carry this label in his or her health record. Concerns that this information may be accessed and used by insurers, employers, or schools have prevented many individuals from pursuing genetic testing for health benefits (Kenan, 1996; Peters, Djurdjinovic, & Baker, 1999).

Use of genetic testing to identify individuals who may benefit from or be harmed by certain medications or treatments is growing. It is known that different people respond to different medicines in different ways. Efforts are currently underway to prescribe medicines based on race and to perform diagnoses considering race as an important biological variable. As an example, race is being used as a pharmacogenomic indicator in glaucoma treatments (Lee, Mountain, & Koenig, 2001).

Two recent studies have been conducted to learn more about the attitudes and perceptions of the public and patients related to prescribing medicine, specifically using individualized genetic testing, race-based prescription, and traditional prescription. Participants in both studies chose genetic testing because it offered individualized attention, even if the costs were high. Overall, participants were highly suspicious of race-based prescription, including both safety and efficacy. Participants expressed that using race as a way to prescribe medication was similar to "racial profiling." The high level of suspicion, if race-based prescriptions were an option, could cause individuals to be less likely to adhere to a drug over the long term. *This would be emphasized if there are negative side effects that might confirm the level of drug safety and if improvement in a condition is not immediately obvious.* In addition, prescribing drugs or treatments based on race is not likely to increase the trust between the medical profession and various racial groups, especially where there has been a strained relationship due to historical and institutional factors. Nurses need to be aware of public attitudes toward use of race as a means of tailoring medications to specific groups or individuals in the application of pharmacogenomics, and the concerns regarding privacy, discrimination, and aversion to race-based prescribing. (Bevan et al., 2003; Condit et al., 2003; Corbie-Smith, Thomas, & St. George, 2002).

Genomics and the Family: Traits and Behaviors

Genetic information has particular meaning for biological family members that most other health information does not. It connects us to our ancestors and descendants and influences our physical, intellectual, and emotional attributes and traits (Grady, 1998). Analysis of an extensive family history is currently the tool used to estimate an individual's risk of disease. Information gained from the family history is increasingly used to estimate a person's risk of disease for the purpose of offering predictive genetic testing (Finkler, Skrzynia, & Evans, 2003). The family history is also being explored at the public health level as a bridge from "genetics to genomics in practice because it reflects the presence of shared genes, shared environments, and complex genetic interactions" (Khoury, 2003, p. 261).

Modern medicine increasingly regards disease as rooted in genetic inheritance. This view, however, may be in contrast to family culture and beliefs about health and illness. The risk inherent in this view is that the family and kinship are "being medicalized as a result of the current emphasis on medical genetics and its clinical application" (Finkler, Skrzynia, & Evans, 2003, p. 404). Medicalization refers to reinterpreting as genetic certain physical characteristics and behaviors that were once viewed in terms of religious, ethical, or moral transgressions. For example, alcoholism, hyperactivity, learning disabilities, obesity, and antisocial conduct are increasingly being looked at through a genetic lens. On the one hand, medicalization of such conditions and traits can lead to a sense of helplessness or lack of responsibility—a preordained trait. On the other hand, it can instill a sense of hope that the situation will be corrected by medical intervention. According to Finkler, Skrzynia, and Evans (2003), medicalization changes people's view of reality and sense of being in the world. When viewed as medical problems, behaviors such as alcoholism are characterized as deviations from the proper levels of functioning, setting the individual apart from the rest (Finkler, Skrzynia, & Evans, 2003).

Kinship and family viewed through a genetic lens are framed in terms of genetic inheritance from parents, grandparents, and other relatives. With the medicalization of the family and kinship, the core patient becomes the family rather than the individual. For example, in order for a genetic diagnosis to be made for conditions such as hereditary breast or colon cancer, family members need to be involved in accessing health care services. In addition, a person's diagnosis can have a profound effect on the health status of other family members outside of the context of their social relationship (Finkler, Skrzynia, & Evans, 2003).

Making the family the central focus can lead to ethical and legal issues. A legal case against Virginia Commonwealth University is one example. In this situation, the concern was whether a family member, in this case the father, becomes a research subject when information about him is revealed by a relative, his daughter, who enrolled in the research study. The questions raised are how much does the family history disclose about the rest of the family, and how much are other family members willing to share when constructing a family tree (Botkin, 2001; Finkler, Skrzynia, & Evans, 2003)? In another example, a woman wished to know her genetic presdisposition to breast cancer, which required that her brother be tested. Her brother refused. This and other family situations highlight the fact that the individual is no longer an independent decision maker in his or her access to health care.

In the 21st century, the traditional family concept of the nuclear household no longer prevails in the United States. As a result of divorce and remarriage, gay marriages, and shared households, there is no one specific pattern of family with diversity of kin in contemporary America. The presence of other cultures, such as Hispanic, Muslim, and Southeast Asian, adds diversity to the concept of family, kinship, and health beliefs. The medicalization of kinship challenges both the traditional ability of the patient to decide about treatment options and how genetic conditions may be perceived in families of diverse cultures.

Nurses need to consider these issues with care as they approach individuals and families with new genetic information, diagnostics, and therapies. Nurses can serve as a vital link between the individual and the family. As genetic information, including family history, becomes an integral component of the practice of everyday health care and disease prevention, "health care providers and public health professionals will need to be equipped with tools of evaluating and communicating complex genomic (and family history) information" (Khoury, 2003, p. 267).

Genomics and Community Identity

Individuals and families belong to specific communities, groups, and cultures. Genomics discoveries are also affecting our understanding of the nature of "difference" among human groups (Lee, Mountain & Koenig, 2001). Use of genetic testing and technologies raises the question of how, as a society, we will understand and consider common conditions associated with socially identifiable populations. For example, how will knowledge of the specific hereditary breast–ovarian cancer gene alterations present in Ashkenazi-Jewish populations affect the treatment of those communities? As another example, research has been conducted on the genetic basis for differences between African Americans and non-Hispanic Caucasians in smoking, and the incidence of lung disease focusing on biomarkers for rates of tobacco metabolism. Research findings suggest that biological differences may account for the differential health status of certain groups, such as higher incidence of lung disease and smoking (Schaeffeler et al., 2001). Such linkages raise the concern that the relationship between "race" and disease occurrence will lead to negative outcomes, despite good intentions. Given the consequences of the association of sickle cell anemia with African Americans and the mid-20th century Nazi science regarding Ashkenazi-Jewish populations, individuals belonging to these groups have reason to fear that they are somehow biologically distinct. The emerging tools of genomics, while having the potential to map biological variation more precisely, may reinforce the concept that human populations can be divided into specific biological entities, while overlooking the social and economic contributions to health and disease (Lee, Mountain & Koenig, 2001). Nurses practicing community medicine need to become participants in policy discussions about how to avoid the potential harms that could evolve from such thinking.

Genomics, Race, Ethnicity, and Research

The changing demographics of the United States is raising questions about the rationale for racial categories, whether they be for census or research purposes. In the 2000 Census Bureau count for example, Hispanic individuals, the largest minority group in the United States, refused to identify themselves by any of the five standard racial categories on the census forms: white, black, Asian, American Indian, or Alaska native. Individuals can mark their ethnicity as Hispanic, but then they are also asked to choose a racial category. In 2000, nearly half of Hispanic respondents (48%) identified themselves as white, while only 2% chose black (Navarro, 2003). The growing number of interracial unions is also changing the way the government accounts for people by racial categories. The issue of racial categories has become a significant issue for genomic and other research debate among scholars.

Some genome researchers argue that race has an important role in biomedical research and should not be replaced with genetic tests to categorize populations of people. In epidemiologic research, Risch et al. (2002) found that racial differences are very much aligned with certain genetic markers, and studies based on race take significantly less time and money to determine useful information (Risch et al., 2002). Many civil rights advocates and government officials also hold that racial categories, although imperfect as they may be, are the only way to measure disparities among groups and provide remedies (Navarro, 2003).

Other researchers hold that what is of concern is the difficulty of translating differences among groups to an individual level and into tests that have a value to help with clinical decisions. Race, they believe, can help to target screening for a disease-associated mutation that is present at a higher frequency in a particular population and very rare in another. Cystic fibrosis in Caucasian populations is an example. However, it is not practical to use race as a surrogate for genetic constitution in medicine or public health. In their opinion, race at a continental level has not been shown to provide useful categorization of genetic information about cause of disease, diagnosis, or response to drugs (Cooper, Kaufman, & Ward, 2003).

The use of genetic technologies in directly determining race and ethnicity has the potential to redirect identity from the social domain into the physical attributes of the body, and affects the way of defining who and what humans are in the arena of biomedicine. Testing for race and ethnicity may be a means of improving the health status of minority populations. However, targeting disease prevention programs to individuals of certain groups supports the concept that disease results from essential characteristics within the individual and denies the role of the social, emotional, and physical environment. Targeting genetic screening for racially identified "at risk" groups also has the potential for stigmatization and discrimination (Lee, Mountain, & Koenig, 2001).

Nurse researchers are also grappling with the concepts of race and ethnicity. A growing body of nursing literature mentions race and associated race–ethnic terms. Similar to scientists in other disciplines, nurse researchers seldom define these terms or relate the terms to health practices, nutritional inclinations, or religious customs. Furthermore, there is little differentiation between biologic differences among health of different groups and the results of living in a racialized society, which may lead to biologic differences in health but do not reside in racial biology. Nurse and other genome researchers need to continue thoughtful dialogue regarding issues of race, ethnicity and culture, and health differences so that well-constructed research that avoids pathologizing particular racial or ethnic groups can be carried out (Drevdahl, Taylor, & Phillips, 2001).

Ethical Boundaries and Use of Genomics

Genomics creates opportunities to explore how we define ourselves, our families, and our communities. Some of the clinical applications currently under development, however, are controversial, with members of the public raising questions about the appropriateness of their scientific exploration. As a society, we will need to come to consensus on the definition of appropriate and inappropriate uses of genomic discoveries. For example, reproductive testing, such as preimplantation diagnosis for couples at risk for having a child with a specific genetic disorder, raises concerns about when life begins and the value of individuals with disability. Individuals from different cultures and religious traditions will have differing views about this and other types of reproductive testing including application of genomics to genetic enhancement. The use of genetic and genomic information for determining life, disability, and long-term care in legal systems, the military, educational institutions, and adoption agencies is a future use of genomics. The appropriateness of the use of genetic information in these arenas needs to be delineated as well. Genomics also offers opportunities for people to understand their ancestry and origins, raising issues about whether genetic information will in the future be used for defining membership in a minority group. Research in all of these areas is needed and will require discussions among participants from both the genomics community and the broader community of stakeholders (Collins et al., 2003). Nurses will need to explore their own views and ethical boundaries in such areas in order to participate fully in this important public dialogue.

Translation of Research to Policy: Integration of Genetics and Genomics

The NHGRI, Ethical, Legal and Social Issues (ELSI) branch funds research that focuses on translation of genomic advances to society. The ELSI branch offers a new approach to scientific research in that it supports research to identify ethical, legal, and social implications of human genome discoveries at the same time that the basic genomic science is being studied. The fundamental concept of ELSI research is that problem areas in the translation of human genome scientific discoveries can be identified and solutions developed before the scientific information is integrated into health care education and practice. The ELSI goals are outlined in **Table 9-1.** Research issues addressed by ELSI are presented in **Table 9-2.**

The Secretary's Advisory Committee on Genetics, Health, and Society (SACGHS) is another important group considering the implications of genomics for society including the nonmedical applications in areas such as education, employment, and law. SACGHS was established to provide a forum for exploration, analysis, and deliberation on a broad range of human health and societal issues raised by new technological developments in human genetics. The Committee consists of a core of 13 members, including a representative from nursing. It is assessing how genetic technologies are being integrated into health care and public health. In addition, attention is being given to exploring the use of genetics for nonmedical purposes such as in bioterrorism, immigration, and forensics. Examining current licensing and patent policies for their impact on access to genetic technologies is another important area that the Committee is analyzing (SACGHS, 2004, *http://www.sacghs.org*). Meetings are held twice a year and are open to the public; annual reports are prepared and posted on its Web Site.

Table 9-1 ELSI Research Goals

- To examine the issues surrounding the completion of the human DNA sequence and the study of human genetic variation.
- To examine issues raised by the integration of genetic technologies and information into health care and public health activities.
- To examine issues raised by the integration of knowledge about genomics and gene-environment interactions into nonclinical settings.
- To explore ways in which new genetic knowledge may interact with a variety of philosophical, theological, and ethical perspectives.
- To explore how socioeconomic factors and concepts of race and ethnicity influence the use and interpretation of genetic information, the utilization of genetic services, and the development of policy.

(ELSI Research Goals 1998–2003 at *http://www.genome.gov*)

Global Genomics

The translation of genomics research to policy that will foster safe and effective integration of genomic technologies into society extends to global health. Globally, heart disease is now the most common cause of death, and late-onset diabetes and obesity are becoming a new pandemic. In some countries, the rate for stroke is four

Table 9-2 Research Program Areas Addressed by NHGRI/ELSI

- **Privacy and Fairness in the Use and Interpretation of Genetic Information**
  Activities in this area examine the meaning of genetic information and how to prevent its misinterpretation or misuse.
- **Clinical Integration of New Genetic Technologies**
  These activities examine the impact of genetic testing on individuals, families, and society, and inform clinical policies related to genetic testing and counseling.
- **Issues Surrounding Genetics Research**
  Activities in this area focus on informed consent and other research-ethics review issues related to the design, conduct, participation in, and reporting of genetics research.
- **Public and Professional Education**
  This area includes activities that provide education on genetics and related ELSI issues to health professionals, policy makers, and the general public.

(http://www.genome.gov)

to five times that in countries that are more prosperous. Many countries grappling with these emerging health care issues are still battling infectious diseases such as tuberculosis and HIV/AIDS concurrently (Weatherall, 2003). In 2001, the World Health Organization (WHO) initiated a report on the role of genomics in world health. The central issue considered in this report was whether there were already genomics advances that could be applied in developing countries to benefit their populations or whether the global community should wait until further progress in genomics research and applications to health care had been made in richer countries (Alberti, WHO, 2001).

After deliberation and exploration, WHO concluded that there is widespread support for the introduction of DNA technology now into developing countries, primarily in the fields of communicable disease and single-gene disorders. For example, it is estimated that 7% of the world's population are carriers of hemoglobin disorders such as thalassemia and sickle cell anemia, with thousands of babies having severe forms of disease. Development of international partnerships has led to understanding the specific mutations in various ethnic groups, and reliable methods have been developed for prenatal diagnosis. The result has been a significant reduction in the frequency of these conditions among Mediterranean populations, and in many parts of the Indian subcontinent and Asia (Weatherall, 2003).

Another example of global integration of genomics and DNA technology by international partnerships is the development of DNA-based diagnostic methods for communicable diseases in individual developing countries. The Swiss Tropical Institute and Tanzania have offered training in DNA technology to identify drug resistance to malaria and for genotyping malarial parasites. Pilot studies conducted in Africa are proving the value of these approaches for typing patients for genetic resistance to drugs used in HIV/AIDS treatment and for community studies of drug-resistant malarial parasites (Djimde, Doumbo, Steketee, & Plowe, 2001; Schaeffeler, 2001).

There is concern and fear that the benefits of genomics will apply only to the richer countries, and this will widen the gap in health care between rich and poor countries and individuals (World Health Organization, 2002). International efforts in the area of hemoglobinopathy and communicable disease demonstrate the possibility of transferring cost-effective DNA technology to developing countries. WHO and other public policy organizations are considering how to mobilize the skills and resources of richer countries for the benefit of the health of the global community. Developments of this nature require that universities and research centers in wealthier countries begin to conceptualize genomic research and applications in terms of the world community. Nurses have an opportunity to be involved in shifting the emphasis in education and research toward a more global view of disease and the development of infrastructure for the development and organization of overseas programs.

Implications for Nursing: Transcultural Nursing

To provide optimum care to individuals, families, and communities in the postgenomic era, nurses must not only have knowledge of genetics and genomics and how they influence health and disease; they must also have a comprehensive understanding of multiculturalism and how to incorporate multiculturalism into practice. Such knowledge and understanding will support nurses in translating complex genetic information to individuals, families, and communities in a culturally competent way. "Multiculturalism refers to an ability to appreciate the values, beliefs and behavior of cultures other than an individual's own culture" (Bassetti, 2002, p. 256). Understanding one's own individual values and beliefs in comparison to those of other cultures is a critical component of multicultural nursing care.

In the United States, nurses increasingly care for diverse individuals from various cultures and ethnic backgrounds. Culture is frequently a significant variable in how families seek, receive, or act on genetic health information. Family interpretations and values associated with genetic information will vary depending on their cultural affiliation. Cultural interpretations of genetic information will in turn affect a family's perceptions of health risk and illness (Bassetti, 2002). A transcultural nursing perspective expands nurses' abilities to provide optimal care for all patients, as it is grounded in the understanding that culture and ethnicity are strong determinants in perception and use of medical care, and that religion, culture, and ethnicity are part of the fabric of each individual's response to interventions *(http://www.culturediversity.org/basic.html)*.

Nursing assessment is a critical step of the nursing process and is important in determining interrelationships between individuals and families. Viewing the individual or family in the context wherein they exist is central to this process. Giger and Davidhizar (1991) propose six cultural phenomena that nurses need to understand so that they can provide effective and culturally safe care for all clients. These phenomena apply to transcultural nursing care in the postgenomic era. They include communication, space, social organizations, time, environmental control, and biological factors. Knowing the communication norm within a particular culture is important to facilitate communication and decrease miscommunication. Having an understanding of a person's comfort level related to personal space and touch practices is another essential assessment. Recognition and acceptance that individuals from culturally diverse backgrounds may have varying degrees of acculturation into the dominant culture, particularly as this relates to age and life cycle factors, will facilitate nurses to provide meaningful care. The concept of time—for example, duration of time, points in time, present or past-orientation—is assessed to gain an understanding of the value of tradition, how new procedures may be received, and how health care measures will be carried out. Assessment of environmental control allows the nurse to gain an understanding of individuals' beliefs on their control of nature, health beliefs, values, and definitions of health and illness. Assessing biologic variation refers to nursing assessment of physical characteristics, genetic variations, susceptibility to disease, nutritional preferences and deficiencies, and psychological characteristics (Giger & Davidhizar, 1991). The Oncology Nursing Society and other nursing organizations have developed multicultural outcomes and guidelines for cultural competence. These resource organizations are listed in **Table 9-3. Table 9-4** lists eight areas of nursing assessment and data gathering that reflect cultural variation.

Practicing transcultural nursing in the genomic era can be complex and requires a high tolerance for ambiguity, since many perspectives within and between individuals must be recognized and appreciated (Baker, Schuette, & Uhlmann, 1998; Bassetti, 2002). Culture strongly influences a person's belief and value systems and inevitably affects the ways in which he or she views health events and conditions such as reproduction, childbearing, pregnancy termination, birth defects, susceptibility testing, and chronic disease. The informed decision-making and consent process becomes especially important in this context. The basis of informed consent is mutual participation, respect, and shared decision making as outlined in the International Society of Nurses *Informed Decision-Making and Consent: The Role of Nursing Position Statement (http://www.isong.org)*. The focus is on the patient and the communication process so that what the patient desires actually occurs. How genetic information is presented in this process, what is said, and how it is said can have a significant impact on a person or family's ability to process, understand, and assimilate such knowledge. The challenge for nursing is to communicate genetic information in a culturally supportive manner and in a culturally sensitive climate that

Table 9-3 Resources for Multicultural Nursing

The Oncology Nursing Society and other nursing organizations have developed multicultural outcomes and guidelines for cultural competence.
http://www.ons.org

The Journal of Multicultural Nursing and Health (JMCNH) Advance the promotion of health and the provision of culturally competent health care.
http://www2.cecomet.net/eestar/jmcnh

Multicultural Nursing, Health, and Medicine
Provides transcultural and multicultural health links.
http://www.edchange.org/multicultural/sites/health.html

Association of Operating Nurses
http://www.aorn.org/SA/multicultural.html

Multicultural Nursing Societies, and Professional Sites
http://www.nursing.umich.edu/oma/Tools%20&%20Resources/societies.html

Table 9-4 Nursing Assessment Reflecting Cultural Variation

1. History of the origins of the patient's culture
2. Value orientations—view of the world, ethics, norms, standards of behavior, attitudes towards time, work, money, education, beauty, and change
3. Interpersonal relationships—family patterns, demeanor, and roles and relationships
4. Communication patterns and forms
5. Religion and magic
6. Social systems—economic values, political systems, education
7. Diet and food habits
8. Health and illness belief systems—behaviors, decision making, and health care providers

(Giger & Davidhizar, 1991; Oncology Nursing Society, 1999)

encourages individual autonomy in decision making. In empowering individuals and families to come to their own decisions, nurses can help facilitate this process by discussing the particular cultural beliefs, traditions, and family values that are viewed as an important part of their identity (Baker, Schuette, & Uhlmann, 1998; Bassetti, 2002).

Chapter Activities

1. Arielle states, "By focusing so much on genetic explanations we may be missing important environmental *causes* like poverty or malnutrition, and we will lose what I have come to see as the richness of cultural beliefs and explanations about how illness occurs in families and what it means." Discuss how you interpret this statement, and how it may influence your interaction with Arielle.
2. Review the *Research Program Areas Addressed by NHGRI/ELSI* listed in Table 9-2. Identify potential research ideas for each of the program areas that could be targeted by nursing research.

References

Alberti, G. (2001). Non-communicable diseases: Tomorrow's pandemics. *WHO Bulletin, 79*(10), 907.

American Anthropological Association. (1997). Response to OMB Directive 15: Race and Ethnic Standards for Federal Statistics and Administrative Reporting. Accessed September 15, 2003 from, *http://www.aaanet.org/gvt/ombdraft.htm.*

Baker, D. L., Schuette, J. L., & Uhlmann, W. R. (1998). *A guide to genetic counseling.* New York: Wiley-Liss.

Bassetti, S. (2002). Cuturally relevant genetic counseling. *AWHONN Lifelines, 6*(3), 254–257.

Bevan, J., Lynch, J., Dubriwny, T., Harris, T., Achter, P., Reeder, A., & Condit, C. (2003). Informed lay preferences for delivery of racially varied pharmacogenomics. *Genetics in Medicine, 5*(5), 393–399.

Botkin, J. (2001). Protecting the privacy of family members in survey and pedigree research. *Journal of the American Medical Association, 285,* 2207–2211.

Collins, F. S., Green, E. D., Gutmacher, A. E., & Guyer, M. S. (2003). A vision for the future of genomics research. *Nature, 422,* 835–847.

Condit, C., Templeton, A., Bates, B. R., Bevan, J. L., & Harris, T. M. (2003). Attitudinal barrirs to delivery of race-targeted pharmacogenomics among informed lay persons. *Genetics in Medicine, 5*(5), 385–392.

Cooper, C., & Denner, R. (1998). Theories linking culture and psychology: Universal and community-specific processes. *Annual Review of Psychology, 49,* 559–584.

Cooper, R. S., Kaufman, J. S., & Ward, R. (2003). Race and genomics. *The New England Journal of Medicine, 348*(12), 1166–1170.

Corbie-Smith, G., Thomas, S. B., & St. George M. M. (2002). Distrust, race, and research. *Archives of Internal Medicine, 162,* 2458–2463.

Department of Health and Human Services. (2004). Developing Cultural Competence in Disaster Mental Health Programs: Guiding Principles and Recommendations. Accessed September 15, 2004 from, *http://www.mentalhealth.samhsa.gov/publications/allpubs/SMA03-3828/introduction.asp.*

Djimde, A., Duombo, O. K., Steketee, R. W., & Plowe, C. V. (2001). *Lancet, 358,* 890.

Drevdahl, D., Taylor, J. Y., & Phillips, D. A. (2001). Race and ethnicity as variables in nursing research. *Nursing Research, 50*(5), 305–313.

Durant, J., Hansen, A., & Bauer, M. (1997). Public understanding of the new genetics. In Marteau, T., Richards, M. (Eds.) *The troubled helix: Social and psychological implications of the new human genetics.* Cambridge, UK: Cambridge University Press.

Ferrante, J., & Brown, P. (1999). Classifying people by race. In F. L. Pincus & H. J. Erlich (Eds.), *Race and ethnic conflict* (2nd ed.). (pp. 14–23). Boulder, CO: Westview Press.

Finkler, K., Skrzynia, C., & Evans, J. P. (2003). The new genetics and its consequences for family, kinship, medicine and medical genetics. *Social Science & Medicine, 57,* 403–412.

Garfield, R., Dresden, E., & Boyle, J. S. (2003). Health Care in Iraq. *Nursing Outlook, July/August,* 171–176, p. 174.

Giger, J. N., & Davidhizar, R. E. (1991). *Transcultural nursing: Assessment and intervention.* St. Louis, MO: Mosby.

Grady, C. (1998). Chapter 9: Ethics, genetics, and nursing practice. In *Genetics in clinical practice.* D. Lea, J. Jenkins, & C. Francomano (Eds.). Boston: Jones and Bartlett.

Guttmacher, A., & Collins, F. (2004). *Genomic Medicine—A Primer: Articles from the New England Journal of Medicine.* In Guttmacher, A., Collins, F., & Drazne, J. (Eds.). Baltimore, MD: Johns Hopkins University Press.

Kenan, R. (1996). The at-risk health status and technology: A diagnostic invitation and the gift of knowing. *Social Science and Medicine, 42*(11), 1545–1553.

Khoury, M. J. (2003). Genetics and genomics in practice: The continuum from genetic disease to genetic information in health and disease. *Genetics in Medicine, 5*(4), 261–268.

Lee, S. S., Mountain, J., & Koenig, B. A. (2001). The meanings of "race" in new genomics: Implications for health disparities. *Yale Journal of Health Policy, Law and Ethics, 1,* 33–75.

Lopez, S., & Guarnaccia, P. (2000). Cultural psychopathology: Uncovering the social world of mental illness. *Annual Review of Psychology, 51,* 571–598.

Navarro, M. (November 9, 2003). Going beyond black and white, Hispanics in Census pick 'other.' *New York Times,* Sunday, p.1, p.21.

Oncology Nursing Society. (1999). Oncology Nursing Society Multicultural Outcomes: Guidelines for Cultural Competence. Pittsburgh: Author.

Peters, J. A., Djurdjinovic, L., & Baker, D. (1999). The genetic self: The Human Genome Project, genetic counseling and family therapy. *Families, Systems & Health, 17*(1), 5–25.

Risch, N., Burchard, E., Ziv, E., & Tang, H. (2002). Categorization of humans in biomedical research: Genes race and disease. *Genome Biology, 3*(7), 1–11.

Schaeffeler, E., Eichelbaum, M., Brinkmann, U., Penger, A., Asante-Poku, S., Zanger, U., & Schwab, M. (2001). Frequency of C3435T polymorphism of MDR1 gene in African people. *Lancet, 358*(9279), 383–384.

Secretary's Advisory Committee on Genetics, Health, and Society. Accessed September 15, 2004 from, *http://www4.od.nih.gov/oba/sacghs.htm.*

Weatherall, D. J. (2003). Genomics and global health: Time for reappraisal. *Science, 302*(24), 597–598.

Wooten, J. (2004). Race Reversal. Accessed 7/28/04 http://abcnews.go.com/sections/Nightline/SciTech/racial_identity_031228-1.html.

World Health Organization. (2002). *Genomics and World Health.* Geneva: World Health Organization.

CHAPTER 10
Connecting Genomics to Society: Spirituality and Religious Traditions

Knowledge is not intrinsically good or evil—it is the use to which it is put. We must collectively make those decisions and set those boundaries.
Francis Collins, MD, PhD, Director National Human Genome Research Institute, 2003

Molly

My daughter Molly is 21 years of age. All through her life, particularly early on, we were referred to neurologists because of her severe developmental delay. She has the clinical signs and symptoms of Rett's (see **Box 10-1**), yet we were never counseled about genetic testing. Most of the neurologists that she saw early on suggested that we should just wait and see how she progressed. They would give her various diagnoses like autism or deafness. At one point we moved so that she could attend a school for the deaf. After a year at the school, the teachers suggested further testing, which showed Molly did not have a hearing problem, she just did not respond to her surroundings.

I first considered a genetic disorder when I attended the Summer Genetic Institute (SGI) at the National Institutes of Health. After I returned from the SGI, a Baylor College of Medicine Rett's syndrome researcher moved in across the street. It was through conversations with her that we began to think about Rett's. Molly has had all the behavioral and physical symptoms of the syndrome. We still have not had her tested.

Underneath all the clinical visits and information from health care professionals there were the underlying and unanswered questions of, "Why me? Why our child? Why our family?" For us, our faith provided the answers to these questions. Both David and I have always had a strong faith in God and believe that He cares and guides us through life. I am often reminded that He will never leave me, and that His values sometimes differ from the values of our society. I know that the ability to love unconditionally, the ability to understand what a person feels like when they are hurt, angry, or brokenhearted, and the ability to be grateful for all the little

Box 10-1 Rett Syndrome: What It Is and How It Is Inherited

General Information

Rett's syndrome is a progressive genetic developmental condition, which occurs almost exclusively in girls. Girls with Rett's syndrome have apparently normal psychomotor development during the early months of life and then begin a rapid regression in language and motor skills. The hallmark clinical symptom of Rett's syndrome is the loss of purposeful hand use and its replacement with repetitive hand movements. Girls with Rett's syndrome typically live into adulthood. An atypical form of Rett's syndrome caused by a gene mutation, MECP2, has been identified in individuals previously diagnosed with autism, mild learning disability, mental retardation, and clinically suspected Angelman syndrome (see http://www.genetests.org for Gene Reviews-Angelman Syndrome)

Clinical Symptoms

In most individuals with Rett's syndrome, there is a predictable pattern of developmental stages. Individuals with Rett's syndrome generally develop normally for the first 6 to 18 months of their lives. Many attain expected milestones such as short words, smiling, and finger feeding. At 5 months to 3 years, head growth begins to slow (acquired microcephaly) and after 18 months other development and physical symptoms begin to appear. The children may be slower to acquire skills and have increasingly diminished muscle tone (hypotonia). Between the ages of 1 and 4 years, children with Rett's syndrome lose previously acquired skills, experiencing development regression. One of the main characteristics of Rett's syndrome is distinctive, uncontrolled hand movements that are performed continually during waking hours. Girls with Rett's syndrome often develop foot and hand deformities as they grow older, and 50% develop seizures. Other clinical features include autistic features, bruxism, and episodic apnea.

Boys meeting the clinical criteria of Rett's syndrome have been identified in association with a 47, XXY karyotype and MECP2 gene mutations that occur after fertilization, leading to somatic mosaicism. Males with a 46, XY karyotype and MECP2 gene mutation may have a severe form of neonatal encephalopathy and die before reaching age 2 years.

Pathology

The diagnosis of Rett's syndrome is based on clinical diagnostic criteria established for the classic syndrome, named after Dr. Andreas Rett who discovered the condition in 1996, and/or molecular testing of the MECP2 gene located on the X chromosome (Xq28). Molecular genetic testing identifies gene mutations in approximately 80% of girls with classic Rett's syndrome.

Box 10-1 continued

Prognosis

Girls with Rett's syndrome have abnormal brain development that causes significant mental retardation. After the period of rapid deterioration, the disease becomes relatively stable. Girls will typically survive into adulthood, but the incidence of sudden, unexplained death is increased. The sudden death is thought to be due to a higher incidence of cardiac arrhythmia and reduced heart rate variability.

Boys with Rett's syndrome may also survive into adulthood with moderate to severe mental retardation, impaired language development, and movement disorder. These boys do not undergo a period of normal development.

Inheritance

Rett's syndrome is inherited in an X-linked dominant manner. Greater than 99% of Rett's syndrome cases are single occurrences in a family, resulting from a new (de novo) gene mutation in the child with Rett's syndrome or from inheritance of the gene mutation from one parent who has somatic or germline mosaicism.

The frequency of the MECP2 mutations in individuals with mental retardation is only now beginning to be understood.

Prevalence

Rett's syndrome appears in an estimated 1 out of every 15,000 female births around the world today, regardless of race.

Diagnosis

The diagnosis of Rett's syndrome rests on clinical diagnostic criteria including:

- Normal development until 6 to 18 months of age
- Acquired microcephaly
- Severe impairment of receptive and expressive language
- Severe mental retardation
- Development of uncontrolled, persistent hand movements
- Impaired ability to coordinate movements required for walking
- Fine tremors of the torso
- Periods of apnea
- Spasticity
- Joint contractures
- Scoliosis
- Chewing, swallowing difficulties, and teeth grinding
- Growth retardation

Box 10-1 continued

MCEP2 mutations should be considered in male infants with severe hypotonia or infantile spasms. MECP2 mutations have been found in girls with a Rett's syndrome variant, mild learning disability, and even in women with no apparent symptoms who have skewed X-inactivation. MECP2 mutations have also been identified in individuals with previously diagnosed autism, mild learning disability, clinically suspected but molecularly unconfirmed Angelman syndrome, or mental retardation with spasticity and tremor.

Relevant Testing

Diagnostic testing for Rett's syndrome involves molecular genetic testing for the presence of the MECP2 gene mutation (located at Xq28). Molecular genetic testing identifies gene mutations in about 80% of affected females. This testing is now available on a clinical basis.

Genetic Counseling

The diagnosis of Rett's syndrome may result in evaluation and diagnosis of the mother and other family members who were previously unaware of the presence of a genetic condition in their family. This discovery may be difficult for families as it has implications for their own health and reproduction. Healthy sisters of a girl with Rett's syndrome could be carriers of the MECP2 gene mutation present in their sister but have few or no symptoms due to skewed X-inactivation. Genetic counseling should include this possibility since these healthy sisters have an increased chance for passing on the gene alteration to their children. Prenatal testing is now available for Rett's syndrome. Preconception genetic counseling should be offered to families for determination of future risk and discussion of the availability of prenatal testing.

Management

There is no known treatment that can improve or alter the neurological outcome of individuals with Rett's syndrome. Current management focuses on supportive and symptomatic therapy. Occupational and physical therapies are important for maintaining function and preventing scoliosis and deformities. Therapeutic activities such as swimming, horseback riding, and music therapy are often beneficial.

Box 10-1 continued

Support Resources

International Rett's Syndrome Association (IRSA)
9121 Piscataway Road
Clinton, MD 20735
Phone: 800-818-7388 or 301-856-3334
Fax: 301-856-3336
E-mail: *irsa@rettsyndrome.org*
http://www.rettsyndrome.org

NCBI Genes and Disease Webpage
Rett syndrome
http://www.ncbi.nlm.nih.gov

Rett's Syndrome Research Foundation
4600 Devitt Drive
Cincinnati, OH 45246
Phone: 513-874-2520
E-mail: *mgriffin@rsrf.org*
http://www.rsrf.org

(http://www.geneclinics.org)

things, doesn't necessarily result from a life without care. I value these abilities more than any of the things I own or the perfect life I dreamed about. The growth that results from the hard work of grieving provides a strength and an understanding that will help me through the rest of my life, and I thank God that I had the opportunity to grow through our experience. It has been difficult and it takes time, but I like the person that I have become.

Today, we embrace whatever God gives us, both now and in the future, as part of an overall plan that we don't always understand, but we have faith that He is in control. The belief that His will directs our lives gives us the strength to face a future that may be uncertain. To trust that we will not only have our needs met, but that we may also be able to experience many things and grow in character and wisdom is a wonderful thing.

We decided early on to accept Molly as a gift from God and to embrace having a different child. We love Molly for who she is and for what she brings to our lives. We enjoy her smiles and her personality, and we love and support each other. The belief that God has a purpose for Molly's life, even Molly who cannot talk or even understand much, is amazing to us. At times we have opportunities to encourage others who grieve over the loss of similar dreams. The best thing that we can give another parent in a similar situation is to let them know that life is still worth living, and that although you still experience pain that will never completely go away, incorporating the loss and pain into a life that God has given us can be the most wonderful experience. When I start to worry about the future, I just look back at the past and see that I have been blessed with so much. I then look forward with the expectation that God's blessing will still be there throughout my life as I live and as I face death and perhaps the reality of leaving Molly behind. I know that He has a purpose, and because of that, my life and Molly's life have a purpose. I can trust Him to provide everything I need to fulfill that purpose.

So, Molly's condition has both challenged and improved our life. The challenge was in working through the loss of a dream of the normal child we did not have, but the struggle to get through that loss has made us a stronger family and better human beings. As a couple, we were always close, but the experience of having an atypical child made us closer and provided us with different values. We value differences in people and people who have faced difficulties more than we did before. Molly is in a group home now, and we value the simple pleasures she has always given us, like her smile and her unconditional love when the world is so hectic.

Lorraine—Molly's mom

Introduction

Knowledge gained from understanding the structure and function of the human genome will bring with it a host of medical and moral options. It is important to recognize that, in addition to cultural and ethical perspectives (see Chapter 9), philosophical and theological perspectives influence decisions made regarding use of new innovations such as genetic information and services. Choices made about utilization of genomic innovations in clinical practice will be influenced by personal and societal values, beliefs, and experiences. Such values and beliefs are often based on the framework of religious teachings and experienced through the lens of spirituality.

How one views life and death is often influenced by religious beliefs. Encountering stressful events such as the birth of a baby, dealing with an illness, or facing diminishing treatment options will often challenge these beliefs. Nurses may also recognize that an individual and family are experiencing spiritual conflicts when faced with making decisions about utilization of genetic diagnosis and treatment options. Nurses may find themselves confronting their own beliefs when encountering individuals considering genetic services (Barnum, 2003). Nurses have a responsibility to assist professional organizations, work associates, and the government to consider spiritual and religious tenets when defining the ethical boundaries for uses of genomic technology.

Western religions (i.e., Judaism, Christianity, and Islam) have similarities in key tenets as well as in the belief of one God or universal spirit (Gerardi, 1989). Eastern philosophical and spiritual thought tends to focus on the person as a spiritual being with a spiritual goal (Martin, 1989). Health care provider awareness of an individual's religious beliefs, practices, rituals, and requirements, which influence personal health care decisions, is not always possible. However, collaboration among health care providers, religious counselors, and the client–family to identify and grasp the meaning of health care options for that individual will facilitate inclusion of spirituality in clinical care. Extensive ecumenical discussions have resulted in position statements available from the various religious entities that may clarify viewpoints on issues of concern for individuals and families based on religious beliefs (**Table 10-1**) (Shinn, 1998). Pastoral counselors and other religious resources are available to guide health care professionals and their clients when ethical questions are prompting theological reflection.

Potential ethical questions resulting from genomic advances as identified by Peters (1998) are detailed in **Table 10-2.** Examples of several of these topics will be briefly presented in this chapter as fictitious stories with an accompanying viewpoint(s) excerpted with permission from presentations provided by theological

Table 10-1 Position Statements

Episcopal Church	*http://www.episcopalarchives.org*
Evangelical Lutheran Church	*http://www.elca.org/docs/humancloning.html*
Islamic Views	*http://www.people.virginia.edu/~aas/issues/cloning.htm* *http://www.biol.tsukuba.ac.jp/~macer/EJ52H.html*
Jewish Law	*http://www.jlaw.com/Articles/cloning.html* *http://www.jlaw.com/Articles/stemcellres.html*
Mormon	*http://www.lds.org*
Religious Action Center of Reform Judaism	*http://www.rac.org/issues/issuebe.html*
Religious Tolerance	*http://www.religioustolerance.org/clo_reac.htm*
Southern Baptist Convention	*http://sbcannualmeeting.org/sbc01/sbcresolution.asp?ID=2*
The Pew Forum on Religion & Public Life	*http://www.pewtrusts.com/pdf/rel_pew_forum_adams_rib.pdf*
United Church of Christ	*http://www.ucc.org/synod/resolution/res30.htm*
United Methodist Church	*http://umns.umc.org* *http://umns.umc.org/backgrounders/cloning.html*
United States Conference on Catholic Bishops	*http://www.usccb.org/profile/issues/bioethic/index.htm*

TABLE 10-2 Ethical Issues Influenced by Genomics Advances

Genetic discrimination	What is normal?
Abortion	When does life begin?
Identity	What does it mean to be human?
Patenting	Who owns God's creation? Marketing of human beings?
Genetic determinism	Is it all in the genes?
Question of sin	If genetic predisposition to a negative behavior is linked, is it still considered immoral?
Gene therapy	When is germline gene therapy justified?
Cloning	What can we do? What should we do?
Playing God	Are we asking scientists to play God? Or refrain from playing God?

(Peters, 1998)

leaders* at an intensive genetics education course offered in 2003 *(http://www.gpnf.org)*. In the following sections, a notation of the voiced religious perspective will be identified next to the quote. If available, a position paper will be discussed. These perspectives highlight the varied belief systems, which may be very different from your own. Take time to reflect on how you can effectively respect and support individuals with these differing viewpoints to promote ethical decision making when facing issues arising during this genomic age (Swaney, 2001).

Genomics and the Concept of *Normal:* What Is Normal?

Jeremy and Kate are interested in having a baby. Jeremy's side of the family has a significant history of cancer, and although currently unaffected, he has been diagnosed with a gene mutation for Hereditary Non-Polyposis Colorectal Cancer syndrome (HNPCC). Jeremy and Kate are exploring options through their local Genetics Invitro Fertilization Clinic to have preimplantation genetic diagnosis that

*Robert Baumiller, PhD, Catholic perspective.
Baher S. Foad, MD, FACP, FACR, Muslim perspective.
Robert Hopkin, MD, Mormon perspective.
Howard Saal, MD, Jewish perspective.

allows them to only have an embryo implanted that is free of the HNPCC family mutation. Jeremy and Kate are also interested in assuring that they have a son. They've asked that during PGD, since they are already paying for the predisposition HNPCC genetic testing, gender selection also be assured.

Technological advances in human genetics and reproductive science have provided the capability to perform genetic tests on embryos produced by in vitro fertilization. Embryos found to be free of a disease-causing gene mutation can then be selectively implanted into the woman to initiate pregnancy. However, a genetic test is not an infallible prediction of future health. Preimplantation genetic diagnosis (PGD) provides parents the option of selectively determining genetic characteristics of their children (Jones, 2002). Significant efforts by the Genetics & Public Policy Center *(http://www.DNApolicy.org)* are in process to gather feedback from stakeholders and experts in preparation for development of policy recommendations regarding "Custom Kids? Genetic Testing of Embryos."

Religious Perspectives

"The testing itself is not morally right or wrong. What one chooses to do as a result of the testing is where the moral decision is made." (Catholic)

"Preimplantation diagnosis poses some problems. There is no official status on when a conception becomes a human being, but genetically it has been recognized that the individual identity is present at the time of fertilization. This implies that the same thoughtful and prayerful consideration should be applied to preimplantation diagnosis as to other prenatal diagnosis, and that the decision will ultimately be left to the couple in question." (Mormon)

A Sample Religious Statement

Jewish law provides a perspective regarding the use of cyropreserved sperm and pre-embryos *(http://www.jlaw.com)*. This statement addresses several topics including assisted reproduction, infertility, and retrieval of sperm from individuals who are brain-dead, or even postmortem. According to this statement, Jewish law recognizes that assisted reproductive technologies are allowed only in the context of an intact marriage, during the life span of the marriage, and with both partners alive. The authors of this statement encourage religious deliberations to generate and contribute to public debate on the issues, with the recognition that religious sources rather than public consensus are crucial to development of moral positions.

Genomics and the Concept of *Soul:* When Does Life Begin?

Paula and Kevin are excited about the news of their upcoming new baby. Paula delayed this pregnancy until her career was established, so at the age of 36 it has been recommended that she have an amniocentesis to determine the health of her baby. Paula and Kevin are struggling with their decision to have the recommended procedure since their religious teachings oppose elective abortions. They are concerned that the procedure may provide greater risks than benefits in their situation. They discuss with the counselor their opposition to having an abortion even if the amniocentesis indicates a genetic disorder, such as Down syndrome. They do recognize that the amniocentesis results could provide to them information that would assist them in preparing for the care of their child. They ask if they can have their pastor attend and participate with them as they hear the information and make this difficult decision about having the amniocentesis.

Genetic testing in the prenatal setting has been possible for many years. However, as greater understanding of genetic contributions to risk for future health problems is possible, an increasing number of genetic tests may be offered and performed. The complexity of the testing methodology, nuances of test outcomes, and the meaning of test results will challenge health care providers to be able to give accurate information and guidance to their clients (Burke, 2002). Multidisciplinary consultation may be required to provide sufficient information to the individual–couple to make an informed decision.

Religious Perspectives

"Abortion is forbidden in Islam unless the pregnancy poses serious danger to the health of the mother. In other words, one of the most common genetic disorders, Down syndrome, having a baby with Down syndrome is not considered an indication or a reason for abortion . . . Every child is a blessing from God, and sometimes the child with Down syndrome is a blessing in disguise because it teaches us patience, caring, and love." (Muslim)

"Now genetic testing causes individuals to make, implicitly or explicitly, certain ethical and moral decisions. While the testing is not necessarily a testing because there is a high probability of a problem, the testing will still uncover difficulties where decisions will have to be made . . . about marrying or not, becoming pregnant or not, continuing a pregnancy already accomplished or not. Genetics is nothing else but those kinds of questions." (Catholic)

"I think that most people make their decisions without consulting because they are not sure what answer they will receive, what support they will receive . . . so that is not the best solution either. People need to be sensitive to the kind of situation that the person is placed in and the pastoral needs that they have." (Catholic)

A Sample Religious Statement

The United States Catholic Bishops in 1996 developed a statement on Genetic Testing and its Implications *(http://ww.usccb.org)*. Using fictitious examples to illustrate the potential uses of genetic testing, this statement offers guidance for pastors and Catholic health care providers. The statement recognizes that genetic testing will create moral issues for the individual, for the family, for racial and ethnic groups, and for society as a whole. Guidance, based on tradition, knowledge, and moral reflection encourages use of genetic testing as an important tool in conjunction with recognition of dignity of the human person.

Genomics and Identity: What Does It Mean to Be Human?

Insights into our evolutionary history and the small variations within individual humans' genomes may affect concepts of race, ethnicity, and even gender. In the light of genetic discoveries, new dimensions of religious or philosophical concepts about identity, responsibility, and what it means to be human may become apparent *(http://www.genome.gov/10001848)*. As increasing numbers of organisms' genomes are determined, identification of similarities to the human genome opens avenues to determining gene function *(http://www.genome.gov/10000375)*. Through sequencing of multiple organisms, such as mouse and chimpanzee, an evolutionary perspective will be obtained. This knowledge may create tension regarding religious philosophy and beliefs about the origin of man. The concept of human identity and genomics was not discussed at the conference, so comments from religious leaders are unavailable.

Religious Boundaries and Use of Genomics: What Can We Do? What Should We Do?

Innovative scientists will be able to find profound uses for biotechnology. The President's Council on Bioethics *(http://bioethics.gov)* released a report, *Beyond Therapy: Biotechnology and the Pursuit of Happiness* (2003), which explores the potential, the limitations, and the implications of the promise of these biotechnological advances. Step one in determining what innovations our society values begins with an awareness of particular developments made possible because of scientific discoveries. A

Table 10-3 Resources

Bioethics Resources on the Web	*www.georgetown.edu/research/nrcbl/nirehg/scope.htm*
Center for Genetics and Society	*www.genetics-and-society.org/perspective/religious.htm*
Council for Responsible Genetics	*www.gene-watch.org*
Genetics and Ethics Page	*www.genethics.ca*
National Bioethics Advisory Commission	*www.bioethics.gov*
National Human Genome Research Institute	*www.genome.gov/100001848*
The Genetics and Public Policy Center	*www.DNApolicy.org*
U.S. Dept. of Energy Human Cloning Fact Sheet	*www.ornl.gov/hgmis/elsi/cloning.html*

survey of 1,003 adults (Virginia Commonwealth University, 2003) reported that a significant number of responders (75%) said that "Sometimes new developments in science seem so complicated that a person like me can't really understand what's going on." Not all of those innovations will be successful, understood, or valued, but all will need careful moral and ethical reflection to assure that through genomics research we move toward improving the human condition.

The human genome is only one component of what makes us human beings. Other factors, such as environment, lifestyle, behaviors, and free will also contribute to the outcomes of whatever interventions may be applied in the future clinical setting. The Web site for the Center for Genetics and Society *(http://www.genetics-and-society.org)* provides excellent background materials, references, and a listing of adopted policies for each of these topics. Other resources that offer insight into these topics are listed in **Table 10-3.**

Gene Therapy: Is It All in the Genes?

Amanda's husband, Craig, has been very successful in sports. In fact, he is a professional football player. They are planning to have a child and have talked about interest in enhancing their baby's chances of growing into a healthy and successful adult. Craig has utilized exercise and diet throughout his life to build a strong body. Along with his dad's assistance and long hours of practice, these interventions have worked together to ensure Craig's stamina and muscular development.

It didn't hurt that Craig's father had also been a successful football player. Craig and Amanda want to make sure that, if there is anything else they can do to jump start their baby's success with sports, they take advantage of it. They certainly have the financial ability to pay for emerging services, such as the designer babies they had heard about on television. Craig and Amanda live next door and have come over to ask what you know about the availability of genetic services called "germline engineering."

Although germline engineering is talked about as a future possibility, no single gene has been identified as contributing to development of an enhanced sports performer. Realistically, even if we did know the genetic contribution to superiority, genetics alone does not determine the final outcome. All the other factors encountered in living (diet, environment, lifestyle, training, equipment) affect the final results. Additionally, the methodology to perform such a genetic enhancement to an embryo is not currently possible. Concerns have been expressed about, when such interventions one day become possible, who will have access to the technology and services? How will a determination of superior performance be made and ensured through such technology? And what implications will that have for future generations of humans?

Religious Perspectives

"The greatest challenge, which is just around the corner, is the one of control. We can define disease rather easily. But does that definition include things like height, intellectual capability, physical structure, and so many other things which one can look at as improvement as health . . . but it swamps the diversity and individuality of humankind. So that the better we get at controlling and adjusting genetic makeup, the greater the temptation to fiddle with it in some way. Where are we going to get the wisdom in the future to do that?" (Catholic)

A Sample Religious Statement

The Religious Action Center of Reform Judaism in 2000 identified issues of concern to include genetic discrimination and genetic engineering *(http://www.rac.org)*. Although this summary did not specifically address gene therapy, the Jewish view expressed may provide guidance when considering future decisions of this possibility. "In the Jewish view, God allows human beings to be partners in creating a better world, and has given us the freedom to do so. We are expected to use our God-given wisdom to help create a better world" (p. 4 of 5).

Cloning

Amy and Ken are experiencing devastating grief as they learn that their only daughter Brenda, at the age of four, was killed in a car accident. Compounding this grief is the fact that, because of health reasons, Ken is no longer able to father another child. Amy and Ken have read about the possibility of cloning in humans. There have already been a lot of success stories of different animals being "cloned." They wonder if perhaps they can save Brenda's DNA as suggested at the funeral home *(http://www.dnaconnections.com)* and explore the possibility of having their daughter cloned. Amy is your sister and asks you what you know about the availability of cloning humans.

Cloning is a form of reproduction in which offspring result from the asexual production (manufacturing) of new human organisms that are virtually identical to an existing individual *(http://bioethics.gov/topics/cloning_faq.html)*. Cloning results from the transfering of DNA material extracted from an existing donor cell, placing the DNA into an egg from which the nucleus has been removed, and inducing the new egg material to develop into an embryo. Reproductive or human procreation cloning has the intention of producing a baby (Evans, 2002). Research or therapeutic cloning is the use of somatic cell nuclear transfer, as just described, with the purpose of using the new material for performing research or treating disease. Seven species have been successfully cloned, and mixed reports on human cloning success have been publicized. Reproductive cloning technology currently has a high rate of morbidity and mortality and elicits strong negative emotional reactions from the public, and some religious and government bodies. Discussion, proposing, and passing of state, federal, and international legislation on cloning involves an active process that changes daily. Web sites that track progress in this area include *http://www.ncsl.org/programs/health/Genetics/03clone.htm*, *http://www.usccb.org/prolife/issues/bioethic/statelaw.htm*, and *http://www.humgen.umontreal.ca*.

Religious Perspectives

"Judaism does not rule out cloning, especially if cloning is done for therapeutic reasons. It really remains very controversial, nonetheless. Cloning for reproductive reasons is not allowed, and again still remains controversial." (Jewish)

"We believe that life is sacred and we do not create life—God creates life. We may participate in bringing a child into the world, but that child is a gift from God. Human life is sacred and should be regarded as sacred and treated with reverence from the beginning to its end." (Mormon)

Religious Statements

Multiple religious responses are available, which express their positions on cloning. Because there is limited knowledge of the long-term consequences of cloning, some religious groups have provided statements banning all cloning (Baptist, Catholic, Methodist, Muslim). Others ban cloning because they consider it biological manufacturing (Baptist, Protestant). Some religions ban reproductive cloning but value therapeutic cloning (Church of Christ, Jewish). Others have developed extensive education materials for their membership (Jewish—Broyde, 1997; Lutheran—Willer, 2001). Evans (2002) provides an excellent summary of the religious voices heard that offer a diversity of perspectives from which society can build an informed public policy.

Stem Cell Research

Janet's husband, Rick, has been diagnosed with symptoms indicating early-onset Parkinson's disease. He is starting to exhibit symptoms, such as shaky hands, slow movements, and impaired balance, despite taking medications to improve movement control. Rick is depressed and expresses concern to Janet that, as his disease progresses, he will become an increasing burden to his family. Janet has been doing Internet searches to seek out the availability of treatment options and has been surprised by the limitations of known interventions. Checking on the National Institutes of Health Web site *(http://www.nih.gov)*, she gains a little glimmer of hope as she reads about the potential for stem cells for the future treatment of Parkinson's disease. This hope is immediately diminished as she reads the fact sheet information that outlines obstacles that currently limit progress in application of human stem cells for treatment of illness. Janet and Rick turn to their Church of Christ minister. The minister is aware of the issues related to embryonic research. He had attended a meeting during which the Church of Christ passed a resolution in 2001 voicing their support of stem cell research *(http://www.ucc.org/synod/resolutions/res30.htm)*. Janet and Rick ask for his assistance to let federal legislators, currently considering a ban on stem cell research, be made aware of the personal implications for a real person as they struggle with the outcomes of this critical decision for Rick's future.

Stem cell research is a promising, yet controversial technological breakthrough. Stem cells maintain the ability to divide. When derived from fetal or placental tissue, these cells are capable of developing into all tissue types. Adult stem cells are capable of developing into only a limited number of tissue types. Researchers are excited about the potential to utilize stem cells to create replacement cells and organs and to treat diseases such as Alzheimer's disease and Parkinson's disease (Nussbaum & Ellis, 2003) (for more information see *http://nih.gov/news/stemcell/primer.htm*). Current legislation and research guidelines regulate federally approved and funded research studies utilizing fetal cells *(http://www.nih.gov/news/stemcell/NOT-OD-00-050.html)*. Updates on proposed federal legislation regarding stem cell research can be located at *http://www.thomas.loc.gov*. Background reports on ethical issues in human stem cell research are available from the National Bioethics Advisory Committee at *http://www.georgetown.edu/research/nrcbl/nbac/pubs.html* (1999) and *http://www.bioethics.gov/reports/stemcell/index.html* (The President's Council on Bioethics, 2004).

Religious Perspectives

"When does life begin? When is a fetus considered a human being? The fertilized egg is biologically alive from the moment of conception. But as Muslims we believe that the spirit is breathed into the embryo at 40 days, according to the interpretation of the saying of the Prophet Muhammad." (Muslim)

"While the first Presidency and the Quorum of the Twelve apostles have not taken a position at this time on the newly emerging field of stem cell research, it merits cautious scrutiny. The proclaimed potential to provide cures or treatments for many serious diseases needs careful and continuing study by conscientious, qualified investigators. As with any emerging technology, there are concerns that must be addressed. Scientific and religious viewpoints both demand that strict moral and ethical guidelines be followed." (Mormon)

A Sample Religious Statements

A summary position paper offered by Eisenberg (2001) provides a perspective on stem cell research based on Jewish law *(http://www.jlaw.com/Articles/stemcellres.html)*. Since stem cell research is a new endeavor, there are no published responses on the topic. Dr. Eisenberg discusses issues raised by stem cell research in conjunction with religious teachings and Jewish law. Issues include the Jewish approach to abortion and pre-embryos. For the first 40 days, the fetus is not yet considered a person, which has implications for interpretation of the law permitting utilization of early embryos for stem cell harvest. Concluding remarks identify the promise of human embryonic stem cell research for the benefit of those suffering and in pain.

Translation of Religion to Policy: Integration of Genetics and Genomics

Translation of religious tenets to policy requires interpretation of doctrine and application to real-life situations. Recognition of human limitations in this process of integration of religious values and beliefs in determining how genetics and genomic knowledge is translated into health care options was expressed at the conference.

"I would just start by saying that we should recognize that, as individuals, we are only interpreting divine guidance; and as human beings we can make mistakes in our interpretation. . . . Life is a series of tests and trials and it is through these tests and trials that we evolve spiritually as human beings and become closer and closer to God." (Muslim)

"One of my concerns is that there cannot only be one interpretation of religious scriptures. We assume these are divinely inspired, however, the interpretation is up to human beings. I do not want to see a limitation of our development of technology for all uses, for all types of health issues based on religious tenets. . . . In Judaism, it really supports development of technology because the basic tenet is God will help those who help themselves." (Jewish)

A Sample Religious Statement

The Archives of the Episcopal Church *(http://www.epsicopalarchives.org)* provides two statements regarding genetics. The first, written in 1982, encourages education of clergy and laity in counseling and recognizes topics, such as areas of genetics and biotechnology, as relevant areas (1982-A064). A resolution in 1985 (1985-A090) encouraged genetic engineering and the study of it implications. The House of Bishops indicated that human DNA is a great gift of God. Basic training in human genetics in continuing education programs is valued by the Church. Specialized training of clergy and ministers in genetics and ethics was highlighted as a way to work progressively with others.

Implications for Nursing: Spirituality in Care

People may have greater spiritual needs when confronted with a health issue. Most studies have indicated that religious involvement and spirituality are associated with better health outcomes (Mueller, Plevak, & Rummans, 2001).

"I frequently recommend to families that they discuss these issues and specifically address religious connotations in addition to family implications and emotional issues. I encourage people to seek out counsel with their religious leaders. Then it is incumbent on those in leadership positions to listen with an open mind and an open heart to hear the perspectives of the family and the whole situation before making a decision or making a declaration of any kind, or they will never understand the story and the situation completely." (Mormon)

How Then Should We Face the Future?

> "I think basically we all believe, as other people in other religions, treat others as you would treat yourself. We have a societal reason to care for everybody. I don't look at God as making mistakes. People who exist here on Earth, for whatever reason we exist, we all have a responsibility as a community, as a religious community, as a society, to care for all people. A lot of these things go beyond religion." (Jewish)

After working on this chapter, as I slept I had a dream that was significant to my perspective of my own responsibilities with issues identified in this chapter. As I remember it, the dream began on a brisk January evening. I was standing with many others along the railway track platform awaiting the train's arrival. I eagerly anticipated the benefits of being in a warm, comfortable environment, sitting quietly reading a book on my trip home. Observing the train light in the distance, I heard the train whistle announcing its approach. Immediately in front of me appeared two beautiful children laughing and slowly walking on the track. All of us on the platform stood there with concern on our faces, frantically looking for someone, maybe a parent, to step forward and rescue the children from the obvious danger of the approaching train. The children were unaware that they were in danger, and as they bent to reach a dropped toy, the sounding horn and noise of the approaching train to the station signaled immediate need for action. I covered my eyes, stepped back, and yelled for help. Someone farther down the track jumped over the railing and pulled the children to safety. Some of the persons with me hadn't even seen the children or sensed any danger. Others talked excitedly about what almost happened.

Jean, author

I approach this genetic age (the train) with such excitement and enthusiasm for what the future holds. I eagerly anticipate the benefits to individuals, families, and society from enhanced understanding of the genomic contributions to health and well-being. But I, like many others, have concerns about the potential signals (horns/lights) of approaching danger (for all) if we do not assume personal responsibility for making some tough decisions. It is important that all are made aware of the issues, the reality of the science, and the options and expectations of scientific advances that are fast approaching. We all have a responsibility to be aware, be alert, pay attention, and take action (not cover our eyes) to understand what's going on here. Only then, can we appreciate the potential hazards of these decisions to the future and for our children. We cannot stand back and let others act on our behalf. These decisions need to be based on reflection about the issues by all, so that we can effectively protect the rights of those yet to come. Religious foundations are a very important component of that reflection and should provide guidance for the future.

Chapter Activities

1. Based on your values and beliefs, present to your colleagues your thoughts about Religious Boundaries and Use of Genomics: What can we do? What should we do?
2. Select one of the religious position statements listed on Table 10-1 and locate it on the Web site. Identify when the statement was written, who authored the statement, and how it was approved to represent the voice of the people. How might you utilize the statement to guide your clinical care?

References

Barnum, B. (2003). *Spirituality in Nursing: From Traditional to New Age.* New York: Springer Publishing Company.

Broyde, M. (1997). Cloning people and Jewish law: A Preliminary Analysis. Found at http://www.jlaw.com/Articles/cloning.html.

Burke, W. (2002). Genetic testing. *New England Journal of Medicine, 347*(23), 1867–1875.

Cole-Turner, R., & Waters, B. (1996). *Pastoral Genetics: Theology and Care at the Beginning of Life.* Ohio: The Pilgrim Press.

Eisenberg, D. (2001). Stem cell research in Jewish law. Found at http://www.jlaw.com/Articles/stemcellres.html.

Evans, J. (2002). *Cloning Adam's rib: A primer on religious responses to cloning.* Pew Forum found at http://www.pewtrusts.com/pdf/rel_pew_forum_adams_rib.pdf.

Gerardi, R. In Carson, V. (1989). *Spiritual Dimensions of Nursing Practice.* Pennsylvania: W. B. Saunders Company, pp. 76–112.

Jones, S. (2002). Reproductive options for individuals at risk for transmission of a genetic disorder. *Journal of Obstetric, Gynecologic, and Neonatal Nursing, 31,* 193–199.

Martin, J. P. In Carson, V. (1989). *Spiritual Dimensions of Nursing Practice.* Pennsylvania: W. B. Saunders Company, pp. 113–131.

Mueller, P., Plevak, D., & Rummans, T. (2001). Religious involvement, spirituality, and medicine: Implications for clinical practice. *Mayo Clinic Proceedings, 76,* 1225–1235.

Nussbaum, R., & Ellis, C. (2003). Alzheimer's disease and Parkinson's disease. *New England Journal of Medicine, 348*(14), 1356–1364.

Peters, T. (1998). *Genetics: Issues of Social Justice.* Ohio: The Pilgrim Press.

Shinn, R. In Peters, T. (1998). *Genetics: Issues of Social Justice.* Ohio: The Pilgrim Press, pp. 122–143.

Swaney, J. Spirituality. In R. Gates & R. Fink, *Oncology Nursing Secrets* (2nd ed.). Philadelphia: Hanley & Belfus, Inc. 2001, pp. 524–533.

The President's Council on Bioethics. (2003). *Beyond Therapy: Biotechnology and the Pursuit of Happiness.* Washington, DC.

The President's Council on Bioethics. (2004). *Monitoring Stem Cell Research.* Washington, DC.

Virginia Commonwealth University. (2003). VCU Life Science Survey. Found at http://www.vcu.edu/lifesci/overview/polls.html.

Willer, R. (2001). Human cloning: Papers from a church consultation. Division for Church in Society. Found at http://www.elca.org/dcs/humancloning.html.

CHAPTER 11
My Hopes and Fears: How I Feel about Genetic Research

Having an understanding of the ethical, legal, and social issues related to genomic research including genetic testing, therapeutics, and the recording of genetic information gained from these studies (e.g., privacy, the potential for discrimination in health insurance and employment) will enable nurses to more fully inform and support individuals and families involved in genomic research.

Introduction

In the spring of 2003, the scientific community celebrated the completion of the sequencing of the human genome. Discoveries and innovations generated by human genome research are continuing to expand understanding of the complexities of human health and disease. Human genome research involving patients and families has contributed greatly to this new knowledge. Nurses will be caring for individuals and families who wish to participate or who are already enrolled in genetic research studies. To more fully inform individuals and families about genome research possibilities and to support their participation, nurses will need to have knowledge of the different types of genetic research available and the attendant ethical, legal, and social issues. In this chapter, we present two stories highlighting the hopes and fears that individuals and families bring to participation in genetic research protocols. Caleb's story offers a hopeful perspective, while Leigh's story presents some of the concerns and fears that patients may have when considering participation in genetic research. Different types of current genetic research and ethical and societal concerns regarding each will also be described. The various nursing roles in genetic research are highlighted.

Caleb's Story: My Hopes about Genetic Research

Before my pregnancy, my husband and I hadn't given much thought to genetics. When I did think about it, I wondered if my child would share my hair or eye color. During our fifth month of pregnancy, genetics became a part of our daily thoughts. At our 20-week ultrasound, the technician noticed our baby boy had shortened and bowed legs. We did not understand how serious this was until we met with a genetic counselor and obstetric specialist. We were told that it was a possibility that our son had a genetic disorder called Osteogenesis Imperfecta (OI). OI is caused by a genetic defect that affects the body's production of collagen. People with OI have bones that break easily, often from little or no apparent cause and have other symptoms, such as loose joints. No one could tell us for certain that this was what our son had, but it was a possibility. We had never heard of OI before, and the more we read, the more we realized that we needed to be prepared if there was a chance our son would have OI (**Box 11-1**).

I read about everything from diaper changes and special clothing for children with OI to experimental treatments and surgeries. We had planned to have a C-section to give our baby the best chance of being born with the fewest number of fractures. We met with a specialist from a major children's hospital nearby. My husband and I had a lot of discussions, and we gave a lot of thought about how to give our son the best and most fulfilling life possible. We knew that if our son did have OI, we would have to work hard to let him be independent and protect him at the same time.

When our son, Caleb, was born, we were overjoyed. Shortly after birth, the geneticist noted that, in addition to what was detected during pregnancy, the white part of Caleb's eyes looked gray and his face was unusually triangular shaped. The signs were pointing to OI, and a skin biopsy confirmed it 4 months later.

We firmly believe that the prenatal testing we had gave Caleb and our family a great advantage. We were able to prepare in some small ways for Caleb—and that made all the difference.

Some parents of children with undiagnosed OI are falsely accused of child abuse due to the fractures that can happen so easily with OI.

Today, Caleb is 22 months old. He is warm, inquisitive, and adventurous. He enjoys putting puzzles together and climbing on anything he can. In the last 3 months he has fractured his thigh twice while trying to learn to stand. He is determined to stand up and walk, and that is a goal we are working toward. Fortunately, we have the privilege of being involved in a research protocol at the National Institutes of Health. Through this protocol our family has received help with physical therapy and other medical issues that are part of living with OI. Being a part of this research program has helped us with Caleb's medical care as well as connecting us with other families facing similar issues. We hope that our involvement will help in some small way to advance the research to help find effective treatments and eventually a cure for OI.

Being the parents of this wonderful little boy has enriched our lives. The genetics and research staff have been an integral part of our endeavor to keep OI in perspective. As tempting as it would be to make Caleb's life all about preventing

Box 11-1 Clinical Trials Research: Osteogenesis Imperfecta Example

Research for many rare genetic conditions is being conducted in national and university research centers nationwide. Funding supports research activities ranging from cutting-edge gene and cell therapies to testing new drug treatments both in animals and in humans. The hope, as Caleb's story illustrates, is for improved diagnosis, treatment, and cure.

OI, as an example, is a rare genetic disorder affecting 20,000 to 50,000 adults, children, and infants in the United States. Individuals with OI have bones that break easily, often from little or no apparent cause (Nussbaum, McInnes, & Willard, 2001). OI is caused by a genetic alteration that affects the body's production of collagen, a major protein of connective tissue, cartilage, and bone. A person who has OI has either less collagen than usual or a poorer quality of collagen than usual, causing weakened bones that fracture easily. At present, there is no cure for OI, but research is ongoing in centers such as the National Institute of Arthritis and Musculoskeletal and Skin Diseases (NIAMS), a center at the National Institutes of Health *(http://www.nih.gov)*.

Funded clinical trials are underway to evaluate the safety and efficacy of new treatments for OI. The study that Caleb is participating in, for example, is an interventional trial using a randomized study design to evaluate the safety and efficacy of Bisphosphonate Treatment of Osteogenesis Imperfecta. The goal of this trial is to evaluate whether the investigational medication is safe and effective and has the ability to increase bone density in osteogenesis imperfecta *(http://www.oif.org/site/PageServer?pagename=Bisphosphonates)*. Children such as Caleb, ages 3 months through 17 years who have OI type I, III, or IV, are being enrolled in this study. Other clinical trials for OI include one study investigating treatment and efficacy of growth hormone therapy in OI, and observational studies looking at genetics, rehabilitation, growth, and natural history of the multiple secondary features of OI. Each of the studies is described on a Web site—*http://www.clinicaltrials.gov*—as a service of the National Institutes of Health and developed by the National Library of Medicine. Adults and parents with OI can access this resource Web site and learn about the various clinical trials for OI. Each of the trial descriptions provides information about the type of study, the study design and goal, eligibility requirements including inclusion and exclusion criteria, and the expected total enrollment. The location of the centers participating in the trials and contact numbers are also provided.

National organizations such as the Osteogenesis Imperfecta Foundation are another important resource for families wanting information about a particular genetic condition and research that is currently underway. These organizations can be located through the Genetic Alliance *(http://www.geneticalliance.org)* or the National Organization for Rare Disorders *(http://www.nord.org)*, two national organizations that serve as repositories for support groups for many genetic conditions.

Reference

Nussbaum, R. L., McInnes, R. R., & Willard, H. F. (2001). *Thompson & Thompson. Genetics in Medicine* (6th ed.). Philadelphia: W. B. Saunders Company.

fractures, we have tried to make our family life about appreciating the good times and coping to the best of our abilities with the challenges.

Melinda, Caleb's mom

Leigh's Story: My Fears about Genetic Research

I had always thought of myself as a healthy person. I ran 2 miles every day, ate well-balanced meals, and got enough sleep. I gave birth to and raised two wonderful healthy daughters, and now I am a grandmother at 60 enjoying my time with my new husband and family. I just wasn't prepared for the abnormal mammogram that showed a lump in my breast. I wasn't prepared to hear the words "You have invasive breast cancer." With those words, I entered the new world of being an oncology patient, and I have been reeling ever since.

I lived through the chemotherapy and loss of hair, and I lived through the radiation therapy. It was when I got to the post-breast-cancer medications that I entered the world of research. Not long after I started taking the tamoxifen, I began experiencing significant depression. I was told that this is one of the side effects, but I was not prepared for the depth of depression. I am a high energy, positive person and this was not the person I became. Because I have a family history of depression—my brother committed suicide and my mother has had depression and mania—my doctor stopped the tamoxifen for a period of 6 months. She recommended starting the tamoxifen again in 6 months, and based on all of the information I had at hand about its preventing future breast cancer, I agreed. This time I started having terrible insomnia and chemically induced hot flashes. I couldn't sleep through the night. I couldn't stop roasting then freezing 20 times or more a day. I didn't have the energy level I was used to. It was horrendous.

My doctor prescribed some other medications to offset the hot flashes—SSRIs. I was tried on three different types of SSRI. Not only did these medications make the insomnia much worse, they also caused a racing pulse and an increase in blood pressure—not fun. I really thought I was going to lose my mind. The insomnia and all of the other side effects were ruining my life. My doctor didn't know what to do with me.

My sister-in-law is a genetics nurse specialist. When I called her to talk about the insomnia and my inability to tolerate the medications, she told me about a new field of research called pharmacogenetics. She explained that there were researchers at the Mayo Clinic and around the country looking at people's genes and their responses to medication. I asked her for the name of the researcher at the Mayo Clinic. After she gave me the name, my husband and I called the research lab and actually spoke with him. He gave us a half-hour of his time listening to the side effects I was experiencing and my concerns. He told us about research being done on the cytochrome P450 system and suggested we contact researchers at this research laboratory.

I didn't waste any time. I called. I was put through to a wonderful laboratory assistant, Annie, who started me on my journey of research participation. I was able to talk with the researcher there too. It was amazing to me how open the researchers were and how generous they were with their time. I had many questions and both Annie and the researcher were there to answer me. The researcher suggested I try cutting the dose of tamoxifen in half when I went on it again as one solution, and take antidepressants to counteract any depression I might have. They also sent me the information I needed to have my blood drawn and to participate in their studies. I learned that they would be studying my alleles to see which ones I had and whether they were involved in my reactions.

I had my blood drawn and waited for results. As it turns out, they tested me for the four most common variant alleles of the CYP2D6 and the three most common of CYP2C19. My test results showed that I am a "wild" (i.e., normal) metabolizer in the CYP2D6 alleles. I am waiting to hear about the CYP2C19. The researcher told me that my reaction to the SSRIs may be due to a "reuptake inhibitor" response and not due to any genetic issue. I told the laboratory that I want to know if there is any research being done on the SSRIs that I could look at. I am pursuing this now.

I cannot tell you how grateful I have been to my sister-in-law and the research laboratory staff. I wrote to the researcher and thanked her, saying:

Working with you, your lab, and Annie has been one of the most interesting experiences I have ever had. I cannot thank you enough for your interest and time that you have spent listening to me and addressing the drug side effects of my post-breast-cancer medications. I know that there is a percentage of women out there who have had the same types of side effects and who either tried to live with them or who have just stopped taking any medication for quality of life reasons. That, in fact, was one of the options suggested to me before I spoke with you.

I have your research table and Web site on "My favorite" Web sites and I have sent them on to several colleagues and breast cancer survivors. I will follow your research and my participation with great interest and I look forward to seeing your patient Web site. If you are going to be speaking somewhere in the northeast part of this country, please let me know as I would come to hear you.

I am now officially a participant in one of their research studies. I am excited about participating, but I told them I was afraid, with all of the genetic databases out there and people able to tap into them, of what would happen to my genetic results. I am already a breast cancer survivor and that is on my insurance. What if they find out that I am a poor metabolizer and cannot tolerate certain drugs, but need others? Will that affect my reimbursement? I am not looking for work right now, but what if I were? Would my employer find out about my genetic issues if I were a part of the company's health plan?

My other concerns are about my family. I still don't know why I am having the reactions I am, but what if they are inherited? What does that mean for my daughters? I also wonder about my mother. She has had a terrible time with certain medications and depression. Maybe that's what has been going on with her too.

The other thing I don't get is why don't my doctors and nurses know about this research? Everyone I talked to is still into the "one-size-fits-all" concept of prescribing medication. My sister-in-law told me that this is new research that is slowly infiltrating the medical practice. She said that I am on the cutting edge. For once, I guess I am a pioneer. My hope is that the research will lead to better treatment of breast cancer and afterward during my lifetime. My fear is that, in being a pioneer, I will somehow be identified as a genetic outsider and lose some of the privileges I am used to.

Leigh

Genetic Research

The history of genetic research dates back to Gregor Mendel, an Austrian monk, who in 1865 outlined the basic laws of inheritance that still stand today. The term "genetics" was not coined, however, until 1906. In 1952, Martha Chase and Alfred Hershey demonstrated that DNA is the substance that transmits inherited characteristics from one generation to the next. Watson and Crick's now-famous research that characterized the physical structure of DNA followed this in 1953. By 1975, DNA sequencing techniques, developed by Gilbert, Maxam, and Sanger, were being used to further understand the structure and function of DNA. During the 1980s, there were many genetic research discoveries and advances. For example, the USDA approved the first genetically engineered drug, a form of human insulin. The first genetic test for a disease, Huntington Disease, was developed in 1983. The National Center for Human Genome Research, now known as the National Human Genome Research Institute (NHGRI), was created in 1989 to oversee the sequencing of the human genome by the year 2005. By 1990, the first gene therapy took place when a 4-year-old girl diagnosed with ADA deficiency (an immune system disorder) was successfully treated. The year 2000 brought the announcement of the completion of a working draft of the sequences that make up the human genome. A new era of genetic research called genomics—the study of the genetic and environmental factors that influence health, disease, prevention, and management—has now begun (Jorde et al., 2000). The many different types of genetic research that are underway today involve individuals, families, and communities. For each, research holds promise and hope on the one hand, and ethical concerns about privacy and confidentiality, discrimination, and stigmatization on the other hand.

Genetic research at the National Institutes of Health involves many of the institutes including the NHGRI. This research involves developing advanced methods for studying the fundamental mechanisms of inherited and acquired genetic conditions. Researchers are working to discover the genetic components of common conditions such as heart disease, cancer, and diabetes as well as more rare genetic conditions such as OI. To achieve their research goals, scientists involve families with a history of inherited conditions to help unravel the complex roles that multiple genes and environmental factors play in disease. The hope is that these studies will lead to the development of new, effective, and reliable gene-based diagnostics and treatments *(http://www.genome.gov)*.

Genetic Testing Research

Genetic tests, as defined in Chapter 2, involve analyses of human DNA, RNA, chromosomes, proteins, or certain metabolites to identify alterations related to a genetic condition or trait. Significant research efforts have occurred in the area of molecular genetics leading to the rapid development and availability of new clinical genetic tests. Research genetic tests are those genetic tests in which specimens are analyzed for better understanding the genetic condition, or for developing a clinical diagnostic or predisposition genetic test. When genetic testing is conducted for research purposes, the cost of the testing is covered by the researcher; however, results are not generally given to patients *(http://www.genetests.org)*. Research genetic testing is ongoing, for example, for hereditary cancers such as melanoma, breast, and ovarian cancer, to identify susceptibility genes that can then be tested for use in clinical settings.

Genetic research may involve finding genes or mapping, learning how genes function, or about diagnosing, treating, or curing genetic conditions. Leigh's story is an example of genetic research to learn how genes function, while Caleb's story highlights genetic research about diagnosing and treating a genetic condition. Individuals, families, and communities participate in genetic research for a variety of reasons. Some individuals and families, such as Caleb's, wish to work with scientists to advance understanding of a particular genetic condition without the expectation of immediate benefit to themselves. Others, such as Leigh, are participating in research as this offers the possibility of developing clinical genetic tests or treatments that have the potential to directly benefit them and their families. Participation in a study can also put individuals such as Leigh and Caleb in touch with specialists who would otherwise not have been accessible to them. Being part of a research project offers individuals and families the opportunity to meet other families with the same condition. For many, such as Caleb's parents, this has provided needed support.

Genetic Mapping Research

Mapping a gene requires finding its specific location or a marker close by in an individual's genetic code. Mapping research may involve family studies and employ linkage analysis or association studies. Linkage analysis is a method that allows researchers to determine regions of chromosomes that are likely to contain a gene alteration of interest. The technique involves using markers—well-characterized regions of DNA—close to a gene of interest. Researchers search for a marker that is consistently present in those individuals who have a condition, such as Huntington Disease or hereditary breast cancer, and is not present in those who do not have the condition. When a marker is found in the presence of the genetic condition, the marker and the disease-causing gene are "linked" and are assumed to be very close together (Jorde et al., 2003).

Linkage analysis research involves families that have both members who have and do not have a particular condition. By looking at markers in large numbers of families, researchers can begin to see that the gene of interest is somewhere close to the marker.

Genetic mapping using linkage analysis for a particular gene allows for study of the genetic basis of the condition and for the development of tests for the gene. This technique was used to identify the gene for Huntington Disease. Now that the specific gene for Huntington Disease has been identified, individuals can choose to undergo direct DNA testing for Huntington Disease (Jorde et al., 2003).

There are some drawbacks to linkage analysis research. First, it requires DNA material from both affected and unaffected biological family members. Linkage analysis also requires DNA from many families. Linkage is a very time and labor intensive research method. An analogy that has been made for linkage analysis by John's Hopkins researchers is as follows. "Consider a person's genome as a telephone book. Researchers are interested in locating a specific person—Bob—in the phone book, but they do not know Bob's last name or address. Researchers can use linkage to take the phone book and narrow the search for Bob to a few pages of names. Bob has still not been located, but researchers are getting closer" (*http://www.hopkinsmedicine.org/epigen/what_is_genetic_mapping.html*).

Genetic Association Studies

Association studies are used after linkage analysis has been able to locate the gene of interest. Researchers test very small genetic regions called candidate genes to determine if they are associated with having a particular condition or trait. Association studies can lead to the specific location of the gene of interest. Like linkage analysis, association studies require the use of DNA from many individuals. Association studies, however, do not involve studying families. This type of genetic mapping investigates DNA from individuals who have a particular condition or trait compared to the DNA of individuals who do not have the condition or trait (called controls). The analogy for association studies is that "linkage studies help researchers to narrow the search for Bob in the phone book to a few pages. They have not yet located Bob. Association studies, in this analogy, would allow researchers to call all of the phone numbers of individuals named Bob until one of the numbers turns up the Bob they are seeking" *(http://www.hopkinsmedicine.org/epigen/)*. Once a gene has been mapped, researchers can begin to focus on how gene mutations contribute to health and disease and ultimately to a treatment or cure.

An example of ongoing linkage analysis research is its use to characterize the genetic (biochemical), developmental, and environmental components of mental illnesses. One study's goal is to determine the genetic components of schizophrenia and bipolar disorder. Researchers are recruiting individuals and families where at least one living relative has been diagnosed with the disorder to help study the components of these disorders. In-person psychiatric evaluations are conducted and blood is drawn from affected and unaffected family members. Linkage analysis is used with these families to get an idea of where the predisposing gene(s) are located. Linkage analysis has helped researchers narrow down the chromosome regions where these genes may be. There is now evidence that there are predisposing genes for schizophrenia, for example, on chromosome 22 (22q11), chromosome 13 (13q32), and chromosome 8 (8p21). With this information, researchers are now moving forward with association methods to help identify the specific genes predisposing for bipolar disorder and schizophrenia *(http://www.johnshopkins.org)*.

Genetics Cohort Research Studies

Cohort studies involve the collection of extensive health information and biological specimens from different groups or populations who are then followed over time to determine the occurrence of a genetic condition and death from that condition. The goal is to understand the causes of a particular genetic-related condition, for example, cancer, and to develop effective interventions. Current genetics cohort studies are being conducted in the area of genetics and health disparities. Ongoing, large cohort studies are being used to understand the determinants of racial and ethnic disparities in cancer incidence and mortality. The focus is on research in two main areas: (1) the effect of racial and ethnic differences alone or in combination with genetic, lifestyle, and other factors on cancer incidence and mortality; and (2) to understand the basis for regional variations in cancer incidence and mortality *(http://epi.grants.cancer.gov/ResPort/Hdoverview.html)*.

Population-Based Genetics Studies

Population-based genetics research involves collection and correlation of detailed information on disease and genetic variation across as large a group of people as possible—a population. A population-based genetics approach is used to locate major genes involved in common diseases that are public health problems such as stroke, heart disease, and Alzheimer's, and diseases that result from the interplay of multiple genes, environmental, and health factors. The deCODE genetics project located in Rehkjavik,

Iceland is an example of one ongoing and intriguing population-based genomics research project. The deCODE project is using Iceland's unique population to learn from research on the genetic causes of common diseases and translate this knowledge into new drugs and DNA-based diagnostics and therapeutics. Iceland offers a special resource for this type of population-based genomics research in that it has a population with all three of the required sets of data—genetic, disease, and genealogical. The deCODE project is bringing all three sets of data together, allowing for population-wide linkage scans for disease-linked genes. Thus far, DNA samples from some 80,000 volunteer participants in more than 50 diseases (approximately 33% for Iceland's adult population and more than 90% of Icelanders over the age of 65) are participating. Data collected from participants will be anonymized and encrypted in a database and will enable users to conduct population analyses of longitudinal health care data including life-style information, disease diagnoses, treatments, and outcomes *(http://www. reykjavikresources.com)*.

Treatment Research: Clinical Trials

Clinical trials research involves human subjects and the study of new and emerging interventions and therapies. Clinical trials help to move basic scientific research from the laboratory "bench" to interventions for people—at the "bedside." Conducting and evaluating clinical trial results allows researchers to determine ways to prevent and diagnose both rare genetic conditions such as OI and more common genetic conditions such as breast cancer. Clinical trials are also designed to find better treatments for genetic conditions. Each clinical trial has a protocol or set of rules describing the type of individuals who are eligible to participate; the schedule of tests, procedures, the intervention such as medications and dosages if applicable; and the length of the study. Clinical trials are a final step in a long research process and contribute to both the knowledge of and progress toward treatment and cure of human disease. Their goal may be to study a drug or medical device to determine whether or not it is safe and effective. This must be done before the Food and Drug Administration can approve them and providers can prescribe them *(http://www.genome.gov; http://www. mdanderson.org; http://www.cancer.gov)*. Or, the study may focus on measuring outcomes of a noninvasive intervention, such as outcomes of education and counseling for hereditary non-polyposis colorectal cancer.

Clinical trials differ by the type of trial and the phase of the trial. Types of clinical trials include treatment, prevention, early-detection/screening, diagnostic, and quality of life/supportive care trials. Caleb, for example, is enrolled in a treatment trial evaluating a new treatment approach to OI (Box 11-1). Leigh, on the other hand, is enrolled in a diagnostic clinical trial answering the question about how new genetic tests can more clearly delineate the appropriate medication and dosage for an individual (**Box 11-2**).

Box 11-2 Prospective Trials Research: CYP2D6 and Tamoxifen Metabolism

Research is currently underway in many centers throughout the United States and worldwide to study the multiple genetic influences on the clinical pharmacology of many classes of drugs. The goal of these studies is to develop a better understanding of the expression of an individual's genetic makeup and effects on drug metabolism. One important class of drugs—Selective Estrogen Receptor Modulators (SERMs)—for which the prototype is tamoxifen is currently being studied to define the multiple genetic influences on the clinical pharmacology of this particular medication. The hope for individuals such as Leigh is that, through understanding the variability in contributions of enzymes such as CYP2D6 to tamoxifen metabolism, more targeted and appropriate treatment, without the extreme side effects such as hot flashes, can be achieved.

Tamoxifen, a SERMs, is used in the treatment and prevention of breast cancer. It is known that there is variability in response to tamoxifen, including side effects. Researchers belonging to the Pharmacogenetics Network (PGRN) at Indiana University are studying tamoxifen pharmacogenetics to define the multiple genetic influences on the action of this drug as a model for other drugs in this class, where multiple genetic variations are likely to alter pharmacologic responses. Tamoxifen has been shown to be metabolized in the human liver by the cytochrome P450 enzymes of which there are many variants. Researchers at PGRN, Indiana University, are investigating the possibility that mutations in the genes that code for these enzymes, specifically CYP2C9, CYP2D6, and CYP 3A, might alter the drug's metabolism in women with breast cancer.

The Division of Clinical Pharmacology has a clinical and research focus on individualized response to drug therapy. Their mission is to carry out research and clinical activities that improve rational therapy for the treatment of disease, regardless of age, gender, or ethnicity. Researchers are particularly interested in differences in response to drug treatment that exist on the basis of age, gender, or ethnic origin. Their research covers a broad range of specific areas of expertise including the effects of drug and herbal interactions and pharmacogenetics on therapy in the elderly, women, and children. With funding from the Food and Drug Administration, National Institutes of Health, and the Pharmaceutical Manufacturers Association, they have established a well-funded research and training program. The Division of Clinical Pharmacology has enrolled 300 patients prescribed tamoxifen to treat or prevent breast cancer in a prospective trial to evaluate the pharmacogenetic variants that influence the efficacy and toxicity of tamoxifen. Participants will be genotyped for variants in the CYP isoforms as well as estrogen receptors and a number of other specific candidate genes. The hope is that these studies, while focused on tamoxifen, will apply to other SERMs including estrogen and their effects on bone, lipid, and coagulation to varying degrees. Data gained from this prospective trial may have considerable relevance to the current controversies surrounding the use of estrogen as hormone replacement therapy.

(http://medicine.iupui.edu/clinical/pg_intro.htm)

Phases of Clinical Trials

There are generally four phases in clinical trials. A Phase I trial involves a small number of individuals. The purpose of a Phase I trial is to identify a safe dosage, to decide how the medication–treatment should be given, and to observe the side effects. Major nursing responsibilities in Phase I trials include the observation for toxic side effects, including early recognition and management of the side effects of treatment. Phase II clinical trials expand the intervention to a larger group of patients, allowing researchers to build on what they have learned from Phase I trials. Phase II clinical trials include less than 100 people with the goal of determining if the medication or intervention has a particular effect on the condition and to see how the medication or intervention affects the human body. Since more individuals are involved in Phase II trials, investigators may discover common side effects and continue to evaluate the safety and efficacy of the treatment. Nursing involvement in Phase II studies includes preparing patients for the studies that monitor disease response and documenting the benefits and side effects of the treatment. Phase III trials usually involve thousands of individuals, often at several medical institutions. During this phase, researchers compare the new medication or intervention, or new use of a treatment, with the standard treatment. A treatment group is given the experimental treatment while a control group is given the standard treatment so that researchers can compare the outcomes. Nursing roles in Phase III studies include monitoring patient eligibility and protocol compliance, documenting therapeutic effects and toxicities, and providing continuity of care. Phase IV trials take place after a treatment has been made available on the market and involve thousands of individuals. The purpose of the Phase IV trial is to further evaluate the long-term safety and effectiveness of a new treatment. In Phase IV clinical studies, nurses provide education and evaluate nursing interventions to decrease morbidity associated with the intervention (Jenkins & Curt, 1996; *http://www.cancer.gov; http://www.genome.gov*).

Pharmacogenetic Research

Pharmacogenetics, as described in Chapter 5, involves the study of how people respond differently to medications based on their genetic makeup. This type of research is not new but has been limited in the past by researchers' lack of knowledge about human genetic variation. Pharmaceutical research, now that the human genome has been mapped and sequenced, is expanding dramatically. The risks to participants in clinical trials will be greatly reduced by targeting only those persons capable of responding to a particular intervention. In addition, it is anticipated that the number of failed drug trials and the time it takes to get a drug approved will be reduced. All of these advances will require research participants such as Leigh. The potential health benefits are great and include personalized treatment and better and safer drugs the first time instead of through trial-and-error or one-size-fits-all methods. The goal of pharmacogenetic research is to speed recovery and decrease adverse reactions while maximizing the therapy's value *(http://www.geneticalliance.org/geneticissues/mediainfo/personalmedicine.html)*.

Prevention, Early Detection, and Screening Trials

Prevention trials involve studying healthy people who have an increased risk for developing a condition such as cancer. Prevention trials may focus on what actions people can take to prevent cancer, such as quitting smoking, or on agent studies, which focus on taking something (e.g., medication or vitamin) to decrease the risk of developing cancer. Early detection and screening trials attempt to discover ways to detect cancer as early as possible. For example, using imaging tests (MRI; virtual colonoscopy), laboratory tests that check body fluids for evidence of cancer, and genetic tests that look for inherited genetic markers linked to some types of cancer *(http://www.cancer.gov)*.

Diagnostic Trials

The focus of diagnostic trials is to identify new tests or procedures that can better diagnose individuals who have a genetic condition. Some diagnostic trials compare two or more techniques, find out how accurate they are, and determine whether they can provide any new and valuable information about the condition. Genetic tests are an example of diagnostic tests that are being evaluated as tools to further classify certain genetic conditions. These test results may then help direct specific treatments to improve clinical outcomes of those who have the specific genetic changes *(http://www.genome.gov)*.

Quality of Life and Supportive Care Trials

Quality of life and supportive care clinical trials aim to evaluate improvements in the comfort and quality of life of individuals who have a genetic condition. Researchers conducting these trials are hoping to help individuals who are having problems such as physical limitations, systemic symptoms such as infection, and other effects from the condition or its treatment. Supportive care trials may involve families and caregivers to help them cope with both their own needs and those of the individual with the genetic condition *(http://www.genome.gov)*.

Clinical Trial Protocols

Each clinical trial follows strict scientific guidelines that clearly outline the study's design and who will be eligible for participation in the trial. The protocol explains the reason for the study, how the study will be conducted, and the reason behind each of the study components. The protocol includes information on: 1) the purpose of the study; 2) the number of people who will be enrolled in the study; 3) requirements for participation; 4) the treatments or interventions that will be part of the study; 5) what medical tests participants will have and their frequency; 6) potential risks and benefits of study participation; 7) information that will be gathered on the study participants; and 8) the endpoints of the study. Endpoints of the study depend on the phase of the clinical trial and range from toxicity to survival and quality of life. All centers or physicians implementing the research use the same clinical protocol. This means that all study participants receive the same treatment or intervention equally regardless of where they are entered into the study *(http://www.cancer.gov; http://www.genome.gov)*.

Clinical trials for many other rare and common genetic conditions are ongoing and are listed on Web sites such as *http://www.clinicaltrials.gov* and *http://ghr.nlm.nih.gov*. Examples of clinical trials focusing on common diseases include those of lung disease, hereditary hemochromatosis (HH), and cancer. One clinical trial is evaluating genetic mechanisms of lung disease. In the clinical trial evaluating genetic mechanisms of lung disease, polymorphic genes involved in respiratory function will be surveyed and gene expression in lung cells examined in individuals with pulmonary diseases such as alpha-1 antitrypsin deficiency, cystic fibrosis, chronic obstructive pulmonary disease, and sarcoidosis as compared with healthy control individuals. Examining the effect of iron buildup in the hearts of patients with HH, a common genetic disease that causes the body to accumulate excess amounts of iron is another ongoing clinical trial. The excess iron can damage the heart, liver, pancreas, skin, and joints. Early treatment with phlebotomy helps to slow organ damage in individuals with HH. The goal of this study is to try to gain a better understanding of the effect of iron buildup in the heart and to determine whether phlebotomy helps to keep the heart healthy *(http://www.clinicaltrials.gov)*.

Examples in cancer research include biological therapy—immunotherapy—a form of therapy that uses the body's immune system, either directly or indirectly, to combat cancer or lessen the side effects of cancer treatments. Immunotherapies are designed to stimulate, enhance, or repair the body's immune system responses. As an example, monoclonal antibodies are being studied, as they may help an individual's own immune system fight cancer by locating cancer cells and killing them or by delivering cancer-killing substances to the cancer cells without harming normal body cells. Cancer vaccines are another form of immunotherpy. Vaccines are being developed to help the body reject tumors or to prevent cancer from recurring *(http://www.cancer.gov)*.

Many more clinical trials are underway for rare genetic conditions such as Duchenne muscular dystrophy, lysosomal storage diseases, inherited eye diseases, and others. Information about the purpose of the clinical trial, the study type and design (e.g., observational, screening, intervention), the type and number of patients eligible to be included in the study, and the location of the study centers is available. A contact person for each study is also listed at *http://www.clinicaltrials.gov.*

Ethical and Social Considerations with Genetic Testing and Treatment Research

Genetic research projects are often located in university medical centers. An Institutional Review Board (IRB) must review all research projects. The IRB is composed of scientists, physicians, nurses, social workers, clergy, lawyers, ethicists, and community members. The task of the IRB is to review all research projects to ensure that the interests of participating individuals (human subjects) are well protected. The IRB members consider not only the physical safety of research participants but also issues of informed consent, family issues, privacy, and confidentiality of genetic information gained from the research. IRBs also monitor the clinical trials. This activity includes monitoring participant safety, data monitoring, and reporting of the results to federal oversight agencies *(http://www.cancer.gov)*.

In the United States, the National Commission for the Protection of Human Subjects of Biomedical and Behavioral Research (1976) developed three basic principles that govern research involving human subjects. These principles still stand today and form the basis for human subject protection. They include:

- Respect for persons—recognition of personal dignity and autonomy of the individual with special protection for those individuals with diminished autonomy
- Beneficence—the obligation to protect study participants from harm by maximizing unanticipated benefits and minimizing possible risks of harm
- Justice—fairness in the distribution of research burdens and benefits *(http://www.cancer.gov)*.

Barriers to Participation in Clinical Trials Research

Although there are numerous clinical trials available to individuals with genetic conditions, many individuals do not participate, which is the case, for example, for cancer clinical trial research. In an analysis of NCI trial enrollment (1998–1999), only 3 percent of adult cancer patients in the United States participate in clinical trials—far fewer than the number needed to answer the most pressing cancer questions quickly (*http://www.cancer.gov/clinicaltrials/facts-and-figures#patients*). An identified barrier to participating is lack of awareness by patients that this is an option. Perceived or real lack of access and the concern of seeking care at a distant trial site that would require additional travel and time are other barriers. Patients may fear and distrust research, having suspicions that they are being experimented upon or that they will be denied insurance coverage if they participate in a clinical trial. The cost of being away from work and home may be deterrents as well. Individuals belonging to certain ethnic or racial groups or who are medically underserved may have concern that the care they receive within the trial and the recruitment strategies are not sensitive to their needs. Finally, patients may be unwilling to go against their personal physician's wishes for their treatment *(http://www.cancer.gov)*. A recent survey by the Harris Interactive (2001) evaluated why so few adults participate in cancer trials. Some of the results, both negative and positive, are outlined in **Table 11-1.**

There are also barriers to clinical trial participation for health professionals. Health care providers are not always aware of available clinical trials. Some may assume that no clinical trial would be appropriate for their patients. Unwillingness to "lose control" of a person's care or concern that the clinical trial will affect the physician–patient relationship are additional barriers for providers. Other providers, as illustrated in Leigh's story, may not adequately understand how clinical trials are conducted or their importance, believing instead that the standard therapy is best. Concern that referring to or having patients participate in a clinical trial may add an administrative burden has also been described as a professional barrier *(http://www.cancer.gov)*. Nurses can be advocates for individuals and families considering participation in clinical trials. Nurses can serve as the liaison between the physician and the patient by fielding questions and identifying concerns (Jenkins & Curt, 1996).

Table 11-1 Reasons for Declining Participation in and Positive Aspects of Clinical Research Trials

Reasons for Declining: Beliefs and Myths

- Medical treatment received in a clinical trial would not be as effective as standard care.
- They might get a placebo.
- They would be treated like a "guinea pig."
- Their insurance company would not cover the costs involved.

Positive Aspects of Clinical Trials

- Participants were treated with dignity and respect.
- Quality of care received was "excellent" or "good."
- Most said their treatment was covered by insurance.

(*http://www.cancer.gov;* Harris Interactive (2001). *Health Care News, 1*(3) (Poll). Available from *http://www.harrisinteractive.com*)

Potential Risks with Genetic Research

A major concern about participation in genetic research is the potential for psychological harm related to participation. Potential psychological harm includes changes in feelings, thoughts, or beliefs that can cause anxiety or depression. Family relationships may be stressed or altered. For example, the risks may relate to identification as an "at risk" family member such as a member of a family with Huntington Disease or hereditary cancer syndrome. There is also the potential for insurance or employment discrimination based on genetic information, as when an insurer inquires about a person's knowledge of a particular disease risk status and the individual must be forthcoming with known information. Individual or group stigmatization based on perceptions regarding genetic information is also a potential harm. As an example, members of the Jewish community have spoken out about the potential for stigmatization by the identification of "Jewish genes," causing a perception of genetic inferiority for their population. Nurses can serve as advocates for ensuring that patients participating in genetic research have access to appropriate psychosocial support throughout the study. Nurses, as links to the patient, the health care system, and the community can take an active role in involving community leaders or the specific population under study in the research plans. Development and implementation of research that includes ongoing communication and education among researchers and community members can improve clinical trial recruitment and address issues of concern (National Cancer Institute, 10/16/03).

Informed Consent

Making sure that individuals considering participation in a research project are adequately informed about the risks, benefits, and limitations of the research and that they understand that their participation is voluntary are critical components of the informed consent process. Nurses providing care to individuals participating in genetic research have a central role in this process. Nurses have the necessary skills to assess and evaluate the patient's understanding of a research study and to determine that the patient's decision to participate has been voluntary. The International Society of Nurses in Genetics has outlined the role of the nurse in informed decision making and consent (**Box 11-3**), highlighting the importance of the communication process and dialogue between the client and the nurse. The nurse has an important role in making sure that individuals and families receive information about the purpose of the research study, as well as the benefits and risks, in language that is understandable to the individual *(http://www.genome.gov)*. Individuals considering participation in a clinical trial should consider and ask a number of important questions to help with their decision making. These are outlined in **Table 11-2.** The Genetic Alliance has created a brochure for families, *Informed Consent: Participation in Genetic Research Studies,* that provides information in lay terms that can help

Box 11-3 International Society of Nurses in Genetics, Inc. (ISONG) Position Statement: Informed Decision Making and Consent: The Role of Nursing

Background

The indispensable initial step in the preparation for genetic testing is the process of "informed consent." A written statement describing the risk and benefits of genetic testing is presented to the client to read and sign before the evaluation and/or test is performed. It is intended as a safeguard to ensure the client's autonomy and to provide an opportunity to learn and understand information with respect to both the positive and negative consequences of genetic testing. In the more conventional use, the informed consent process focuses on the provider conveying information about the genetic test and the client signifying acceptance of the testing by a signature on the written consent form.

ISONG supports a more active selection process with an emphasis on the informed decision-making authority of the client to choose either to accept or reject genetic testing. Pivotal to accomplishing this process is a dialogue between the client and the providers in a joint endeavor to facilitate informed decision making and consent. This educational and informative dialogue should occur at the level of language and comprehension of the competent client.

Nurses should encourage clients to seek information and identify concerns prior to giving informed consent. The nursing process can be universally utilized to assist clients contemplating any type of genetic testing and to ascertain whether essential elements of informed consent are present in the decision-making process.

Informed Decision Making and Consent for Genetic Testing: The Role of the Nurse

Genetic testing can now be used for diagnosis, management, treatment, or health and reproductive decision making. The benefits of genetic testing are many and range from early detection for treatable disorders to prevention by health planning before the onset of symptoms for those who are at risk for a genetic disorder. Genetic testing should be carried out within the context of voluntariness, informed consent, and confidentiality. Nurses, as the omnipresent health care providers, have a central role in providing information and support to individuals, families, and communities in the multiphase processes of genetic testing. With genetics knowledge, nurses can advocate, educate, counsel, and support patients and families during the informed decision-making and consent process.

Box 11-3 continued

It is the position of the International Society of Nurses in Genetics that:

- All professional nurses are responsible for alerting clients to their right for an informed decision-making and consent process prior to genetic testing.
- All professional nurses should advocate for client autonomy, privacy, and confidentiality in the informed decision-making and consent process.
- All professional nurses should ensure that the informed decision-making and consent process includes discussion of benefits and risks including the potential psychological and societal injury by stigmatization, discrimination, and emotional stress, in addition to, if any, potential physical harm.
- All professional nurses should be aware of the criteria that delineate research versus clinical uses of genetic tests, and advise clients of the status of a specific test.
- All professional nurses who have an established relationship with and are providing ongoing care to a client contemplating genetic testing should augment the informed decision-making and consent process by assisting the client in the context of the client's specific circumstances of family, culture, and community life.
- All professional nurses should integrate into their practice the guidelines for practice (e.g., informed consent, privacy and confidentiality, truth telling and disclosure, and nondiscrimination) identified by the American Nurses Association.
- All advance practice nurses in preparation for providing genetic services should receive appropriate education that includes knowledge of the implications and complexities of genetic testing, ability to interpret results, and knowledge of the ethical, legal, social, and psychological consequences of genetic testing.
- All health care professionals should collaborate to maximize the potential for the client to make an informed decision.

References

1. International Society of Nurses in Genetics (ISONG) (January 27, 2000). ISONG Testimony to Secretary's Advisory Committee on Genetic Testing (SACGT).
2. Scanlon, C. & Fibison, W. (1995). *Managing genetic information: Policies for U.S. nurses.* American Nurses Association: Washington, DC.

Approved 9/30/00

TABLE 11-2 Questions to Ask When Considering Participation in a Research Trial

- Why is this trial taking place?
- Why do the doctors who designed the trial believe that the treatment being studied may be better than the one being used now? Why may it not be better?
- How long will I be in the trial?
- What kinds of tests and treatments are involved?
- What are the possible side effects or risks of the new treatment or intervention?
- What are the possible benefits?
- How could the trial affect my daily life?
- Will I have to travel long distances?
- Will I have to pay for any of the treatments or tests?
- Does the trial include long-term follow-up or care?
- What are the other treatment choices, including standard treatments?
- How does the treatment I would receive in this trial compare with the other treatment choices in terms of possible outcomes, possible side effects, time involved, costs to me, and the quality of my life?

(Adapted from *http://www.cancer.gov*, Weighing Decisions About Participating)

individuals make an informed decision about whether or not to participate in a genetics research project *(http://www.geneticalliance.org/geneticissues/informedconsent.html)*. The Centers for Disease Control and Prevention has information for individuals needing to make a decision about participating as well *(http://www.cdc.gov)*. Questions to ask researchers are one component of each of these resources. Nurses can use these resources to guide their discussion with patients and further support their decision-making process.

Monitoring, Documenting, and Data Management

Nurses coordinating research studies are often responsible for monitoring patients during the study. Monitoring includes various kinds of assessments, planning, and documentation, as well as ensuring that the patient fits the study eligibility criteria. Monitoring responsibilities include assessing for side effects of interventions on a frequent basis and documenting these observations in the patient's record.

Family Issues

A person's genetic makeup or "genetic self" is linked to biological family members. Genetic information gained from participation in research may affect a person's self-identity and perceptions of that person by the family. The family may also experience alterations in their identity. For example, genetic research may reveal unexpected information, such as nonpaternity. Genetic research involving family members may challenge previously held family beliefs about the origins of a particular condition or trait in a family. Such information may be disabling or require redefinition of the individual and family participating in the research (Baker, Schuette, & Uhlmann, 1998; Peters, Djurdjinovic, & Baker, 1999; Williams, 1999; Williams & Schutte, 1999). Nurses have a key role in assessing individual and family beliefs about genetic conditions and in properly informing and counseling individuals and their families about the potential aspects of research.

Privacy and Confidentiality Concerns

As Leigh expresses in her story, privacy and confidentiality of her genetic information are of great concern. The concern expressed is the potential for discrimination or stigmatization—by family, friends, employers, and society—based on genetic information gained from research. Assurance that these concerns are addressed in the informed consent process is an important activity of the IRB. Individuals and families participating in genetic research need to understand that, although researchers will try to guarantee protection of personal genetic information gathered through participation in genetic research, such information may become known to other persons outside of the research team. Questions that individuals will need to have addressed regarding privacy and confidentiality of their genetic information include:

- What will happen to the stored DNA sample or genetic information once the research is complete?
- Will any participant's DNA or genetic information be distributed, such as to biotechnology companies, genetic laboratories, or government agencies?
- Who will get the results of the research if they are to be shared?
- How will confidentiality of research records be maintained? *(http://www.geneticalliance.org/geneticissues/informedconsent.html)*

Nurses, along with other health care providers, share the responsibility to ensure privacy and confidentiality of genetic information, including genetic research information. The role of the nurse in ensuring privacy and confidentiality is outlined in the International Society of Nurses in Genetics position statement *Privacy and Confidentiality of Genetic Information: The Role of the Nurse* (2002) (**Box 11-4**). In addition to professional and institutional measures to protect privacy and confidentiality, many states now have laws against discrimination based on genetic information. In 2003, the Senate passed the Genetics and Discrimination Bill, but this federal bill has not yet been passed by the House (American Society of Human Genetics, 2004 at *http://www.ashg.org/genetics.ashg/news/012.shtml*).

Box 11-4 International Society of Nurses in Genetics, Inc. Position Statement: Privacy and Confidentiality of Genetic Information: The Role of the Nurse

Background

An increasing volume of genetic information about individuals is becoming available as a result of genetic advances. While this information has the potential to provide health benefits, it may also increase risk of harm. Of major concern is the potential for misuse of genetic information resulting in any kind of discrimination or stigmatization. All nurses must advocate for and ensure strict maintenance of the privacy and confidentiality of genetic information, educate clients and the public about the nature of genetic information and protection of client privacy, and collaborate with other members of the health care team to protect individuals from any improper use of their genetic information.

Genetic information refers to any information about a person that identifies heritable contributions to a person's biological self or genetic alterations that are acquired during a person's lifetime. There is a broad spectrum of sources of genetic information. Genetic information may be identified before birth, at any point during the life span, and after death. Genetic information can be determined from a simple blood test or a complex evaluation of a genetic condition. Sources of genetic information include family history information, genetic analyses of tissues and cells or other genetic-related test results and treatments, findings on physical examination, or medical records. Genetic information includes an identified genetic variation or trait, a predisposition to a genetic condition, a presymptomatic condition, or the diagnosis of an inherited condition. In some cases, genetic information can be established for a person based on testing another family member. Sensitive information such as obligate carrier status or nonpaternity may become known.

Genetic information is considered private and linked to a person's identity. It also has particular meaning for biological family members that most other medical information does not. The availability of genetic information creates legal, ethical, and social responsibilities for all health professionals. These responsibilities guide the collection, recording, storage, and reporting of genetic information. Maintaining privacy and confidentiality and protecting a person's right to privacy are major ethical considerations in the management of genetic information. Privacy refers to a person's right to control his or her own body, thoughts, and actions. Confidentiality refers to the health professional's obligation to protect and not to disclose personal information provided in confidence by a client.

Box 11-4 continued

Nurses in all practice settings are involved in managing genetic information and share in responsibility with other members of the health care team to safeguard clients and their families against misuse of genetic information. This can be accomplished by a careful consideration of the significance of genetic information, by creating a practice environment in which clients can be assured that information will be shared in a professional manner and restricted to those who require it to provide care.

Privacy and Confidentiality of Genetic Information: The Role of the Nurse

All nurses must be mindful of the management of genetic information, follow the American Nurses Association Code for Nurses, and guard against invasion of client privacy. The International Society of Nurses in Genetics, Inc. (ISONG) recognizes that nurses play a critical role in managing genetic information and in maintaining an individual's privacy. The nurse's role is central in managing genetic information and in maintaining privacy because essential to the relationship between the nurse and a client is the trust that implies a duty to hold in confidence personal and family genetic information. A nurse is obligated to use genetic information in a confidential manner and to prevent unauthorized access.

Nurses must maintain a central role in educating clients, other health care providers, and the public about the nature and protection of genetic information. The role of the nurse includes informing clients about their right to privacy and confidentiality of genetic information, and guarding against a violation of a client's right to privacy. A nurse should not release genetic information without a signed release from the client or client's legal surrogate that reflects that the client has been fully informed regarding the possible consequences of such a release. For clients considering participation in genetic research, the nurse makes clients aware that the research needs proper review and approval by the appropriate Institutional Review Board. The nurse seeks ethical and legal counsel in instances where it may be necessary to release genetic information if significant harm could be caused to a third party if the genetic information is withheld.

Nurses are present in all practice settings throughout the health care system, and therefore have an ongoing responsibility to recognize the unique nature of genetic information and its many sources. Nurses collaborate with other members of the health care team and health facilities to ensure ongoing maintenance of the privacy and confidentiality of genetic information.

It is the position of ISONG that professional nurses will:

- safeguard a client's right to privacy.
- adopt into their practice, guidelines for ethical practice, identified by the American Nurses Association (1) regarding privacy and confidentiality, informed consent, truth telling and disclosure, and nondiscrimination.

Box 11-4 continued

- become familiar with legislation in their own state with regard to the nurse–client relationship, confidentiality of medical information, and privileged status.
- obtain a client or their designee's written consent prior to releasing genetic information to any third party.
- understand that family culture, values, traditions, and relationships influence the sharing of genetic information.
- recognize that each individual in the family has autonomy with respect to genetic matters that may be compromised by the decisions of other family members.
- collaborate with all other health professionals to ensure that clients receive the highest level of genetic health care by:
  1. advocating for the creation of practice guidelines that assure privacy and confidentiality of genetic information.
  2. keeping informed regarding legal and ethical issues associated with the use of tissue samples in genetic research.
  3. educating about the various ways in which (a) abandoned tissues and cells might be used as a source of genetic information and (b) genetic information might be used (positively or negatively) by employers or insurance companies.
  4. Becoming aware of the potential for stigmatization and discrimination as a consequence of linking genetic information with ethnicity, race, gender, or other social variables.
- In addition to the above, it is the position of ISONG that advanced practice nurses will be prepared at an advanced level to integrate knowledge of privacy and confidentiality issues and psychological consequences of the use of genetic information into health care.

In summary, privacy and confidentiality of all health information is of great concern to nurses in all practice settings. Assuring privacy and confidentiality of genetic information, in particular, demands continued vigilance on the part of all nurses as genetic technologies and discoveries are translated into clinical application and practice.

References

1. American Nurses Association (2001). Code of ethics for nurses with interpretive statements. Kansas City, MO: Author.
2. International Society of Nurses in Genetics, Inc. (1998). *Statement on the scope and standards of genetics clinical nursing practice.* Washington, DC: American Nurses Association.
3. International Society of Nurses in Genetics, Inc. (Winter 2000). Position Statement: Informed decision-making and consent: The role of nursing. *International Society of Nurses in Genetics Newsletter, 11*(3), 7–8.

Box 11-4 continued

4. National Coalition for Health Professional Education in Genetics (2001). Committee Report: Recommendations of core competencies in genetics essential for all health professionals. *Genetics in Medicine, 3*(2), 155–158.
5. Scanlon, C., & Fibison, W. (1995). *Managing Genetic Information: Implications for Nursing Practice.* American Nurses Association: Washington, DC.
6. Secretary's Advisory Committee on Genetic Testing (2000). *A public consultation on oversight of genetic tests.* Bethesda, MD: National Institutes of Health.

Approved on 10/9/01

Genetic Research and Racially or Ethnically Diverse Populations

Ethnically diverse and minority communities face many health challenges including lack of access to genetic services. These challenges extend to minority participation in clinical research trials, particularly by individuals from certain ethnic or racial backgrounds and those who are medically underserved. Diverse populations in the United States include minority ethnic and racial groups as designated by the U.S. government: American Indian or Alaska Native, Asian American, Black or African American, Hispanic or Latin American, Native Hawaiian or other Pacific Islander. Diverse populations are growing rapidly in the United States, with 25% of the U.S. population reporting their race as other than White in the 2000 Census. Researchers also consider medically underserved populations in their working definition of diverse populations. Medically underserved populations include those that lack easy or any access to or do not make use of prevention, screening, early detection, treatment, or rehabilitation services. These are individuals who live in rural areas or who have low income or literacy levels. Concerning cancer, for example, medically underserved populations are often characterized as having higher cancer mortality rates and insufficient participation in cancer control programs *(http://www.cancer.gov)*. Specific barriers identified in diverse populations to participation in genetics and other research include differing cultural beliefs about health and disease that Western medicine cannot address and language and literacy barriers that make it difficult for people to understand and consider participating. Other barriers to participation are outlined in **Table 11-3.**

Table 11-3 Specific Barriers to Genetic Research Participation in Diverse Populations

- Long-standing fear, apprehension, and skepticism about medical research because of abuses that have happened in the past (e.g., the outcome of the Tuskegee syphilis study)
- Physician avoidance of offering clinical trials as an option out of concern that they would seem insensitive
- Differing cultural beliefs about health and disease that Western medicine cannot address (e.g., fatalism, family decisions about treatment, use of traditional healers)
- Language or literacy barriers making it difficult for people to understand and consider participation
- Additional access problems such as transportation and time off from work
- Lack of health insurance or a source of health care
- Costs associated with clinical trials—concern that a participant's insurance company will not cover their participation in a clinical trial
- Minority groups are disproportionately over-represented in the uninsured population.

(*http://www.geneticalliance.org,* Minority Communities and Healthcare Access Problems)

Genetic Testing and Community Research

Advances in genetic technology have raised new issues concerning human subjects research in terms of harm to identifiable communities (Kegley, 1996). Newer studies of human genetic variation raise concerns about risks to all members of a particular social group, not just the individuals participating in the research. One of the concerns raised, for example, is that findings from research that associate a particular ethnic group with a predisposition to a genetic condition could lead to group stigmatization and discrimination. Risks of this nature have been discussed widely with regard to the studies of hereditary cancer genes—BRCA1 and BRCA2—as some mutations in these genes are more common among persons of Ashkenazi Jewish ancestry (Streuwing et al., 1997). These research findings raise the possibility that all individuals of Ashkenazi Jewish ancestry may be asked to pay higher insurance premiums or face other more subtle forms of discrimination on the basis of this association between genetic variation and the risk of developing breast cancer. Until recently, current federal regulations for human subjects research have not addressed this issue or required ethics review boards to consider the consequences to communities (Sharp & Foster, 2002).

The National Bioethics Advisory Commission (NBAC) is a presidential commission on the protection of human subjects research. NBAC has proposed that regulatory oversight now include the protection of social groups (National Bioethics Advisory Commission, 1999). The Commission recommends that individual research volunteers, investigators, and review boards consider how to minimize group harms such as working directly with community representatives to develop study methods that minimize the potential for harm and discuss collective risks as a part of the informed consent process. With reporting of results, the Commission recommends that researchers and journal editors carefully consider the potential implications for social groups.

Nurses in all practice settings serve as a vital link between individuals, families, communities, and health care systems. Nurses implementing genetics research involving communities can build on this foundation and work with researchers in several ways. One way is to involve community representatives at the earliest possible stage in a research process. The early involvement of the community has the potential to lay the groundwork for partnership between the community and the researcher (Weijer, Goldsand, & Emanuel, 1999). Such communication can help researchers improve understanding of how communities and social organizations evaluate well-being and risk. Participating in communal discourse is another nursing role. Communal discourse involves working with researchers to seek out representatives of a particular community to learn about the ways in which individuals and families are accustomed to making health care decisions. For example, is it the individual, the family, or a community leader who makes health care decisions? Nurses can help with communicating genetic research information in social gatherings including discussion of the risks and benefits. Further, communicating information from these gatherings to family elders, community leaders, and elected officials can improve the research experience. This approach helps to support a diversity of social and ethical arrangements and identifies decision-making patterns that are a function of collective identity (Foster, Eisenbraun, & Carter, 1997).

At present, there is little known about how members of underserved or marginalized communities understand individual research versus community risk. It is not known how members of various communities may weigh individual research risks against group risks; how important collective risks are in relation to other risks faced in their daily lives; or how individuals may try to reconcile potential conflicts that exist between personal interests in research participation and collective opposition to proposed research trials. Nursing research can contribute to further understanding these dynamics.

Summary: Implications for Nurses

Many nurses are actively coordinating genetic mapping and association research and genomic clinical trials research. At the basic level, nurses are involved in collecting medical histories, explaining the purpose of the testing or clinical trial, obtaining informed consent, and obtaining the blood or tissue specimen for analysis. At an advanced practice level, genomic nurse researchers are participating in study design, developing clinical trial protocols, and overseeing quality and accuracy of data obtained. At the clinical research level, advanced practice nurses have many activities and roles. These are outlined in **Table 11-4.**

Nurses in primary and tertiary care clinical settings can play an important role in making sure that those patients with genetic conditions or genetic-related concerns are aware of the possibility of participating in clinical research trials. Knowledge of available studies for individuals and families is a first step. Most people would like to have all of the information about options presented to them with discussion of the risks and benefits before making important health-related decisions. Nurses can support patients in making informed decisions about participation in genetic-related research by knowing the important questions to ask concerning informed consent, privacy of genetic information, and costs of participation. **Table 11-5** provides a listing of clinical trial resources.

Nurses also play an essential role in ensuring privacy and confidentiality of genetic information. In caring for individuals and families who are considering participation in a research trial, the nurse can review with them information from the study protocol about how human subjects and genetic information will be protected. In caring for individuals who have participated in clinical trials, the nurse ensures that any genetic information he or she receives regarding the clinical trial will be handled responsibly. Having an understanding of the types of hopes and fears individuals and families may bring to considering participation in a clinical trial, and the ethical, legal, and social issues related to genomic research will enable nurses to more fully inform and support individuals and families considering participation in genomic research. This proactive role will also facilitate that study outcomes can be achieved more effectively and efficiently with the potential to improve health benefit for all.

TABLE 11-4 Advanced Practice Nurse in Genetics: Clinical Research Activities

- Conduct health assessments
- Collect medical and family histories
- Confirm family histories of genetic condition (e.g., breast cancer, colon cancer)
- Conduct physical assessments
- Educate and provide pre- and postintervention genetic counseling
- Conduct clinical procedures and interventions
- Follow and support individuals who are living with their particular condition
- Oversee follow-up in the clinical research center and as links to the primary care setting

TABLE 11-5 Clinical Trials Resources for Nurses and Their Clients

ClinicalTrials.gov—*http://clinicaltrials.gov/*
A comprehensive database of more than 5,700 current clinical studies sponsored by the federal government and pharmaceutical industry. This site provides regularly updated information about federally and privately supported clinical research in human volunteers. *ClinicalTrials.gov* offers information about a trial's purpose, who may participate, locations, and phone numbers for more details. Includes a link to Genetics Home Reference. Open protocols at NHGRI can be found at this Web site listing of NHGRI investigators' clinical studies that are currently recruiting participants.

Search the Studies *http://clinicalstudies.info.nih.gov/*
This is a database of all clinical studies being conducted at the NIH Clinical Center, Bethesda, MD. Providers, individuals, and families can search by diagnosis, sign, symptom, or key words or phrases.

Centerwatch—*http://www.centerwatch.com/*
This site offers a wealth of information about clinical research, including listings of more than 41,000 active industry- and government-sponsored clinical trials, as well as new drug therapies in research and those recently approved by the FDA. This site is designed to be an *open resource* for patients interested in participating in clinical trials and for research professionals. Please visit the *'Your Privacy'* page to learn more about its privacy and confidentiality policies.

Cancer.gov—*http://www.cancer.gov/search/clinical_trials*
On the PDQ (Physicians Data Query) database, more than 1,800 active cancer clinical trials are listed. Providers and families can do a broad search, for example, "breast cancer," or a narrow search specifying other criteria such as the type of trial and location. This site is hosted by the National Cancer Institute.

GeneTests—*http://www.genetests.org*
A Web site that offers information about availability and location of genetic testing including the availability of research studies for an array of genetic conditions.

Chapter Activities

1. Leigh, in this chapter, describes significant side effects experienced while taking tamoxifen. She expresses frustration when she states, "*The other thing I don't get is why don't my doctors and nurses know about this research? Everyone I talked to is still into the "one-size-fits-all" concept of prescribing medication.*" Do you feel that it is worthwhile to invest your time to learn more about pharmacogenetics? Why or why not? How would you explain this to Leigh?
2. Privacy and confidentiality of genetic information is an important concern to many individuals considering genetic testing. Go to the THOMAS Legislative Information site on the Internet at *http://thomas.loc.gov* and search for the proposed legislation that was passed by the Senate and is under consideration by the House to protect the public from genetic discrimination. Do you feel such a bill is needed?

References

American Society of Human Genetics. (2004). At http://www.asgh.org/genetics/ashg/news/012.shtml.

Baker, D. L., Schuette, J. L., & Uhlmann, W. R. (1998). *A guide to genetic counseling.* New York: Wiley-Liss.

Foster, M. W., Eisenbraun, A. J., & Carter, T. H. (1997). Communal discourse as a supplement to informed consent for genetic research. *Nature Genetics, 17,* 277–279.

Harris Interactive (2001). *Health Care News, 1*(3) [Poll]. Available from http://www.harrisinteractive.com.

International Society of Nurses in Genetics, Inc. (Winter 2002). Position statement: Privacy and confidentiality of genetic information: The role of the nurse. *International Society of Nurses in Genetics Newsletter, 13*(1), Insert.

International Society of Nurses in Genetics, Inc. (Winter 2000). Position statement: Informed decision-making and consent: The role of nursing. *International Society of Nurses in Genetics Newsletter, 11*(3), 7–8.

Jenkins, J., & Curt, G. A. (1996). Implementation of clinical trials. Chapter 25. In *Cancer Nursing: A Comprehensive Textbook* (2nd ed.). McCorkle, R., Grant, M., Frank-Stromberg, M., & Baird, S. (Eds.) Philadelphia, PA: W. B. Saunders Company.

Jorde, L. B., Carey, J. C., Bamshad, M. J., & White, R. L. (2003). *Medical Genetics* (3rd ed.). St. Louis: Mosby.

Kegley, J. (1996). Using genetic information: The individual and the community. *Medicine and Law, 15,* 377–389.

National Bioethics Advisory Commission. (1999). *Research Involving Human Biological Materials: Ethical Issues and Policy Guidance.* Rockville, MD: U.S. Government Printing Office.

National Cancer Institute (10/16/03). *Cancer Clinical Trials: The Basic Workbook.* At http://www.cancer.gov.

OPRR Reports. (1979). National Commission for the Protection of Human Subjects of Biomedical and Behavioral Research. The Belmont Report: Ethical Principles and Guidelines for the Protection of Human Subjects. Accessed September 15, 2004 from *http://ohsr.od.nih.gov/guidelines/belmont. html.*

Peters, J. L., Djurdjinovic, L., & Baker, D. (1999). The genetic self: The Human Genome Project, genetic counseling and family therapy. *Families, Systems & Health, 17*(1), 5–25.

Sharp, R. R., & Foster, M. W. (2002). Community involvement in the ethical review of genetic research: Lessons from American Indian and Alaska Native populations. *Environmental Health Perspectives, 110*(2), 145–148.

Streuwing, J., Hartge, P., Wacholder, S., Baker, S., & Berlin, M. et al. (1997). The risk of cancer associated with specific mutations of BRCA1 and BRCA2 among Ashkenazi Jews. *New England Journal of Medicine, 336,* 1401–1408.

Weijer, C., Goldsand, G., & Emanuel, E.J. (1999). Protecting communities in research: Current guidelines and limits of extrapolation. *Nature Genetics, 23,* 275–280.

Williams, J., & Schutte, D. (1999). Chap. 26 Genetic counseling in Bulechek, G., & McCloskey, J. (Eds.). *Nursing Interventions: Effective Nursing Treatments* (3rd ed.). Philadelphia, PA: W. B. Saunders Co.

CHAPTER 12
Braving New Frontiers: What Is My Future with My Genetic Condition?

Introduction

It's tough to make predictions, especially about the future. But with the advent of the genomic era, it is certainly the end of the world of health care as we know it. The possibilities are simultaneously exciting and frightening. The expansion in genetic services challenges resources: professional, financial, and political. Unless steps are taken soon, the profound advances in biomedical research today may never be effectively transformed into quality outcomes in health care for society (Bloom, 2003). And as a provider and consumer of health care services, that is unacceptable. You, too, are at a crossroads. We hope that you will embrace the challenges associated with integrating genomics into your foundation for nursing in the 21st century (Porter-O'Grady, 2001). Consider what role you will play in moving this new frontier forward, and take active steps to make it a reality. Only then can we ensure that the boundaries that surround nursing practice, education, and research are pushed past the artificial limitations of our minds to a reality that we create.

Profound Change

For some reason as we began to think about what the future offers, lots of "P' words seem to come to mind. This chapter will highlight those "P" words as we discuss the potential future for those with a genetic condition. We all have genetic errors—some are more obvious than others (physical manifestations), some are predictive but not absolute (predisposition testing), while others have no obvious negative outcomes. Genomic technology will provide methodologies to be able to determine those risks, potentially even identifying that information before or at birth. Living with the knowledge of the "potential" or "actual" manifestations of a genetic error(s) will push the limits of our current models of health care delivery. Profound change in preparation of health care professionals is needed to provide care in a predictive and preventive model of care rather than in a model of intervention and treatment. Additionally, financial coverage of preventive services is not always considered a priority in this country. Potential discrimination associated with knowing predictive risk characteristics may limit consumer interest and utilization of such information. Each of these factors will affect access to potentially life-saving information. Importantly, balancing this knowledge with the realization that genetics is not destiny is essential; genetics interfaces with environmental, lifestyle, and behavioral influences. Having this knowledge may also open up opportunities for interventions never before considered. Parker's story highlights an example of one of many individuals braving the way to trailblaze the new frontiers of genomic-based health care services.

Parker's Story

Our son, Parker, has SCID, severe combined immunodeficiency (for more information on SCID, see Chapter 13, Box 13-3). He appeared perfectly healthy for the first 7 to 8 months. He caught pneumonia at 8 months old. Doctors thought it was bronchitis. He would turn red in the face and stop breathing. This went on for about 4 weeks. He would recover quickly, so at the time we were puzzled. We were trying to get answers from his pediatrician. It got so bad that we finally took him to the emergency room on New Year's Eve and told them that we weren't leaving until they figured out what was wrong with him. It took 2 weeks at the hospital, but they finally discovered his condition, SCID.

Once Parker was diagnosed, we then had to find a place to help him. We were told that he needed a bone marrow transplant, but it became evident that usually meant a blood transfusion. We are Jehovah's Witnesses and that would not be an acceptable solution. In addition, I had a 30-day window that my insurance had been dropped before I was being picked up by another company and this was the time period that Parker was diagnosed with SCID. One of the Infectious Disease doctors (Dr. G.) listened to our case and wrote Medicaid about us, telling them that the only solution was a medical center where Dr. R. was one of the foremost specialists in SCID. Within a week we were on our way and remained there for the next 8 months. We were incredibly fortunate and blessed by Dr. G. and by Dr. R. She knew we were Jehovah's Witnesses, and she had previously had two Witness babies so she wasn't concerned with the blood issue. She was and continues to be very supportive. Parker had a total of four bone marrow transplants, one each year of his life from 1 to 4 years old.

Since none of them worked, Dr. R. referred us to a clinical study at NIH (National Institutes of Health). She is a close associate with Dr. P. and Dr. M. They have been seeing Parker for approximately 4 years, and he has just recently had his gene therapy treatment. We have had no positive results as of yet, but it takes months to see any. NIH has been absolutely great, and again we have been incredibly fortunate and blessed by the care Parker has received there and the wonderful people that have taken care of him.

Nick, Parker's Dad

Power

The emerging field of gene therapy promises to yield exciting new treatments for serious illnesses, including SCID (Hacein-Bey-Abina et al., 2002). Parker's family had the power to make a decision related to his care. Frustrated by a lack of diagnosis, they sought answers and pushed the health care system to make it happen. Once a diagnosis was made, treatment options became available. Some of these options created ethical dilemmas for the family. Knowledge can be a powerful tool. But genomic knowledge currently has limitations. Techniques for gene therapy have been found to potentially have serious side effects, such as leukemia (McCormack & Rabbitts, 2004). Sometimes the power to make decisions related to the care of a loved one is overwhelming. When braving new frontiers, the answers to questions about risks and benefits may be clouded by the hopes and expectations of all involved. You have the power and responsibility to help families balance the realities with the possibilities.

Possibilities

One of the things the doctors told us when Parker was first diagnosed was that our lives would change drastically from that point forward. To an extent they were right, but Parker is the only child I've ever had, and the way we have had to live our lives because of it is the only way I've ever known. So to me it's a normal way of life. We used to check into the hospital like most people check into hotels. I would call ahead for admittance, bypassing the admittance desk, walk up to our floor, say hi to all of the nurses, put Parker in bed, start the IV, get the VCR and settle back for 2 to 3 days of IV treatment. His hospital stays have decreased tremendously. He continues to take many medications.

It is hard to predict all the possible outcomes of a life touched by a genetic illness. Interacting with nurses and physicians has the potential to improve upon what can be a tremendously stressful time for the individual and his or her family. The spiraling effects of encountering uncaring or uninformed health care providers in this foreign territory can be frightening and frustrating. Try to remain open about what the experience means to your clients, make no assumptions, and offer support and guidance as requested. You have the power and responsibility to integrate genomic information into practice and offer programs of care that ensure that all possibilities are considered.

Policy

Sometimes, those possibilities are limited by current policy and regulation. Because genomic health services may require new technology, new lab tests, or new information that has implications for family and society, decisions about safety and quality control have not yet been adequately addressed. Guidelines for reimbursement (insurance coverage), privacy standards, and quality control oversight are items of major discussion at many levels of society including the government (Food and Drug Administration, Senate, House). You have the power and responsibility to contribute to policy setting. Only then will creative avenues utilized by Parker's family be unnecessary.

Once my insurance started, Parker was covered and all his medicines and infusions have always been covered completely. The NIH covers all of his expenses related to the protocol and this has been wonderful. My monthly payments used to start at about $400 to 500 per month and would slowly creep up to over $1000 per month. I would then switch companies and start over. Because we live in a state that has no pre-existing clause, I was able to do this.

Patience

I feel that what they have done to break the genetic code is good and can be very useful. Obviously there can be some possible bad uses in the future, but I think for the most part it is a very good thing and will be very helpful at curing many diseases and conditions of people. I hope anything that they discovered while working on Parker can be used to help others.

One of the concerns expressed by many consumers is that it takes so long for new discoveries to really make a difference to them. It may take many years to be able to understand the function of our genes, and perhaps several more years to translate those findings to improving population health. Suggestions for conducting genomics and population-based health research, developing evidence on the value of genomic information, and integrating genomic information in practice and programs are included in a report by the Centers for Disease Control and Prevention *(http://www.cdc.gov/activities/ogdp/2003.htm)*. These activities will help to ensure rapid translation of clinical applications. Although it is expected that this rapidly expanding field will have tremendous benefit for each of us, leadership is needed to ensure that progress is made. Thinking genomically is necessary in guiding the future education of all health care professionals to ensure implementation of these discoveries immediately into clinical care (Gebbie, Rosenstock, & Hernandez, 2003).

Persuaded

I would probably use treatments that are developed in the future that would improve on what Parker has received so far and help him achieve a better quality of life. Even though he has had his treatment, there is a good chance that he will need further treatments of some kind. I don't expect this to fully cure him for the remainder of his life, though it may be possible.

As for myself, I would subject myself to gene therapy of some kind in the future if I thought it could cure me of allergies or other health conditions I have. Generally speaking I am very healthy, but I'm very open to this type of treatment should it become more commonplace.

Parker's Dad is representative of the way many consumers already view the potential of genomic research. The promise of benefit captures the imagination and because of this may open up avenues for deception, or at least for the opportunity for entrepreneurs to sell their products. Direct-to-consumer marketing of tests, pills, profiles, and multiple unproven services is already expanding. Even the media, through video, Web sites, and advertising, is significantly shaping the public's perception of scientific capabilities (Varmus, 2004). Helping the public identify what is credible and of value will require interpretation by knowledgeable professionals.

Professionals

Health care professionals will continue to be challenged by the fast pace of genomic research discovery. One of the untapped resources that we attempted to capture in this text is the voice of the individual and their family. Taking advantage of the voices of those families you meet daily can teach you. Oftentimes, because they are living the experience, they can be our best teachers. Listen to their compelling stories and learn about the consequences of your care (both good and bad). Sometimes we forget how it feels to be on the other side of the stethoscope. Know that in addition to learning through curriculum, continuing education, and Internet sites, people are our best teachers. And they are more than willing to offer their insights.

I'm happy to write down more detailed information on Parker's life and how we have dealt with his situation if you desire. *I hope I have been of help with your work.*

Nick, Parker's Dad

Public Perception

My family's genetic condition is Neurofibromatosis type 2. I have had cysts removed all my life, but it wasn't until I was 17 that the diagnosis was made. My parents died when I was very young so they were never tested. I have four sisters who (for reasons unknown to me—probably fear) refuse to be or have not shown any interest in being tested.

Jayne

It is equally important to recognize that not all persons will have any interest in knowing about genetic information. Multiple reasons exist for why persons do not wish to learn about information that may influence life choices. Health care professionals will feel ethically challenged by individuals in high-risk families who do not want to know about tests or screening recommendations that could be life-saving. Family dynamics may inhibit communication of disease occurrence influencing accurate interpretation of family history. Public education will be important in order to be able to provide a sufficient foundation for the consumer about the benefits, limitations, and risks of knowing as well as not knowing genetic information. They are equal partners in braving this new and unknown territory. How can we as health care professionals responsibly assist them in having enough information to make informed choices?

Public Health

Despite the reservations by some, there is great public expectation that genomic information will make a difference to health outcomes. From a young individual with breast cancer,

---

I would like to see more about patient education and patient empowerment in using this information to improve their lives. Helping them to have a sense of control by actively seeking monitoring and screening tests at earlier ages than suggested for the general population will help to identify disease earlier so that the disease can be treated and more probably cured.

Patty

Preventive

Recognizing signs of high-risk individuals also opens up other options for preventive care that can be life altering.

Our family has Long QT Syndrome. This is a genetic condition (autosomal dominant) that predisposes affected people to cardiac arrhythmias, which can be fatal. In our family's case, we learned of our LQT problems when my then 14-year-old daughter collapsed at a junior high track practice and suffered a cardiac arrest.

The process was complicated due to the fact that our daughter suffered a brain injury secondary to the anoxia during her arrest. Before I could concentrate on the genetics of the condition, we first had to worry about her living in a coma or on a ventilator. At the time, LQT was a fairly obscure entity, but our physician was extremely interested in the syndrome, so we always had his help in understanding LQT. We eventually got connected to the Sudden Arrhythmia Death Society (SADS) and they have been a wealth of information.

The impact has been minimal upon my life as I take a beta-blocker, fish oil, and pay attention to eating a diet high in potassium. To Kara, the affected daughter, the impact includes all of the above plus an implanted pacer–defibrillator. As to our oldest daughter, an EKG revealed she also has LQT but is totally (and thankfully) asymptomatic. She takes a beta-blocker daily prophylactically. I am trying to convince her to take daily fish oil.

I hopefully counsel my children to consider never having biologic children. There are many beautiful children of many beautiful colors who need to be adopted.

Sharing genetic information scares me at this time. I don't think that there is adequate protection for those of us with pre-existing conditions. I would like more information about the law and when our rights might be in jeopardy insurance-wise. More counseling for adult children so they can hear facts from unbiased, knowledgeable professionals would be a service that I would like to see.

Kara's Mom

Predictive

In the following situation a genetic test would have been predictive of Alzheimer's; the individual outcome illustrates the value of the intervention for her.

I wanted to thank you for the time we spent together discussing the genetic testing for Alzheimer's. I really appreciated processing the hopelessness I was feeling about life and figuring out how I was going to live with hope if I did have the gene. I realize now just how suspended I remained with all my hopes and dreams for the future, though I did maintain some hope in general just about life, but I dared not to hold onto specific dreams lest I would need to let them go. I recognize this so clearly because I do not have the mutation and I have walked around for 2 weeks just picking up all of the hopes and dreams I had dared not hold on to. Bill and I will grow old together; I may get to be a grandmother! There is such a sense of relief for my kids and others that love me, siblings, friends. This reality has slowly sunk in but the reality has been truly a relief. Just yesterday morning I lost the keys to my car, I emptied my purse, I went back upstairs to check in the shorts I had on the day before. I retraced my steps from the moment I walked through the door and could not find them. Bill joined me in the search because I was running late and I needed to get to work. I was frustrated and mad but Bill was walking around looking for those keys with a smile on his face like he was walking on air saying "but at least we know you don't have Alzheimer's." It was humorous only after I found my keys and was on my way to work. The relief not only for me but also for those around me is an amazing observation to watch. Thank you once again!

Julie

Perplexities

Genomic research will change the world as we know it forever. The choice we make today as individuals and as part of society will have implications for our future generations. One example that is perplexing but beginning to offer a glimmer of understanding is in behavioral genetics. There is a lot yet to be learned about the role genetic changes play, not only to the occurrence of health and disease, but also for human behavioral aspects (Baker, 2004). Determining the interrelationships of all the variables influencing behaviors such as learning disabilities, violence, and addictions will challenge researchers for many years. Sorting through and letting go of long-held beliefs of causation for antisocial behaviors will be challenging to personal and professional views. An excellent resource for learning more about behavioral genetics is available from the American Association for the Advancement of Science (Baker, 2004).

Performance

An additional area, which will challenge society, is the contemplation of future choices of when and how to use genomic technology in nontherapy situations. One of the emerging possibilities with improved understanding of genomic science is the ability to intervene beyond health interventions and attempt to improve upon other facets of our lives (President's Council on Bioethics, 2003), for example, improved physical strength, faster athletes, or decreased obesity. The public is often looking for ways to improve upon their personal performance. Genomics may be an additional tool that will be considered in improving our lives. What role will health care professionals play in offering consideration of these choices?

Promise

The promise of understanding the DNA blueprint that guides the development, structure, and functioning of our bodies is to improve upon our ability to predict, prevent, treat, and manage health outcomes. The promise brings with it the potential of overexpectations, which many identify as hype. It's important that health care professionals become informed about current knowledge and assist their clients to have realistic expectations. Others express hope that these promises will indeed become reality and create a brighter future, opening up new options of care. You have the power and responsibility to help make these promises become reality.

Helen's Story

As I began to wake, all I could feel was the pain in my back. I tried to turn over but that is when I realized that I was not in a bed. It's impossible to lay on your stomach in a recliner. As I opened my eyes, the sun was beginning to creep over the horizon. It was breathtaking and chilling all at the same time. Breathtaking because in my lifetime I have seen very few sunrises and chilling because it was a déjà vu moment.

So here I sit looking out the window of the intensive care unit at Maine Medical Center as the sun rose on a new day. Suddenly, I remembered why I was there and jumped out of the recliner to look at my son. He was seventeen years old and he had ruptured his spleen and bruised his kidney in a skiing accident the day before.

He was resting comfortably after a long night of excruciating pain. It was a huge relief to see him sleeping peacefully. So I climbed back into the chair and stared at the sunrise. It was at this moment I realized the implications of my earlier chills and déjà vu feeling. . . . I had been in this intensive care wing before and I had seen a sunrise that time too. . . .

As I recall my labor was going very well, faster and less painful than my previous one. I remember thinking to myself, as the doctor said, "One more push," that indeed I was fortunate to have this short, quick, easy labor. Today was going to be my lucky day!

As I tried to center myself after the last grip of pain, something seemed odd. The silence in the room was deafening! I noticed whispering and people moving around quickly. I tried desperately to see what was going on. All I saw was a glimpse of a tiny baby being whisked away with the doctors and nurses. Shouldn't they be giving the baby to me? (was all that was going through my head.) Why are they leaving? I looked into my husband's eyes for answers to my unspoken questions. But he looked in shock. He was pacing back and forth. His eyes were glazed over and he was rubbing his hands together. It was clear he would have not been able to answer my questions even if I had asked them out loud.

Now, how can I make sense of this? It was a drug-free labor and delivery so I knew that I wasn't hallucinating . . . I needed to focus . . . I looked around the delivery room and besides myself and my husband there was only an older woman who appeared to be a nurse's aide. She was quietly cleaning up the room. Suddenly I heard the words "cleft palate." It came from either my husband or the aide. I tried to process what I knew about that term. Being a speech therapist, I was more than familiar with it. I thought to myself, "Oh that's not so bad, I can deal with that." Then the nurse's aide went on and on about having a second cousin whose baby had that and the wonderful things they could do these days with plastic surgery.

I was raised to always be kind and polite but what I really wanted to do was to scream at this woman to shut up and go get my baby!!!

I don't know how much time lapsed, but it felt like an eternity! At this point my husband can pace no longer . . . he says he needs to find out what is going on and he leaves me alone with this woman telling cleft palate stories.

Finally the delivery doctor, who I barely know, and who just happened to be "on call" that night re-enters the delivery room. She looks pale and nervous and very serious. She tells me I have a "girl" but that there are "problems." The neonatology specialist has just arrived at the hospital. He will evaluate her and then come and speak to me. She quietly backs out of the room.

Finally, a nurse arrives. She has very kind eyes and she tries to smile and reassure me that everything will be ok. I think to myself that this person will help me figure this out. But she is busy getting me ready to move to a regular room. I can't seem to engage her in a conversation concerning the "everything" that was going to be ok. She stays focused on getting me situated comfortably in a room in the maternity ward.

Suddenly, my husband enters the room. He now has fear in his eyes. He says he needs to go to the hospital library and he'll be back by 10:00 P.M. because that is when the neonatologist will be coming to speak to us. Before I have a chance to protest, he's gone. Now I know for sure that this is actually going to be my most "unlucky day"!

The neonatologist and my husband both arrive at my room at 10:00. He is a very soft spoken, kind looking man. He had in his hands a photograph taken with an instamatic camera and a book. He told us exactly what we did not want to hear . . . our baby had Trisomy 13. I had never heard of it, but from his tone I gathered it was not good. He proceeded to give us the sad details. I tried to listen carefully to all the components. I forced myself to keep my emotions in check until all information was absorbed.

At this point he said that she was in the neonatal intensive care and on life support. He asked if we wanted to see a photograph of our baby. The first wave of shock passed over me. How can this be? There must be some mistake. As I looked at the photograph, I saw a gaping hole in her precious face. I remember wiping tears from my face but I kept myself from crying out loud, because there were still too many questions to be asked! I heard . . . cleft palate and lip, heart anomalies, mental retardation, life expectancy was a few hours to 3 months. After reviewing the photograph and the information on Trisomy 13 in the book and asking every possible question I could think of, reality was setting in big time! When it was clear that we had run out of questions and had a good grasp of the situation, we asked the ultimate question. What do we do now?

He laid out our options, starting with, do you want us to keep her on life support or not? I can recall only two things that helped me form this ultimate decision. 1. There was no sucking reflex due to the cleft palate damage, therefore there would be no way for her to suck. My firstborn could always be soothed and comforted with a nipple, a bottle, a pacifier, and a thumb! How does a baby sooth itself without any of these. The answer. She doesn't! 2. When asked how will I feed her when I take her home? The answer. She can be fed intravenously, but she would not be going home.

What was the purpose of life support in this situation? I saw life support as a means of keeping someone alive until they could make life livable. But from what I was hearing, it was merely to keep her alive with no hope of making her life livable in the future. At this time someone asked us if we had a name. We did. She was now a real person with a name: Kristen.

At this point we decided to put the situation in God's hands rather than a machine. If she survives on her own, we will deal with what comes. If she does not, it was not meant to be.

We needed to go to the neonatal intensive care in order to take her off life support. As I neared the nursery, I remember watching hospital workers look at this poor creature and then quickly look away, in the same way that I had glanced at the photograph. But for me this was now different. I no longer saw a deformed baby the way the nurses and doctors saw her or the way I had seen her in the photograph. Now she was my baby and there is nothing more beautiful than your own baby and I asked if I could hold her. They brought her to my hospital room and put her in my arms. My husband sat quietly beside me.

I immediately began to pray, "Oh God please don't make her suffer, if she cries and I can't comfort her it will rip my heart out." And God heard my prayers.

She never did cry. She just lay peacefully in my arms. I could hear a slow whispered rattle that was her breathing. Tears again rolled down my cheeks but there was no way I was going to cry loudly. If I cried out loud, I would no longer be able to hear her quiet breathing and it might disturb her. So I held in my sadness and listened to each and every breath she took. Time was hard to fathom. I was in the hospital all of about eight hours but it felt like eight days.

Part of me was praying that God would take her so that she would not have to experience hunger, pain, needles, and sadness. Another part of me was outraged that I would be praying to God to take away my little girl that I so desperately wanted.

I continued to listen to the rhythmic breathing and then it became more and more faint. I held my breath and listened. As much as I wanted to scream and call in the nurses and doctors and yell, "Save her!" I also knew that I wanted her to pass from this life in a warm, calm environment. So I fought my desire to scream and forced my arms to continue to hold her tight and keep her warm and the room quiet.

I clearly remember her last breaths. I felt her spirit leave her body, but I said nothing so that I could continue to hold her. It was this letting go that was going to be difficult. My husband asked if I was ok and I had to tell him that she was gone. He went to tell the staff. A doctor and nurse came to take her away and I did not want to let go! When she left the room, I sobbed so hard that my body shook. How will I ever go on without her?

I looked out the window and the sun was rising. It was bright and beautiful and magical and it reassured me that there was a heaven and that Kristen was there.

Of course, I have gone back over these moments in my mind hundreds of times, thinking maybe I should have done this or that. Maybe she'd be here today if I'd done this differently or that differently. Here it is 19 years later and I am still plagued by "what if's" . . . and I guess I always will.

Helen, Kristen's Mom

Helen's story provides a reminder of our limitations with genetics knowledge—old and new—and the need for ongoing translation of genetics knowledge and technologies into clinical care. Through her story, we can see that communication and comfort are sometimes still the best tools health care professionals have when assisting clients to cope with the manifestations of a genetic condition. You can make a difference.

Priorities

If all the promises from genomic science for improved options in clinical care are fulfilled, society will indeed be faced with prioritization of resources. Even today, expensive interventions are not always available to those in need. The goal of ongoing genomic research is to reduce the cost of technology, such as genomic sequencing. Additional goals of the National Human Genome Research Institute (Collins et al., 2003) include reaching out to educate health care professionals and the public. Only then can we be sufficiently prepared to prioritize utilization of these services and assist in the design of future health care delivery systems that integrate genetics knowledge. Working with payers of care will be essential to ensure their understanding and willingness to cover health care costs based on this new technology.

Passionate

Obviously, we believe in the promise of genomics to make a difference. We have already been persuaded of the potential for improved health because of genomic research. We are passionate about the need for all health care professionals to have a foundation of basic genetics understanding from which to build their clinical practice. We are cautious in recognizing that there will be significant ethical and social challenges facing us as we cross this new frontier. We highly recommend that, as you stand at the crossroads of determining whether to advance into the unknown or turn back to what feels comfortable, you consider both personal and professional ramifications. You have the power and responsibility to make a difference to the health and well-being of those living with a genetic condition.

Chapter Activities

1. Visit the CDC report at *http://www.cdc.gov/activities/ogdp/2003.htm.* Identify at least two activities that you could design to advance the goals of genomics and population health and ensure rapid translation of clinical applications.
2. Visit the Web site discussed by Varmus (2004) at *http://www.godsend.com.* What is your reaction to the Web site? Does this type of information have implications for you and your clients?

References

Baker, C. (2004). *Behavioral Genetics.* Washington, DC: American Association for the Advancement of Science.

Bloom, F. (2003). Science as a way of life: Perplexities of a physician-scientist. *Science, 300,* 1680–1685.

Collins, F. S., Green, E. D., Guttmacher, A., & Guyer, M. S. (2003). A vision for the future of genomics research: A blueprint for the genomic era. *Nature, 422* (24, April, 2003), 835–847.

Gebbie, K., Rosenstock, L., & Hernandez, L. (Eds.). (2003). *Who will keep the public healthy? Educating the public health professionals for the 21st century.* Washington, DC: The National Academies Press.

Hacein-Bey-Abina, S., LeDeist, F., Carlier, F. et al. (2002). Sustained correction of x-linked severe combined immune deficiency by ex vivo gene therapy. *New England Journal of Medicine, 346,* 1185–1193.

McCormack, M., & Rabbitts, T. (2004). Activation of the T-cell oncogene LM02 after gene therapy for x-linked severe combined immunodeficiency. *New England Journal of Medicine, 350*(9), 913–922.

Porter-O'Grady, T. (2001). Profound change: 21st century nursing. *Nursing Outlook, 49,* 182–186.

President's Council on Bioethics. (2003). *Beyond Therapy. Biotechnology and the pursuit of happiness.* Washington, DC: Author.

Varmus, H. (2004). Lost your little boy? We'll make another. *New York Times,* May 2, 2004.

Chapter 12 Interdisciplinary Commentary

Kevin Lewis

Chairman of the Board, Colon Cancer Alliance

Six years ago, through genetic testing, several members of my family learned that we carry a gene mutation that causes Hereditary Non-Polyposis Colorectal Cancer or HNPCC and gives us an 80% chance of getting colon cancer as well as at high risk for getting other cancers, including stomach, renal, ovarian, and uterine cancers. We learned early enough that we were designed to die from colon cancer; but more importantly, we learned about the screening techniques that could prevent the disease and extend our lives.

In just four generations, genomics has contributed to the transformation of my family members from cancer death at age 40, to two generations of trying to survive this terrible disease, and finally to a generation of cancer prevention fighting we hope to win. Quite simply, genomics and genetic testing are one of the tools that has saved or at least significantly extended my life.

And this tool can help manage conditions that many families are facing, but it is a tool that is extremely complex. This tool requires a tremendous amount of information to understand while generating even more data during the course of a patient's journey with managing their genetic condition. Families must be supported in this process, and it falls on the nursing staff to help to fill the gaps that are developing in the health care system.

The hardest part about genetic conditions is that they are extremely rare conditions, and our understanding of them is just emerging. One in 740 people carry the

HNPCC gene mutation that I carry. That means that most doctors and most nurses will not see an HNPCC patient and will not know how to deal with one.

That fact has a profound impact on the way that care and information is delivered to consumers. Both of my doctors know little about my condition, and these are doctors who work at Harvard-affiliated institutions. I proscribe my screening regimen and my primary care physician orders it. My doctors ask *me* about the advances that are being made in genetics and colon cancer screening, not the other way around.

The only people that I have found that know much about my condition are the genetics specialists at the National Cancer Institute (NCI) and at the MD Anderson Cancer Center and a few other people. I was forced to seek out other alternatives to my physicians to be informed about my condition. The way I have learned about genetics, genetic screening, and managing my condition is through the Internet and the doctors and nurses at the NCI.

I was fortunate to be tested within a clinical trial setting with genetic counselors from NCI. Not everyone has access to the NCI, and many patients have doctors who don't know how to help them when it comes to genetics. I believe that genetic counseling and genetic information are crucially important to patients. I hope that nurses can help bridge the gap of information that exists, or at least help patients find credible sources of information.

I find that some of the best information for patients is provided by the companies and, especially, the support organizations that specialize in this area. This is much better than it is possible to have provided by the general health care delivery process. This area of distributing new genetic information is a difficult one, but it is important to understand the limitations of general practitioners to deliver the information. Patients that are concerned with the area of genetics are going to have to manage their health care and seek out information on their own.

When I found out that I carried a colon cancer gene, I reached out to find any resources that I could to help me understand my condition and to understand colon cancer, the disease that I was destined to get. In my search I found little about my condition, a situation that is now drastically different, but I did find a colon cancer support group. This Internet support group on ACOR.org, called the Colon Cancer List, provided me the support to better understand the terrible nature of colon cancer.

Along with a dedicated group of listserv members, we found that there was no patient group representing the voice of colon cancer patients, so we set out to form a patient advocacy group to support patients and their families, while advocating and communicating about the importance of colon cancer screening. I was given information to extend my life, and I have found a lot of fulfillment in trying to inform others of this special gift of knowledge. I hope that, in helping to found and grow the Colon Cancer Alliance, I have been able to pass some of this special gift on to others.

In my volunteer work at the Colon Cancer Alliance, I have realized another aspect of my orphan condition. In the case of HNPCC, more people ask about and get tested for HNPCC than have the disease. This amplification of the necessary services provided places even more burden on a health care system already stretched to provide treatment. What if orphan conditions are most of the genetic conditions? This chapter has mentioned five genetic conditions, but there are many more, and the magnitude of the services required is only starting to be understood. Maybe, every patient is going to need to be somewhat of an expert on his or her own genetics.

In order for consumers to fully realize the potential of genetics, we still have issues to solve in terms of eliminating the consumer fear in using these tools. Genetics offers to us knowledge about our destiny and gives us an opportunity to reshape that destiny. At the same time, this personal information can and will be used in a way that discriminates against the affected individual. The key for consumers is to be able to receive genetic testing without the threat of discrimination in critical areas, like employment, that affect an individual's livelihood.

While I see a possible threat to consumers with genetic information, I have not seen any discrimination in reality, and I believe that the advantages of learning about your genetic makeup far outweigh the risks. In my case, this information has extended my life. What more can you ask for from your medical system?

It's important to remember that genetic testing is a very personal issue. I didn't share my results with my doctors for several years after my tests. I don't know why I did this but I felt more comfortable this way. What I believe is that it is the patient's right to know genetic information and to make decisions about whether or not to get tested. As much information as possible should be provided to the patient, but ultimately the decision to test is a very personal decision that some people feel more comfortable making on their own.

In this context, nurses can readily fill the gap in the ability of the health care system to deliver solid genetic care, but an investment in training will be the only way to keep up with all of the information that will develop in this field.

Getting back to my future with a genetic condition, I know that I will have to get a colonoscopy every year for the foreseeable future. I also know that I will have to continue to learn about my condition so that I can continue to beat the odds. I imagine that someday a blood test will come along, and I will participate in another clinical trial to help replace the colonoscopy.

And sometimes I wonder if I will get cancer anyway, hoping that if I do I will catch it early. Lastly, I hope that, when I have children, their experiences in trying to prevent cancer will be so much easier than mine. And just maybe there will be reproductive services that will help me keep from passing on my gene to my children—a blessing that I hope to see in my lifetime even if it is too late to help me.

CHAPTER 13
Integration of Genomics to Improve Health Care

Gene-based diagnostics and therapeutics will soon be routinely integrated into and commonplace in today's health care. Nurses will be among the many health care providers to use these new interventions to improve the health of all populations. To participate as knowledgeable members of health care teams, nurses will need to become educated in the principles and applications of new genetics and genomics and participate in professional and public education about genetics and genomics.

Introduction

The first phase of the international research project to determine the entire sequence of the human genome was completed 2 years ahead of schedule in 2003. During the last quarter of the 20th century, research has focused on deciphering first genes and then entire genomes, thereby ushering in the era of genomics (International Human Genome Sequencing Consortium, 2001). This last phase of basic research has generated a massive, continuous stream of complex genomic data that have transformed the study and treatment of health and disease (Collins et al., 2003).

One result of genome discoveries is an increasing array of gene-based diagnostics and therapeutics that are becoming integrated into all aspects of health care. Nurses need to be prepared for the transition of genetic tests and treatments from research to the clinical arena. Nursing roles will include assessing patients, for administering to, and monitoring the effects of a range of therapeutic treatment approaches (Collins et al., 2003; Lea & Monsen, 2003). A central role for all nurses will be learning how to assist individuals in making decisions about genetic information (Grady & Collins, 2003). To do this, nurses worldwide must understand the genetic contribution to health and disease, be fluent in the language of genetics, and be competent in communicating complex genetic information as a routine component of clinical practice (Guttmacher, Jenkins, & Uhlmann, 2001; Khoury, 2003;

Lea & Monsen, 2003; Williams & Lea, 2003). Expanding nursing roles in the areas of education, clinical care, and research will ensure that the health and social needs of the public are met, and will begin to address the technological and economic inequities that exist in accessing genomic advances across the world.

In this chapter, a vision for progressing the role of nursing in the postgenomic era is put forth. New role considerations for nurses and the attendant educational implications will be explored. Nursing research focus and direction in genomics is also discussed. Strategies for integrating genomics into all areas of nursing practice and care to improve health care outcomes are suggested. Our vision is founded on the International Council of Nurses (ICN) Code for nurses, which states that the "nurse shares with society the responsibility for initiating and supporting action to meet the health and social needs of the public, in particular, those of vulnerable populations" (ICN, 2000). This vision is also built on the World Health Organization's Proposed Guidelines on Ethical Issues in Medical Genetics and Genetic Services (1997), which emphasize the paramount importance of education about genetics for the public and all health care professionals and the "profound economic and technological inequities that exist between nations" that have an impact on health care practice *(http://www.who.int/ncd/hgn/hgenethic.htm)*. Our vision also recognizes that the "paradigm of genetic services" will always apply for a small proportion of individuals and families that will need support and services for the management and understanding of genetic disorders. However, a threshold between delivery of genetic services for "genetic disorders" and communication of "genetic information" as a routine component of practice will have to be delineated and established (Khoury, 2003).

Research Advances and Integration of Genomics into Health Care

The current phase of genomics research focuses on cataloguing, characterizing, and understanding the entire set of functional elements—the 'parts list'—encoded in the human genome and other genomes (Collins et al., 2003). The single nucleotide polymorphism (SNP) consortium research is one effort toward achieving this goal. Sites in the human genome where the DNA sequences of many individuals differ by a single base are called SNPs. As an example, some people may have a chromosome with a T at a particular location, while others have a chromosome with an A at that same site. Each form is called an allele. The term genotype refers to the SNP alleles that a person has at a particular SNP site or for many SNPs across the human genome *(http://www.hapmap.org/abouthapmap.html)*.

It is known that about 10 million SNPs exist in human populations. Rare SNPs occur in 1% of the population. The SNP consortium is a group of scientific research centers that has mapped 125,000 to 250,000 SNPs. In 2003, the Human Genome Project announced a collaboration with the SNP consortium to generate a new set of human DNA sequence that will come from 24 anonymous, unrelated donors from diverse geographic origins. The creation of a high-density map of SNPs is becoming a research tool that is helping scientists pinpoint genetic differences that predispose some individuals to disease or are the basis of variable responses to treatment. The ultimate goal of SNP research is to develop individualized therapies to treat and prevent disease (see *http://www.genome.gov/10001551*).

HapMap

The search for the genetic cause of complex and common diseases such as heart disease, cancer, and diabetes is moving forward with the development of a map of the genetic variations between individuals, the HapMap. Most chromosome regions have only a few common haplotypes that account for much of the variation between individuals in a population. Haplotypes are large segments of DNA that contain the single-letter variations (SNPs) that distinguish individuals from each other. The HapMap describes common patterns of genetic variation in humans and includes chromosome regions with sets of SNPs. A completed haplotype map will help to decrease the size of genetic information that needs to be deciphered in order to find individual differences. The HapMap is helping researchers to discover the genes that affect common diseases such as diabetes *(http://www.hapmap.org/abouthapmap.html)*.

Genomics and Future Applications

Scientists speculate that genome research will have significant and widespread influence on people living in the 21st century. Research efforts such as the SNP consortium and the HapMap hold promise to understand common diseases, including infections such as tuberculosis and HIV/AIDS. The hope is that, in understanding the underlying causes of disease, new directions for diagnostics, interventions such as drug development, and environmental interventions will develop (Goldstein, Tate & Sisodiya, 2004; http://www.genome.gov/10001551).

Genomics and Nursing Practice: Expanding Roles for Nurses

Nurses practice at the interface between the individual, family, community, and health care system. In this position, nurses will be using new gene-based diagnostics and therapeutics that take into account the health manifestations resulting from combinations of genes in the human genome and environmental influences. Nursing roles integrating genomics are expected to expand in prenatal, pediatric, and adult health care settings. New roles for nurses are already being developed by specialty nursing organizations such as the International Society of Nurses in Genetics (ISONG, 1998), the Association of Womens' Health and Neonatal Nursing (2004), and the Oncology Nursing Society (ONS) (2000a).

Communicating Complex Genetic Information

The "threshold between delivery of genetic services for 'genetic disorders' and communication of 'genetic information' as a routine component of practice" (Khoury, 2003) is one of the first and most important areas for nursing role development. Khoury suggests that the family history can be used as the bridge between genetics and genomics. Family history assessment as a routine component of all health assessments is familiar to nurses. Building on the knowledge that family history can elucidate health risks and help with the appropriate choice of genetic test(s) and treatment(s), nurses can expand their use of family history to encompass communication of complex genetic information. Complex genetic information, such as the role of family history in disease risk, implications for reproduction, and treatment options, can be shared with individuals and families in the context of family history. Sara's story illustrates the nursing roles of family history risk assessment, analysis, client education, and referral.

Sara's Story

I had just met John and our relationship was becoming serious. I decided I needed to go to Family Planning for some help. I met with a nurse who reviewed my medical and family histories. During our discussion, I shared with her that my mother had had a blood clot in her leg when she was pregnant with my brother, and that one of her brothers had a blood clot in his leg when he was in his 40s. My mother's mother and one of her sisters had both died in their 50s from a stroke. My nurse explained to me that this family history suggested that there might be an inherited predisposition to blood clots and strokes in my family. She pointed out how, in my mother's family, there were multiple relatives in several generations with similar problems and that made her think it could be an inherited characteristic. She recommended that I have a blood test called a Factor V Leiden test to find out whether I had a gene change that would predispose me to early clotting too, and suggested that it could be a risk factor for me should I become pregnant. (**Box 13-1**)

To be thorough the nurse also asked me about John's side of the family as well. I told her that John was born with a congenital heart defect that had been re-

Box 13-1 Factor V Leiden

General Information

Factor V Leiden is the most common inherited blood abnormality that results in a predisposition to forming blood clots. This mutation involves the single substitution of glutamine for arginine at position 506 in the gene. This single point mutation results in a form of Factor V, known as Factor V Leiden, that is resistant to degradation by activated protein C and therefore leads to a relatively hypercoagulable state.

Clinical Symptoms

Diagnosis is suspected in patients with a history of venous thrombosis, especially women with a history of thrombosis during pregnancy or in association with oral contraceptive use and families with a high incidence of venous thrombosis. The clinical expression of Factor V Leiden thrombophilia is extremely variable. Many individuals with Factor V Leiden never develop thrombosis. Although most individuals with Factor V thrombophilia do not experience their first thrombotic event until adulthood, some have recurrent thromboembolism before age 30 years. The clinical expression is influenced by the number of Factor V Leiden alleles, the presence of coexisting genetic abnormalities, and circumstantial risk factors. Risks are particularly increased in patients with coexisting predispositions for thrombosis, such as advanced age, surgery, use of oral contraceptives, hyperhomocystinuria, and deficiencies of protein C and protein S.

Prognosis

People who are homozygous for Factor V Leiden (i.e., have two copies of the mutated gene), are at a much greater risk for venous thrombosis than those who are heterozygous (i.e., have only one copy of the abnormal gene), having close to a 100% lifetime risk of a thrombotic event.

The factor V prothrombin G → A 20210 gene mutation/polymorphism is another risk factor for thromboembolism. This mutation is found in 18% of individuals with a personal and family history of thrombosis. Coinheritance of both Factor V Leiden and the prothrombin gene mutation occurs in approximately 2% of persons with venous thromboembolism.

Inheritance Pattern and Recurrence Risks

Individuals who have one Factor V Leiden mutation (heterozygotes) have a 50% chance to pass on the gene mutation to each of their children. This is autosomal dominant inheritance. Individuals who have two copies (homozygotes) have inherited the mutations in an autosomal recessive manner with each of their parents

Box 13-1 continued

having one copy of the gene for Factor V Leiden. These individuals will always pass on one Factor V Leiden mutation to each of their children who will be heterozygotes. Family members of individuals who are heterozygous or homozygous for Factor V Leiden are at increased risk and should be offered genetic counseling.

Prevalence

The Factor V Leiden mutation is found in 4% to 6% of the U.S. population. It is the most frequently found genetic risk factor, found in up to 40% of patients with familial predisposition to venous thrombosis.

Relevant Testing

Because of the high prevalence of the Factor V Leiden mutation in the general population, the genetic status of both parents of an affected individual and/or spouse of an affected individual should be evaluated first before information regarding potential risks to siblings or children is assessed. Prenatal testing is not routinely available or used since Factor V Leiden thrombophilia is a relatively mild condition with available and effective therapy.

Management and Treatment

Management of patients who are found to have Factor V Leiden may include daily aspirin or thrombosis prophylaxis such as Coumadin therapy that is usually instituted when clinical circumstances warrant. In some individuals, long-term anticoagulation therapy might be considered following a serious or idiopathic thrombotic event.

Adapted with permission: D. H. and Smith, 2003. The Genetics Resource Guide: A Handy Reference for Public Health Nurses. Scarborough, ME: Foundation for Blood Research. Written under a grant from the State of Maine Department of Human Services Genetics Program. Grant BH-01-166 Band C. pp. 91–92. Scarborough, ME: Foundation for Blood Researcher Publisher.

paired as an infant, and that he had had some speech delays as a child. His sister had been born with a cleft palate, and she had a daughter who also was born with a congenital heart defect and had some developmental delays. John's father had a history of bipolar disorder. As we talked about John's family, the nurse explained to me that, since there were several family members with birth defects, these might not have happened by chance alone and could be related to each other. She suggested that John and I consider a more detailed evaluation of his family history with a geneticist to find out more and to learn about what the chances were for our children to have a similar condition (**Box 13-2**).

Box 13-2 22q11 Deletion Syndrome

General Information

The 22q11 deletion syndrome represents a spectrum of disorders associated with deletions within the 22q11.2 chromosomal region. This syndrome is commonly referred to as both velo-cardio-facial syndrome (VCFS) and DiGeorge syndrome (DGS).

Clinical Symptoms

Clinical findings of 22q11.2 may include cardiac anomalies, hypoparathyroidism, T-cell immunodeficiency, cleft palate, and unusual facial features. The face is characterized by a flat mid-face and broad, bony base of the nose, often with a squared-off nasal tip. The ears may be prominent and/or small. The palate in people who have 22q11.2 is usually long and narrow, and some people affected by the syndrome have an obvious cleft of the palate. Speech is often hypernasal. Congenital heart defects typically are of a conotruncal etiology. However, not all individuals who have 22q11.2 have a cardiac abnormality. Other features include long, tapered fingers and long hands and feet and low muscle tone. Approximately 40% to 50% of individuals have some sort of developmental delay–mental retardation, although this is usually not of a severely profound type. Learning difficulties tend to be in the verbal–language areas, and there is often speech delay. There are occasionally ocular, auditory, and renal abnormalities, and there is an increased incidence of psychiatric disturbance in older affected individuals.

Prognosis

The 22q11.2 syndrome is a genetic condition that varies widely even within families, with some affected people having only the facial features and mild speech problems, while others may have severe congenital heart defects and cleft palate as well as the other typical physical features. Some people with VCFS/DGS may have learning problems, either as the sole manifestation of the syndrome or in combination with other findings. VCFS/DGS is not progressive.

Inheritance Pattern and Recurrence Risks

The 22q11.2 deletion syndrome is inherited in an autosomal dominant manner. Most cases of VCFS/DGS occur sporadically, but approximately 10% of cases will have a parent who carries the deletion. Children of an individual who has the 22q11.2 deletion have a 50% chance of inheriting the deletion.

Box 13-2 continued

Prevalence

Estimate of prevalence ranges from 1 in 4000 to 1 in 6395. Given the variability of expression of 22q11.2 deletion, the incidence is likely higher than estimated.

Relevant Testing

Diagnostic and prenatal testing are available. The 22q11.2 deletion syndrome is diagnosed in individuals with a submicroscopic deletion of chromosome 22 detected by a fluorescent in situ hybridization (FISH) test. This testing is also widely available for prenatal diagnosis.

Treatment

Treatment is multidisciplinary involving health care providers from the following professionals as needed: medical genetics, plastic surgery, speech pathology, otolaryngology, audiology, dentistry, cardiology, immunology, child development, child psychiatry, neurology, and general pediatrics. Some patients require the expertise of gastroenterology, orthopedics, urology, hematology, and psychiatry providers.

Prevention

Genetic counseling is available to individuals and couples who have a 50% chance for having a child with 22q11.2 deletion syndrome for discussion of recurrence risk and available options such as prenatal diagnosis.

Patient Resources

Further information about 22q11.2 deletion syndrome can be found through the following organizations. These support groups can be very helpful for affected individuals and their families, and for providing the most up-to-date information.

- Velo-Cardio-Facial Syndrome Educational Foundation
  Phone: 315-464-6590
  Web site: *www.vcfs.org*
- Chromosome Deletion Outreach, Inc.
  P.O. Box 724
  Boca Raton, FL 33429-0724
  Phone: 888-236-6680; 561-391-5098 (family helpline)
  Fax: 561-395-4252
  E-mail: *cdo@att.net*
  Web site: *www.chromosomedisorder.org*

After I had the Factor V Leiden blood test, I went back to talk with the nurse. The testing showed that I did have a change in the gene that would predispose me to blood clots. The nurse recommended further evaluation with a hematologist and advised that, when John and I were considering pregnancy, I have an evaluation with a perinatologist to make a plan for treatment during my pregnancy. She also told me that my other family members, my mother and her brother, probably had the same thing and that I should talk with them about testing. John and I have an appointment together with the geneticist in a few weeks; we plan to go back to talk with the nurse again after we find out more about our family.

Sara

Learning How to Assist Individuals in Making Decisions about Genetic Information

Increasingly, screening, diagnosis, and treatment will involve a genomic component and client decision making about genetic information. Nurses participate as members of multidisciplinary teams in all practice settings and will participate in this process. With additional education and training, nurses will need to provide the counseling and teaching that are integral to client decision making about utilization of genetic information. Provision of care for women and newborns is one area where decision making about genetic testing and treatment is increasing rapidly (see Chapter 4). Expanding roles for nurses in this area of practice include supporting women and families in their decision-making process, being prepared to understand the implications of genetic test results, fostering patient understanding, and assisting with adaptation and integration of new genetic information regardless of the outcome of the testing. The Association of Women's Health and Neonatal Nursing has created a position statement regarding the role of the registered nurse as related to genetic testing that serves as a starting point for basic nursing care in helping women make decisions about prenatal and newborn screening, and prenatal diagnosis *(http://www.awhonn.org/awhonn)*.

Emerging genetic technologies and applied genetics in the areas of gene therapy, stem-cell research, and human cloning have the potential to influence client care and the profession of nursing. In response to the growing debates regarding these therapeutics, the American Nurses Association (ANA) plans to expand the study of the ethical, moral, clinical, psychosocial, cultural, and spiritual implications involved in the care of clients receiving such services (2003). The ANA also wants to ensure that nurses become knowledgeable about genetic and genomic science so that they can appropriately educate, counsel, and support clients *(http://www.nursingworld.org)*. As Mary's story illustrates, nurses will have an important role in helping clients sort through their emotional, ethical, and spiritual beliefs when trying to decide on a path for new genomic treatment options.

Mary's Story

I had a brother who died from Duchenne muscular dystrophy when he was 16 years old. I remember so much about all of the doctors visits, the trips to Shriners, when he got a wheelchair, and of course, all of the Jerry Lewis telethons to raise money for a cure. I wish there were a cure. His illness took such a toll on my parents; they looked older than their years when they were in their 40s. My brother was so sweet and kind and he tried so hard to be up. Needless to say, it was devastating for all of us when he died. Before he died, my parents, on the advice of a doctor, talked with my brother about saving some blood for future genetic testing. They were concerned about me and my future. My brother agreed, so some blood was drawn and banked "for my future."

I went on to college and met my husband while we were in school together. I didn't know when to bring up what had happened with my brother. Finally, when we were getting serious, I told him. I told him I didn't know what it meant for me and for our children should we have them. My brother had died 10 years ago, so I hadn't done much research since then. I went on the Internet and learned that Duchenne muscular dystrophy is an X-linked condition and that I could be a carrier. I learned that there was DNA testing out there and new prenatal tests including something called pre-implantation diagnosis. My fiance and I talked a lot about what we should do. We decided to go visit my nurse practitioner and talk with her first.

My nurse practitioner listened to my concerns and said that it would be a good idea if we went to see a geneticist who could help us find out more about my chance of being a carrier and whether we could use my brother's blood for DNA testing. We asked her about the new pre-implantation diagnosis. She was so very kind to us. She said that there were many more options available today for prenatal diagnosis than in the past, but that this was a highly personal decision. She explained that decision making of this type involved our religious, philosophical, and social beliefs. She told us that, for some couples, it is helpful to talk to their pastor or minister to help sort out the issues. I cannot tell you how much we appreciated her sensitivity and thoughtfulness, and her recommendations. We are still in the process of trying to decide what we will do. Our minister and the geneticist have given us a lot of information and food for thought. My nurse practitioner is always available to us to hear where we are in our decision making.

Mary

Using Genomics Throughout the Health–Disease Continuum

The use of genomic technology and information is expanding rapidly in the area of population-based screening. Genetic advances in the public health arena are increasing opportunities for disease prevention through services and practices that improve disease risk assessment; make earlier, more accurate diagnoses; and enhance medical care and public health policies (French & Moore, 2003). This is especially evident in the area of newborn screening where new laboratory technologies increase the capacity to screen for many more genetic conditions in the newborn *(http://genes-r-us.uthscsa.edu/)*. State and national public health departments are creating genetics–genomics plans to expand genetic screening to adult populations in the area of chronic diseases such as heart disease, cancer, and diabetes *(http://genes-r-us.uthscsa.edu/)*. This means that health professionals will increasingly use genetic tests and family histories to assess risk for disease in individual patients, families, and populations. With this information in hand, public health professionals can recommend preventive measures such as increased frequency of preventive health services (e.g., cancer screening or lipid tests) for persons with a specific genetic variation. They can offer immunizations to populations that are especially susceptible to infectious diseases with genetically improved vaccines. Customized treatments can be designed based on an individual's genetic makeup. Removal of harmful exposures at home and in the workplace and public spaces can be implemented to decrease the potential gene-environment interactions that contribute to common diseases (French & Moore, 2003).

Public health nurses will be central to the process of educating patients and families about new genetic screening and treatments and in coordination of care. However, the number of registered nurses identified as "public health nurse" in the National Sample Survey of Registered Nurses has decreased from 39% in 1980 to just 17.6% in 2000. In the meantime, the demands on this group of nurses and the entire public health structure have continued to increase, including issues of uneven access to and disparities in genomic health for minorities and rural populations (Office of Minority Health, *http://www.omhrc.gov;* Ross, 2001; Weatherall, 2003). The American Nurses Association has called for the "acknowledgment of the critical nature of the public health nurse's role in promoting and protecting the health of individuals, families and communities" (p. 2, *http:/www.nursingworld.org/pressrel/2003/pr0708.htm*). This includes advocacy for technology, training, and funding to health departments to support and enhance the role of public health nurses who will help with the integration of genomics into programs and policies that seek to protect individuals and reduce behavioral, environmental, and other health risks.

Access to genomic technologies and applications is also a significant issue in developing countries. As an example, it is already possible to transfer cost-effective DNA technologies to developing countries that can be used for diagnosing and treating chronic genetic conditions such as thalassemia and communicable diseases such as tuberculosis and HIV/AIDS. Expanding this approach can provide a base whereon more sophisticated techniques can be added as more information about genomics advances for world health become available. Developments of this kind will require a major shift in emphasis in teaching of a more global view of disease and in developing the infrastructure and professional education for the organization and implementation of such programs (Weatherall, 2003).

To achieve this aim, a nursing focus on equity and public health measures must be adopted by the nursing profession. Nurses can take a leading role working with state, federal, and international health agencies to provide guidance to health systems to help nurses and other health professionals make appropriate decisions about referring clients for genetic counseling and services. This role will involve nursing participation in the core public health functions—assessment, policy development, assurance—to ensure equal access to genomic technologies and services for individuals and populations throughout the lifespan. Nurses can participate in assessment when they assess individuals, families, and communities, including the global community's health and well-being, from a genomics perspective. This assessment must also include consideration of the resource needs and recommended surveillance and monitoring strategies. Nursing participation in policy development will encompass consideration of alternatives for the best possible use of shared resources, including equal access to genomics services, which results in quality use of genetic testing. The core function of assurance includes ensuring access to and quality of genomic care, and informing populations about relevant genomic health issues and services (Khoury, Burke, & Thomson, 2000).

Pediatric Nursing Roles

The important role of genetics in pediatric illness is increasingly recognized (McCandless, Brunger, & Cassidy, 2004). In an annual review of pediatric mortality data, nearly one quarter of infant deaths are due to genetically determined disorders (Hoyert et al., 2001). As long ago as 1978, a study of more than 4,000 admissions to a pediatric hospital revealed that just over 50% were for a genetically-determined condition, while almost 1 out of every 20 admissions were due to genetic disorders. Twenty-two percent were due to multifactorial conditions and 27% to developmental or familial disorders (Hall et al., 1978). A recent study of more than 5,000 pediatric admissions at an urban hospital showed that more than two thirds of the admissions were attributable to conditions with a genetic component, confirming Hall's earlier study. Children with these genetically determined conditions also remained in the hospital longer than those without a pre-existing chronic medical condition (McCandless, Brunger, & Cassidy, 2004).

The two studies, although 25 years apart, indicate that in a children's hospital, genetic conditions are common, not rare. Individuals, families, and communities have a strong interest in understanding the implications of this new knowledge and application of genomics to health care for their children and themselves. Nurses practicing in pediatric hospitals or units can expect to be asked more questions about the genetic component to health and illness and thus need to be fluent in the complexities of genetic information and competent in explaining these complexities to anxious parents and families. Nurses are already involved with providing gene-based therapeutics to pediatric patients such as DNA-based factor replacement, for example, in the treatment of hemophilia. Nurses are already explaining genetic tests that are used to determine the specific dose of a medication, for example, TMPT testing to determine dosage of 6-mercaptopurine in children with leukemia (Weinshilboum, 2003). In the near future, it will not be uncommon for nurses to be discussing SNPs and gene-based interventions in the same way that germs, infections, and antibiotics are currently discussed (McCandless, Brunger, & Cassidy, 2004). In the rapidly changing health care arena, pediatric nurses will be applying the important new genomic developments as a routine component of their practice and helping parents, children, and families understand the nuances of medicine in the genomic era. Leslie's story of dealing with her daughter's severe combined immunodeficiency (SCID) provides an example of the expanding roles for pediatric nurses.

Leslie's Story

My son was diagnosed with severe combined immunodeficiency (SCID) (**Box 13-3**), within the first few months after his birth. I knew that this condition ran in my family because my aunt had a son who had died from SCID as an infant. My husband and I knew that there was a risk for us because women are carriers of the gene. I am my parents' only child, so we didn't know whether my mom was a carrier or not. When my son was diagnosed, we were devastated. Our knowledge of SCID was limited to what had happened to my cousin 30 years ago. In the hospital, we stayed with our son in the intensive care unit and worried. The pediatric nurse assigned to us listened to our concerns and shared with us that many things had changed with treatment of children with SCID. She told us that the doctor would be talking with us about a treatment that has been used and has been successful, called bone marrow transplant, to help with our son's immune system. We were encouraged. That night we went on the Internet to find out more. We saw that there was another treatment called gene therapy and we wondered whether this would be better for our son. The next day we came back and told our nurse what we had found. She took the time to explain to us how gene therapy is performed using autologous bone marrow stem cells that are treated with a virus and then the good gene is introduced. She told us that this has been successful, but two young infants recently developed leukemia, so this type of treatment is not used unless bone marrow transplant doesn't work. We were so thankful for her explanations and time because with this knowledge we were able to listen and better understand all that our son's doctor was telling us. Our son has since had a bone marrow transplant and is doing well right now. We will never forget our nurse's kindness and the information she gave to us that helped us to understand and get through such a difficult time.

Leslie

Box 13-3 X-Linked Severe Combined Immunodeficiency (X-SCID)

General Information

X-linked severe combined immunodeficiency (X-SCID) is a combined cellular and humoral immunodeficiency resulting from a lack of T and natural killer lymphocytes and nonfunctional B lymphocytes. Since it is X-linked, males are affected. Most males come to medical attention between 3 and 6 months of age. Infections that usually appear ordinary, such as otitis media, do not respond to the usual treatment. Treatment must be instituted immediately to restore the immune system.

Clinical Symptoms

Affected males appear normal at birth. The universal clinical features during the first year of life include failure to thrive, oral–diaper candidiasis, absent tonsils and lymph nodes, recurrent infections, infections with opportunistic organisms, and persistence of infections despite the usual treatment.

Prognosis

X-SCID can be fatal unless early intervention and treatment to provide a functional immune system are instituted.

Inheritance Pattern

X-SCID is inherited in an X-linked manner. More than one half of affected males have no family history of early deaths in male relatives. A father of an affected male will not have the condition, nor be a carrier. In a family with more than one affected individual, the mother of an affected male is an obligate carrier and has a 50% chance to pass on the gene mutation to a son who would be affected, and a 50% chance to pass the gene mutation on to a daughter who would be a carrier.

Prevalence

The incidence of X-SCID is not yet known. It is estimated that X-SCID occurs in 1 in 50,000 to 1 in 100,000 live births. All ethnic populations are affected with equal frequency.

Relevant Testing

Diagnosis is made by lymphocyte testing and molecular genetic testing. IL2RG is the only gene known to be associated with X-SCID. Sequence analysis is used for diagnosis and detects mutations in more than 99% of affected individuals and is available on a clinical basis. Testing for known family-specific mutations is the best approach for carrier testing. Testing is available for prenatal diagnosis when the mutation in the family is known.

Box 13-3 continued

Treatment

Diagnosis of X-SCID requires emergent treatment to provide a functional immune system. Interim management involves treatment of infection and use of immunoglobulin infusions and antibiotics. Prompt immune reconstitution is required for survival of children with X-SCID. Bone marrow transplant (BMT) is the standard means of immune reconstitution. It is estimated that over 90% of infants with X-SCID can be successfully treated with BMT. The best timing for BMT is immediately after birth, because young infants are less likely to have had serious infections or failure to thrive than older infants. Gene therapy performed using autologous bone marrow stem–progenitor cells retrovirally transduced with a therapeutic gene has also been successful in reconstituting the immune system in patients with X-SCID; however, the youngest two of the first 10 infants treated in a French study have developed leukemia due to retroviral insertional mutagenesis. Thus, gene therapy is only a consideration for those individuals who are not candidates for BMT or have failed BMT.

Patient Resources

Immune Deficiency Foundation
40 Chesapeake Ave
Suite 308
Towson, MD 21204
Phone: 410-321-6647
FAX: 410-321-9165
E-mail: *idf@primaryimmune.org*
www.primaryimmune.org

Jeffrey Modell Foundation
747 Third Avenue, 34A
New York, NY 10017
Phone: 1-800-533-3844; 212-819-0200
Fax: 212-764-4180
E-mail: *info@jmfworld.org*
www.info4pi.org

Chronic Disease: Cancer and Nursing Roles

Genomic discoveries have had an immense impact on cancer. Genetic information is the basis for understanding the biology of cancer and is now routinely used to characterize malignancies, develop new therapeutic interventions, and identify individuals and families at risk to develop cancer. In short, genetic information and genomics applies throughout the cancer care continuum (Tranin, Masny, & Jenkins,

2003). The ONS has responded to this revolution by making a commitment to prepare its members to meet the challenges of genomic medicine (ONS, 1997). The commitment includes publication of two position statements that outline the role of the oncology nurse in cancer genetic counseling and in cancer predisposition genetic testing and risk assessment (ONS, 1997). These statements provide guidance to all oncology nurses, helping them respond by informing patients, families, and the public about the implications of these developments for cancer prevention, risk reduction, early detection, and intervention. A specific role for the advanced practice oncology nurse with genetic expertise is outlined encompassing pre- and postgenetic test counseling and follow-up. The publication of a genetics text specifically for oncology nurses is another ONS-sponsored initiative that supports and prepares oncology nurses to incorporate genomics into daily practice (Tranin, Masny, & Jenkins, 2003). The ONS' initiatives and commitment to progressing oncology nursing roles serve as an example for other nursing societies nationally and internationally. John's story about his battle with chronic myeloid leukemia (CML) (**Box 13-4**) and the miracle of his remission following Gleevec© therapy illustrates the important role of the oncology nurse in education, support, and follow-up.

John's Story

My diagnosis of CML (chronic myeloid leukemia) couldn't have come at a worse time, right after my daughter announced her engagement. What should have been a happy time for us all became a nightmare. I learned early on that CML is generally a fatal disease. I was so depressed sitting there in the hospital waiting to hear what would be next. My oncology nurse came in to get some more admission information before more testing and sensed my depression. She asked me what my doctor had told me so far. I said I remembered he said something about testing me for some Philadelphia chromosome thing and then we would determine my treatment. My nurse stopped the interview at that point and explained to me what the Philadelphia chromosome is, in language I could understand. She explained how all cancer involves genetic changes and that CML was the first cancer to be linked to a genetic abnormality—the Philadelphia chromosome—named for the city where it was discovered. She described how this Philadelphia chromosome happens when there is an exchange of genetic material called a translocation between portions of chromosome numbers 9 and 22. The testing that I was going to have was to see if I had the Philadelphia chromosome, which apparently most patients with CML do have. If I did have it, then I would be a candidate for a new treatment called Gleevec© that can help me with improved treatment of my symptoms and, hopefully, survival. I cannot tell you how encouraging this news was for me. I know that my doctor had reviewed a lot of information with me, but I was overwhelmed. My nurse took the

Box 13-4 Chronic Myeloid Leukemia

General Information

Chronic myeloid leukemia (CML) is a malignancy of the myeloid lineage—monocytes, macrophages, dendritic cells, and granulocytes. The granulocytes and this series of cells in all stages of maturity become massively expanded in the bone marrow and peripheral blood in CML. Leukemia, such as CML, generally results from a genetic abnormality such as a point mutation within a gene, loss or duplication of genetic material, or inappropriate recombination of genetic material (chromosomal translocation). These events may cause overexpression of a normal cellular gene, proto-oncogene, that promotes cellular proliferation or survival or a new formation of an abnormal gene (oncogene), which leads to malignant transformation of normal cells.

Clinical Symptoms

After one half of individuals with CML are asymptomatic when they are diagnosed, the CML being discovered during routine laboratory blood testing. Common symptoms in the chronic phase of CML include fatigue, weight loss, abdominal fullness, bleeding, and sweating. A palpable spleen is present in more than 50% of patients. As CML progresses, bone pain and pain from splenic infarction may occur.

Prognosis

CML progresses through three stages that are characterized by worsening clinical features and laboratory findings. Most patients (85%) are diagnosed when they are in the chronic phase of CML. Symptoms may initially be mild and worsen as the chronic phase progresses. The chronic phase of CML is the longest, with an average duration of 5 to 6 years. Optimal therapeutic outcome is achieved during the first year after diagnosis. In the second or accelerated phase, symptoms are generally worse than in the chronic phase and include fever and bone pain. The median duration of this phase is 6 to 9 months. The final stage or blast phase represents the most advanced and terminal stage of CML with a median survival of 3 to 6 months.

Genetic Cause of CML

CML was the first cancer whose cause was linked to a genetic abnormality—the Philadelphia (Ph) chromosome. The Philadelphia chromosome was named for the city where it was discovered. The Ph chromosome is an abnormal and shortened chromosome caused by a chromosomal rearrangement (translocation) of genetic material between the long arms of chromosomes 9 and 22. This is designated as follows: t(9:22)(q34;q11.2). The Ph chromosome is present in approximately 95% of persons who have CML and is considered the genetic hallmark of this disease. The

Box 13-4 continued

Ph chromosome involves a breakpoint cluster region (BCR) of chromosome 9 and the gene for the ABL tyrosine kinase on chromosome 9. The combination of these two genetic segments causes an abnormal hybrid BCR-ABL gene and a continuously activated BCR-ABL fusion protein.

The Ph chromosome is also seen in other forms of leukemia such as childhood and adult cases of acute lymphoblastic leukemia.

Prevalence

Globally, CML has an incidence of 1 or 2 cases per 100,000 population and is the cause of 15% to 20% of all adult leukemia. Approximately 30% of individuals with CML are older than 60 years of age. The average age at diagnosis is 53 years.

Relevant Testing

The Ph chromosome can be detected by standard chromosome analysis in most patients. In patients who are Ph negative, molecular techniques such as fluorescence in situ hybridization (FISH) can be used to detect the expression of the BCR-ABL oncogene.

Treatment

Therapeutic options for CML include stem cell transplantation, interferon-α, and chemotherapy. Recently imatinib mesylate—Gleevec©—has become commercially available for treatment of CML. Gleevec© is an oral agent that selectively blocks cellular proliferation and induces apoptosis in the Ph chromosome. Gleevec© is indicated for the treatment of newly diagnosed adult patients with Ph-positive CML. Recent clinical data show that when given as first-line therapy in de novo Ph-positive CML, Gleevec© achieves a complete cytogenetic response in 54% of patients in early chronic phase and delays disease progression to advanced phases. The discovery of Gleevec© has ushered in a new era of targeted drug therapy and set a new standard for drug development in oncology.

time to explain what would be happening and why I was having the testing in a way that made sense to me and gave me hope. Since that day, I made it through treatment and was given Gleevec© because they found that I did have the Philadelphia chromosome. I made it to my daughter's wedding 2 years later, and I am now expecting my first grandchild.

John

Genomics Nurse Coordinator Role

Case management, counseling, and informational support associated with the many gene-based tests and therapeutics are expected to be time-consuming and complex. Lea and Monsen envision a new role for midlevel nurses—Genomics Nurse Case Coordinator—a role that is expected to become increasingly important as the demand for help from individuals and families who qualify for testing and treatment increases (Lea & Monsen, 2003). Nurses with counseling and support abilities can expand their role in supporting individuals and families in managing their own genomic health. Genomics Nurse Case Coordinators can participate as interdisciplinary team members and conduct family history risk assessments; participate in the informed consent process, providing information about risks, benefits, and limitations of genetic testing and therapeutics; prepare clients for further genetic counseling and evaluation; coordinate care involving gene-based diagnosis and therapeutics; and provide ongoing case management and support. There are several clinical settings where Genomics Nurse Case Coordinators might function. One important area is the primary care setting where risk assessment and referral would be conducted. Specialized care settings such as prenatal, cancer, or psychiatric are another venue where the Genomics Nurse Case Coordinator would receive referrals, review past medical records, and conduct health status assessments. Other settings where the Genomics Nurse Case Coordinator could function as a midlevel coordinator include biotech and commercial laboratories conducting genetic testing and pharmacogenomics testing and insurance companies determining reimbursement for genetic testing, treatments, and services (Lea & Monsen, 2003).

Educating Nurses to Face the Challenge

Collective action is required from nurses worldwide to ensure that all individuals, families, and populations have access to the fruits of genome discoveries. This includes efforts to improve education and training of the 2.7 million nurses nationwide and nurses worldwide. Although nursing as a profession is recognizing the importance of genetics and genomics to nursing practice (American Association of Colleges of Nursing, 1998; Lea, 2002), multiple surveys have demonstrated that nurses receive limited genetics and genomics in their basic nursing programs (Lea, Feetham, & Monsen, 2002; Mertens, Hendrix, & Morris, 1984; Monsen, 1984; Williams, 1985). In 2002, at an ANA convention, more than half of the nurse attendees surveyed said that they did not feel competent enough to provide care that involved genetics, but did agree that genomic science will affect nursing practice (Jenkins & Collins, 2003). Nursing education in genomics must become a top priority for all nursing communities. Training the nursing workforce in genomics must become a top priority in health care, including public health.

National leaders in genetics and nursing urge all nurses to become competent and fluent in genetics (HRSA, 2000; Jenkins & Collins, 2003). This requires diverse approaches and significant changes in the nature and process of nursing education at all levels of training. One way to prepare to meet this challenge is for nurses to become actively engaged in attempting to understand the changes and implications of genomics medicine. Such engagement must take the form of discussion, debate, reading, and thoughtful consideration throughout the nursing community. This in itself will serve as an educational process essential to integrating genomics into daily nursing practice (HRSA, 2000).

ISONG has engaged in such dialogue and discussion since its incorporation in 1988. ISONG recognizes that the hallmark of nursing education is the acquisition of knowledge and skills needed to continuously integrate and apply scientific and technologic advances into the art and science of professional nursing practice. In support of this recognition, ISONG, in collaboration with the ANA, developed a *Statement on the Scope and Standards of Genetics Clinical Nursing Practice* (1998). This document identifies and describes the integration and applications of genetics at the basic and advanced levels of nursing practice. ISONG has also created and adopted multiple position papers that address the multiplicity of issues that confront nurses practicing in today's genomic health care, including informed decision making and consent, privacy, and confidentiality of genetic information, genetic counseling for vulnerable populations, and access to genomic health care *(http://www.isong.org)*.

ISONG members have also taken a leading role in educating and supporting all nurses in the integration and application of genomic knowledge and science. Programs established by ISONG members, now available to nurses worldwide, include the Summer Genetics Institute, based at the National Institute of Nursing Research (Tinkle & Nichols, 2001); the Genetics Program for Nursing Faculty and the Web-Based Genetics Institute, led by Cindy Prows at the University of Cincinnati (2004) *(http://www.cincinnatichildrens.org/ed/clinical/gpnf/default.htm)*; and the Practice-Based Genetics Curriculum for Nurse Educators (Lea & Thomas-Lawson, 2001).

A genomics nursing educational agenda for the future is proposed that includes the development of new models for health care provider and public education (Wilkinson & Targonski, 2003). This model includes an interdisciplinary focus that has its roots, in part, in national nursing efforts during the 1990s. These efforts led to the creation of a basic genetics core curriculum for nurses and the development of the National Coalition for Health Professional Development in Genetics (NCHPEG) recommendations for core genetic competencies for all health professionals (NCHPEG Working Group, 2001). The NCHPEG competencies are broad and inclusive and can serve as a framework for global nursing education (Lea, 2003). The NCHPEG competencies can be used as the foundation for educating all nurses in genomics and health care to (1) recognize when genetic factors are or could be playing a role in health; (2) use family history, genetic screening, and diagnostic tests effectively; (3) communicate complex genomic concepts to patients, families, and the public; and (4) participate in gene-based treatments and therapeutics. (See addendum for the full NCHPEG document.) Using the competencies to ground nursing education models will include development of educational and decision-support systems that provide easy access to reliable information and guidelines (Wilkinson & Targonski, 2003).

Research Directions

Khoury et al. have noted that "the rate of new gene discoveries seems to be outpacing the ability to use genetic information to benefit health" (Khoury, Beskow, & Gwinn, 2001). These leaders in genomics advocate for population studies in epidemiology, health services research, policy, and communication sciences to fully integrate genomics into clinical practice. In 2004, ISONG held an annual conference dedicated to the "assimilation and synthesis of the state of the science for genetic nursing and the establishment of a research agenda that is strongly focused on the delivery of and access to genetic health care services for or by individuals, families, groups, and populations" (Jones, 2003 SACGHS testimony). This research agenda supports nursing research in genomics in multiple areas, including epidemiologic, health services, communication of genomic information, pharmacogenomics, and nursing workforce research.

Nurses can participate in population-based research to help determine which environmental factors and lifestyle factors are the most influential in disease development, and for which specific genotypes they are significant. Knowledge of the prevalence of specific genetic variants within populations and how the diseases are expressed can help improve screening and prevention techniques. Nurses can also have a central role in health services research by participating in the creation of better strategies for using genetic tests, and in health promotion by elucidating factors that may result in better compliance.

Communication of complex information to the public requires innovative strategies in which nurses can take a leading research role. Creative approaches, including Web sites that can help with family history analysis, information about genetic tests, and therapeutics including clinical utility and validity, can be developed with nursing input. Family history analysis and discussion can have a significant impact on families and societies, depending on the definition of kinship and views of family and inheritance. The focus on family history as a bridge for communicating complex genomic information has the potential to stigmatize and "medicalize" kinships (Khoury, 2003; Melzer & Zimmern, 2002). Nurses can collaborate in research on the impact of genomics, including family history, on families and communities. Nursing research in perceived risk and special population groups regarding ethical issues and fears about genetic information can help determine the most sensitive and effective ways to use family history in clinical practice.

Pharmacogenomic research is yet another direction for nursing researchers. Nurses are already participating in laboratory-based research and in developing outcome measures and disease markers that can help with the design of appropriate pharmacogenomic interventions (Frazier et al., in press). Increasing nursing research in this area contributes to understanding disease mechanisms and predictive markers, and monitoring of outcomes of targeting therapeutics to the individual.

A national effort has been underway for the past 3 years to identify and describe the issues confronting the development of an appropriately educated genetics workforce (Lea, 2004). The profession of nursing was the focus of the 2003–2004 effort to assess the roles of nurse specialists and generalists. ISONG was actively involved in this effort, with many members sitting on the advisory board of this federally funded project. Results of the survey are being used to guide development of service responses to new genomics and as the initial phase in assessing nursing activities and roles in genomics at all levels of practice (Lea, 2004). Such research efforts will help to better inform current and future nursing roles in genomics health care. This research will allow nurses to participate in policy development with regard to workforce issues and access to genomic technologies for all populations.

Moving Forward

Nursing has made many strides in "riding the new wave of genetic nursing" (ISONG, 2003). The challenge is for nurses to continue dialogue, exploration, education, and research efforts to progress the role of nursing to meet the demands of the genomic era. Nurses must continue to strive to integrate and apply genomics science to clinical applications so that they can truly practice the art and science of nursing care in today's health care. Nurses must also make efforts to collaborate with all health professionals and communities worldwide so that the fruits of genomic discoveries will be made available to all populations. Using the NCHPEG competencies in genetics for all health professionals to create new and diverse educational strategies to educate nurses worldwide is a beginning step. "Nursing cannot be viewed in isolation; instead the profession must be viewed within the social, cultural, economic, and political context of the health care system and wider society" (Garfield et al., 2003). In gaining an understanding of other nurses' and populations' realities, genomic health care challenges and the challenges these present for nursing can be more holistically and globally addressed now and in the future.

Chapter Activities

1. Nursing roles in family risk assessment, analysis, client education, and referral are expanding as new genome information is applicable in clinical practice. Read the following case history:

Mark is a 2-year-old boy who has speech delay and who is not yet walking. He is currently in foster care. It is known that Mark's biological mother has a history of learning issues and that he has a maternal uncle (his mother's brother) who has a history of significant learning issues. This uncle is currently in a group home. Mark has a maternal half sister who is 5 years old and who has learning issues.

Create Mark's family pedigree. Discuss your assessment and analysis of Mark's family history. What steps would you take to further evaluate the cause of Mark's developmental delays?

2. New roles for nurses are evolving rapidly in the area of prenatal care and services. Read the following case. Identify and discuss potential nursing roles in caring for Shirley.

Shirley is an 18-year-old primagravida. At her first clinic visit, the nurse provides Shirley with information about the maternal serum screening test for neural tube defects, Down syndrome, and cystic fibrosis carrier screening. Shirley tells the nurse that she plans on having the maternal serum screening test. She says that she does not want to have the cystic fibrosis screening test as "my first cousin has cystic fibrosis, and I would never want to be in a position to make a decision about the pregnancy because of cystic fibrosis." The maternal serum screening test reveals that Shirley has an increased risk for neural tube defect in her baby. A follow-up ultrasound evaluation shows that the baby has a lumbar spina bifida. Shirley expresses shock and feelings of guilt, saying "I never expected this to happen. Do you think it is because I was so sick in the beginning of my pregnancy and couldn't eat right that this happened?"

Expanding roles for nurses in prenatal care include supporting women and families in their decision-making process, being prepared to understand the implications of genetic test results, fostering patient understanding, and assisting with adaptation and integration of new genetic information regardless of the outcome of the testing. What role(s) would you feel most comfortable in performing? Where may you need additional training to feel competent in your practice?

References

American Association of Colleges of Nursing. (1998). *The essentials of baccalaureate nursing education for professional nursing practice.* Washington, DC: American Association of Colleges of Nursing.

American Nurses Association. (July 8, 2003). *Nurse leaders take action on a range of workplace and patient care issues at ANA's house of delegates.* At http://www.nursingworld.org/pressrel/203/pr0708.htm.

Association of Women's Health and Neonatal Nursing. (2004). The role of the registered nurse as related to genetic testing. *Clinical Position Statement.* At http://www.awhonn.org/awhonn/?pg=873-6230-6990-4730-5400-7430.

Collins, F. S., Green, E. D., Guttmacher, A. E., & Guyer, A. (2003). A vision for the future of genomics research: A blueprint for the genomics era. *Nature, 422,* 835–847.

Expert panel on genetics and nursing. (2000). Report: Implications for education and practice. Maryland: Health Resources and Services Administration.

Frazier, et al. (in press) article on cardiovascular genomics.

French, M. E., & Moore, J. B. (2003). *Harnessing genetics to prevent disease and promote health.* Washington, DC: Partnership for Prevention.

Garfield, R., Dresden, E., & Boyle, J. S. (2003). Health care in Iraq. *Nursing Outlook.* July/August, 171–176.

Goldstein, D., Tate, S., & Sisodiya, S. (2004). Pharmacogenetics goes genomic. *Nature Reviews Genetics, 5*(1), 76.

Grady, C., & Collins, F. S. (2003). Genetics and nursing science: Realizing the potential. *Nursing Research, 52*(2), 69.

Guttmacher, A. E., Jenkins, J., & Uhlmann, W. R. (2001). Genomic medicine: Who will practice it? A call to open arms. *American Journal of Medical Genetics, 106*(3), 216–222.

Hall, J. G., Powers, E. K., McIlvaine, R. T., & Ean, V. H. (1978). The frequency and financial burden of genetic disease in a pediatric hospital. *American Journal of Medical Genetics, 1,* 417–436.

Hoyert, D. L., Freedman, M. A., Strobino, D. M., & Guyer, B. (2001). Annual summary of vital statistics: 2000. *Pediatrics, 108,* 1241–1255.

HRSA. (2000). *Report of the Expert Panel on Genetics and Nursing: Implications for Education and Practice.* Washington, D. C.: HRSA.

International Council of Nurses. (2000). *The ICN code of ethics for nurses.* Geneva, Switzerland: International Council of Nurses.

International Human Genome Sequencing Consortium. (2001). *Nature, 409,* 860–921.

International Society of Nurses in Genetics, Inc. (1998). *Statement on the scope and standards of genetics clinical nursing practice.* Washington, DC: American Nurses Association. 1–37.

International Society of Nurses in Genetics, Inc. (2003). 16th Annual International Conference: *Riding the New Wave of Genetic Nursing.* November 1–4, 2003, Los Angeles, CA.

Jenkins, J., & Collins, F. S. (2003). Are you genetically literate? *American Journal of Nursing, 103*(4), 13.

Jones, Shirley (October 22–23, 2003). *Public comment section. International Society of Nurses in Genetics.* At the Secretary's Advisory Committee on Genetics, Health and Society. At http://www4.od.nih.gov/oba/SACGHS/meetings/October2003/SACGHS_Oct_2003.htm.

Khoury, M. J. (2003). Genetics and genomics in practice: The continuum from genetic disease to genetic information in health and disease. *Genetics in Medicine, 5*(4), 261–268.
Khoury, M. J., Beskow, L., & Gwinn, M. L. (2001). Translation of genomic research into health care [letter]. *JAMA, 285,* 2447–2448.
Khoury, M. J., Burke, W., & Thomson, E. J. (2000). Chapter 1: Genetics in public health: A framework for the integration of human genetics into public health. In *Genetics and Public Health in the 21st Century: Using Genetic Information to Improve Health and Prevent Disease.* Khoury, M. J., Burke, W., & Thomson, E. J. (Eds.). New York, NY: Oxford University Press.
Lea, D. H. (2002). Position statement: Integrating genetics into baccalaureate and advanced nursing education. *Nursing Outlook, 50,* 167.
Lea, D. H. (2003). Public health genetics—the nursing role of the future? Presented at the 16th Annual International Conference, International Society of Nurses in Genetics. Plenary Session: International Genomic Nursing: A Kaleidoscope of Practice. November 3, 2003. Los Angeles, CA.
Lea, D. H. (2004). Genetics Service Delivery Models in the United States and Emerging Nursing Roles. Presented at the 17th Annual International Conference of the International Society of Nurses in Genetics. October 24, 2004, Toronto, Canada.
Lea, D. H., Feetham, S., & Monsen, R. B. (2002). Genomic-based health care in nursing: A bi-directional approach to bringing genetics into nursing's body of knowledge. *Journal of Professional Nursing, 18*(3), 120–129.
Lea, D. H., & Monsen, R. B. (2003). Preparing nurses for a 21st century role in genomics-based health care. *Nursing Education Perspectives, 24*(2), 75–80.
Lea, D. H., & Thomas-Lawson, M. (2001). Bringing genetics into the classroom: a practice-based approach. *Nursing and Health Care Perspectives, 2293,* 146–151.
McCandless, S. E., Brunger, J. W., & Cassidy, S. B. (2004). The burden of genetic disease on inpatient care in a children's hospital. *American Journal of Human Genetics, 74,* 121–127.
Melzer, D., & Zimmern, R. (2002). Genetics and medicalization. *British Medical Journal, 324,* 863–864.
Mertens, T. R., Hendrix, J. R., & Morris, M. M. (1984). Nursing educators: perceptions of the curricular role of human genetics/bioethics. *Journal of Nursing Education, 23*(3), 98–104.
Monsen, R. B. (1984). Genetics in basic nursing program curricula: A national survey. *Maternal-Child Nursing Journal, 13*(3), 177–185.
NCHPEG Working Group of the National Coalition for Health Professional Education in Genetics. (2001). Recommendations of Core Competencies in Genetics Essential for All Health Professionals. *Genetics in Medicine 3*(2), 155–158.
Office of Minority Health. http://www.omhrc.gov (Accessed 5/3/04).
Oncology Nursing Society. (1997). *1996–1997 annual report: Expanding our horizons.* Pittsburgh, PA: Author.
Oncology Nursing Society. (2000a). Cancer predisposition genetic testing and risk assessment counseling. *ONF, 27*(9), 2.
Oncology Nursing Society. (2000b). The role of the oncology nurse in cancer genetic counseling. *ONF, 27*(9), 1.
Prows, C. (2004). The Genetics Program for Nursing Faculty and the Web-Based Genetics Institute, University of Cincinnati. At (http://www.cincinnatichildrens.org/ed/clinical/gpnf/default.htm.

Ross, H. (2001). *Office of minority health publishes final standards for cultural and linguistic competence. Closing the gap.* Washington, DC: Office of Minority Health, U.S. Department of Health and Human Services.
Tinkle, M., & Nichols (2001). *Summer Genetics Institute: Report of external curriculum review panel.* National Institute of Nursing Research: Bethesda, MD. 1–9.
Tranin, A., Masny, A., & Jenkins, J. (2003). *Genetics in Oncology Practice: Cancer Risk Assessment.* Pittsburgh, PA: Oncology Nursing Society.
Weatherall, D. J. (2003). Genomics and global health: time for a reappraisal. *Science, 302*(24), 597–601.
Weinshilboum, R. (2003). Inheritance and drug response. *New England Journal of Medicine, 348*(6), 529–537.
Wilkinson, J. M., & Targonski, P. V. (2003). Health promotion in a changing world: preparing for the genomics revolution. *American Journal of Health Promotion, 18*(2), 157–161.
Williams, J. K. (1985). *Surveys of nursing knowledge and roles in genetics. Education in genetics II: Nurses and social workers.* Washington, DC: National Maternal and Child Health Clearing House. 81–92.
Williams, J. K., & Lea, D. H. (2003). *Genetic Issues for Perinatal Nurses* (2nd ed). R. R. Wieczorek (Ed). White Plains, New York: March of Dimes Birth Defects Foundation.
World Health Organization. (1997). *Proposed international guidelines on ethical issues in medical genetics and genetic services.* Geneva, 15–16, 1997. At http://www.who.int/ncd/hgn/hgnethic.html.

Chapter 13 Interdisciplinary Commentary

Summary of the Contribution of the Credentialing Process to Assurance of Quality Nursing Care That Integrates Genetics/Genomics

Rita Black Monsen, DSN, MPH, RN
Interim Executive Director and Immediate Past President
Genetic Nursing Credentialing Commission

In the past 30 years, certification has become the standard for recognition of competence in the health care industry. Licensure, while managed by governmental bodies and dependent upon exam-based validation of knowledge and skills of the provider, is designed around the minimum requirements for safety of clinical practice. Certification is included in the concept of credentialing, the process of verifying trust and belief in a person or object that has met a standard or received a testimonial of confidence in quality (Schoon & Smith, 2000). Certification denotes a formal attestation of the quality of the work of a person or the characteristics of an object (*Webster's New Collegiate Dictionary,* 1980). Certification is optional for licensed nurses, but has gained such respect over the past quarter century as to be an expected qualification for advanced practice nursing and many lesser roles in health care today.

Certification in Genetics

In nursing, certification is available to every qualified nurse clinician, with over 150,000 nurses in over 40 specialty and advanced practice areas certified by the American Nurses Credentialing Center (American Nurses Association, 2003) alone (many nursing specialty organizations offer certification as well). ISONG created the Genetic Nursing Credentialing Commission (GNCC) in 1999. This development offered a new opportunity for nurses to demonstrate their abilities to provide comprehensive care based upon the ISONG Scope and Standards of Clinical Genetics Nursing Practice published by ISONG and the ANA (1998). The GNCC grants the APNG, the Advanced Practice Nurse in Genetics credential to qualified clinicians with a minimum of the Masters in Nursing (or equivalent) and the GCN, the Genetics Clinical Nurse credential to those with a minimum of the Baccalaureate in Nursing (or equivalent). Moreover, the achievement of each of these credentials is based upon a portfolio of verifiable evidence that includes (in part) education in basic and clinical genetics, supervised clinical practice, a 50-case log, four case studies, and demonstrated teaching efforts in the areas of genetics and/or genomics by the applicant. The APNG is designed for the nurse who provides genetic counseling, exhibits leadership and research activities, and practices independently and collaboratively with interdisciplinary teams of clinicians in specialty health care settings. The GCN is designed for the nurse who provides supportive counseling about genetic health concerns, uses research findings in practice, and functions collaboratively with interdisciplinary teams of clinicians in a variety of health care settings.

It is likely that the role of the Genomic Nurse Case Coordinator described in this chapter would most appropriately be assumed by nurses in a variety of health care settings and would lead the nurse to apply the GCN credential. While the Genomic Nurse Case Coordinator is not expected to be knowledgeable and prepared to provide the genetic specialty services in a tertiary care agency (as would the APNG), he or she would use a foundational knowledge base of genetics and genomics to assist a wide variety of patients and families who would benefit from evaluation to specialist providers. And, he or she would provide support, follow-up care, and additional resources that are indicated when patients and families require services for genetic concerns.

Title and Rights of the APNG and GCN

The APNG and GCN certifications entail certain rights for nurses who use these letters after their name, degree(s), and license(s). These certifications attest to the nurse's knowledge, attitudes, and skills that are closely akin to the Core Competencies supported by the National Coalition for Health Professional Education in Genetics. In addition, they affirm the credentialed nurse's abilities among his or her colleagues in genetic health care and other settings. Indeed, the APNG credentialed nurses are included in the prestigious clearinghouse for clinical genetics, genetics professionals, and laboratory resources. GeneTests *(http://www.genetests.org)* Licensure by state boards of nursing (or their equivalent) and reimbursement by third-party payers (or their equivalent) for advanced practice nurse services are governed by laws and statutes in effect in the jurisdiction(s) of clinical practice. Most state boards of nursing require certification by bodies such as the American Nurses Credentialing Center for advanced practice nurse licensure. As of today, the credentials offered by ISONG and the GNCC are not recognized for licensure in most areas of the United States. In California, the governing board for nurse licensure does grant recognition to nurses with the APNG as clinical nurse specialists.

Why Certification?

The strengths and benefits of the APNG and the GCN credentials rest upon evidence of competence and clinical expertise based upon the Scope and Standards of Genetics Clinical Nursing Practice widely accepted by the profession and by all of the disciplines in health care today. They provide recognition by professional colleagues and consumers of the ability to provide quality genetic health care. And lastly, they instill pride in accomplishment of advanced practice knowledge and skills in genetic health care or generalist genetic practice knowledge and skills that every qualified nurse can attain.

References

American Nurses Association. (2003). American Nurses Credentialing Center—Certified nursing excellence [Online]. Available at http://nursingworld.org/ancc/indside.html. Accessed 04-15-2004.

International Society of Nurses in Genetics, Inc. & American Nurses Association. (1998). *Statement on the scope and standards of genetics clinical nursing practice*. Washington, DC: American Nurses Publishing.

Schoon, C. G., & Smith, I. L. (2000). *The Licensure and Certification Mission: Legal, Social, and Political Foundations.* New York: Forbes.

——. (1980). *Webster's New Collegiate Dictionary.* Springfield, MA: G. & C. Merriam.

CHAPTER 14
Genomics Resources for Nurses

Throughout this book, the reader has learned from client stories of the increasing relevance of genomics—to biology, to health, and to society in general—and the associated issues and concerns inherent in genomics information. All nurses must be prepared with knowledge and skills to provide quality genomics-based care to individuals, families, and communities. To do this, nurses must know of the resources available to assist clients seeking genetic information or services, including the types of genetics professionals available and their diverse responsibilities and areas of expertise. Nurses must also be able to effectively use new information technologies to obtain current information about genetics. This chapter provides a sampling of resources for nurses and the clients that they care for; these resources will support the development of genomics knowledge and skills needed to practice in today's health care environment.

The Nursing Workforce

In the United States, the nursing workforce consists of over 2.2 million nurses currently practicing as registered nurses. Of these 2.2 million practicing nurses, over two thirds work in inpatient hospital settings, with the vast majority of nurses involved in direct patient care. In the inpatient setting, the nurse is the health care professional who spends the largest amount of time in direct contact with the patient. The nurse is often the health care professional who provides the patient with education about the nature of a newly diagnosed chronic condition; who dispenses and coordinates medications and other interventions to treat the condition; who answers the patient's questions about the meaning of the illness for themselves and family members; and who deals with the entire spectrum of the human response to health and illness (Lewis, 2003 Testimony SACGT *http://www4.od.nih.gov/oba/,* March 1, 2004).

The National Genetics and Health Workforce Research Center has conducted a 3-year national study to assess the provision of genetics-related services in various practice settings in four metropolitan areas across the country. Nurses practicing in genetics specialty and primary care settings are a significant workforce included in the study. The project, based at the University of Maryland, Baltimore, is funded by the Health Resources and Services Administration (HRSA) and the ELSI (Ethical, Legal and Social Implications) Program of the National Human Genome Research Institute (Lea, 2003).

Nurses and Genomics Resources

Nurses are present in all communities and serve as a vital link between individuals, families, and the medical care system. Nurses are often the first health professionals to whom community residents turn with health concerns or questions. Thus, nurses serve an important role in improving the public's knowledge and literacy in genomic health care.

Nurses, by virtue of their training, have a holistic approach to health care that takes into account biological, physical, environmental, social, cultural, and spiritual aspects of each individual. The nursing workforce holds great potential for caring for individuals, families, and communities as medical practice moves toward realizing the benefits of genomic advances in prevention and disease management, including individualized medicine. Specifically, nurses have much to contribute in educating patients, families, other health care professionals, and the public about new genomic health care; advocating for safe and fair use of genomic technologies; and managing and coordinating personalized health care throughout the life span.

Professional Nursing Societies: Genetics Resources for Nurses

Professional nursing societies provide education, support, and guidance to nurses in all areas of nursing practice. Increasingly, professional nursing societies are embracing new genomic applications to education and clinical practice such as targeted genomic educational programs, position statements, and genomics nursing role delineations and credentials. The following resources are available for all nurses and serve as a foundation for nursing professional growth and development in genomics-based health care.

International Society of Nurses in Genetics, Inc.

http://www.isong.org
The International Society of Nurses in Genetics (ISONG) is an international nursing specialty organization dedicated to fostering the scientific and professional growth of nurses in human genetics. ISONG members represent the United States,

Canada, England, Japan, Israel, Korea, Ireland, New Zealand, and Australia, and practice in diverse health care settings including reproductive, prenatal, pediatric, and adult. ISONG provides a forum for education and support for nurses providing genetic health care and promotes the integration of the nursing process into the delivery of genetic health care services. Incorporating the principles of human genetics into all levels of nursing education, promoting the development of standards of practice for nurses in human genetics, and advancing nursing research in human genetics are major ISONG goals.

Documents created by the International Society of Nurses in Genetics *Scope and Standards of Genetics Clinical Nursing Practice* (published in 1998 and currently being revised), was developed, written, and published in conjunction with the American Nurses Association. This document delineates competencies expected for nurses practicing at the basic level, as well as enhanced competencies for the advanced practice nurse.

Position Statements (available at http://www.isong.org)

- Informed Decision-Making and Consent: The Role of Nursing (approved September 30, 2000)
- Privacy and Confidentiality of Genetic Information: The Role of the Nurse (approved October 9, 2001)
- Genetic Counseling for Vulnerable Populations: The Role of Nursing (approved October 10, 2002)
- Access to Genomic Healthcare: The Role of the Nurse (approved September 9, 2003).

Oncology Nursing Society

http://www.ons.org
The Oncology Nursing Society (ONS) is a professional organization of more than 30,000 registered nurses and other health care providers dedicated to excellence in patient care, education, research, and administration in oncology nursing. It is also the largest professional oncology association in the world.

Position Statements (available at http://www.ons.org)

- Cancer Predisposition Genetic Testing and Risk Assessment Counseling
- The Role of the Oncology Nurse in Cancer Genetic Counseling

Cancer Genetics Resources

- ONS Genetics Short Course for Cancer Nurses
- ONS Genetics and Cancer Care Tool Kit (available at http://www.ons.org)
- Internet Resources in Genetics for Cancer Nurses (includes evidence-based practice resource center available at http://www.ons.org)

- Tranin, A., Masny, A., & Jenkins, J. (2003). *Genetics in Oncology Practice.* PA: Oncology Nursing Press. Awarded "First Place Winner" American Medical Writers Association, 2003.

Association of Women's Health and Neonatal Nurses

http://www.awhonn.org
AWHONN is the leading professional association for nurses who specialize in the care of women and newborns. Members include neonatal nurses, APRNs (Advanced Practice Registered Nurses), women's health nurses, OB/GYN and labor and delivery nurses, nurse scientists, nurse executives and managers, childbirth educators, and nurse practitioners.

Position Statement

- The Role of the Registered Nurse as Related to Genetic Testing (available on AWHONN Web site)

American Nurses Association

http://www.ana.org
On-Line Journal of Issues in Nursing: *The Genetic Revolution: What, Why, How?* (available at http://nursingworld.org/ojin/admin/topics.htm)

Sigma Theta Tau International: Honor Society of Nursing

Journal of Nursing Scholarship Series will be published in 2004–2005 and is modeled after the *New England Journal of Medicine* Genomics Series. Articles are being written by genetics nursing experts.

Centers for Disease Control and Prevention

http://www.cdc.gov/genomics/activities/fbr.htm
Model System for Collecting, Analyzing and Disseminating Information on Genetic Tests Office of Genomics and Disease Prevention, CDC

In September 2000, the Office of Genomics and Disease Prevention, CDC, funded a new cooperative agreement with the Foundation for Blood (FBR). FBR is a nonprofit research organization with expertise in clinical and laboratory investigation. It has carried out a wide array of studies evaluating test performance, quality control, and effectiveness. The current project is to develop a model system for assembling, analyzing, disseminating, and updating existing data on the safety and effectiveness of DNA-based genetic tests and testing algorithms. It is called the ACCE Project—Analytic validity, Clinical validity, Clinical Utility, and Ethical, legal, and social implications.

TABLE 14-1 Genetic Information on the Web
www.genome.gov/11510197

Genome Research Resources

Resource	URL	Content
BLAST	*http://www.ensembl.org/Data/blast.html*	Searches of protein or DNA sequence against metazoan genomes
Cancer Genome Anatomy Project	*http://cgap.nci.nih.gov*	Access to all CGAP data and biological resources
Chromosomal Variation in Man	*http://www.wiley.com/legacy/products/subject/life/borgaonkar/*	A catalog of chromosomal variants and anomalies
Ensembl	*http://www.ensembl.org*	Access to DNA and protein sequences with automatic baseline annotation
Human Genome Maps	*http://genome.wustl.edu/projects/human/index.php*	Links to clone and accession maps of the the human genome
National Center for Biotechnology Information	*http://www.ncbi.nlm.nih.gov/genome/guide/*	Views of chromosomes, maps, and loci; links to other NCBI resources
Oak Ridge Genome Channel	*http://compbio.ornl.gov/channel/*	Java viewers for human genome data
Online Mendelian Inheritance in Man (OMIM)	*http://www.ncbi.nlm.nih.gov/Omim/*	Information about human genes and disease
The SNP Consortium	*http://snp.cshl.org/*	A variety of ways to query for SNPs in the human genome
UCSC Genome Bioinformatics	*http://genome.cse.ucsc.edu/*	Reference sequence for the human and C. elegans genomes and working drafts for the mouse, rat, Fugu, Drosophila, C. briggsae, Yeast, and SARS genomes.

TABLE 14-1 continued

Clinical Genetics Resources

Resource	URL	Content
Gene Tests	*http://www.genetests.org/*	Information for health professionals about hundreds of genetic tests
The Genetics Resource Center	*http://www.pitt.edu/~edugene/resource/*	Clinical and educational information related to genetic counseling
Human Genome Epidemiology Network (HuGENet)	*http://www.cdc.gov/genomics/hugenet/default.htm*	Network for sharing population-based human genome epidemiologic information
INFOGENETICS	*http://www.infogenetics.org/*	Clinical practice tools
National Birth Defects Prevention Network	*http://www.nbdpn.org*	Network of birth defect care providers
National Newborn Screening & Genetics Resource Center	*http://www.genes-r-us.uthscsa.edu/*	National Newborn Screening & Genetics Resource Center
Online Mendelian Inheritance in Man (OMIM)	*http://www.ncbi.nlm.nih.gov/Omim/*	Information about human genes and disease

Support and Advocacy Groups

Resource	URL	Content
Coalition for Genetic Fairness	*http://www.nationalpartnership.org/Content.cfm?L1=202&TypeID=1&NewsItemID=381*	Advocacy group for federal legislation regarding genetics discrimination
Family Village	*http://www.familyvillage.wisc.edu/index.html*	Disability-related resources
The Genetic Alliance	*http://www.geneticalliance.org/*	Wide array of genetic-related information

Support and Advocacy Groups

Resource	URL	Content
National Organization for Rare Disorders	*http://www.rarediseases.org/*	Wide array of genetic-related information
Rare Genetic Diseases in Children	*http://mcrcr2.med.nyu.edu/murphp01/homenew.htm*	Focuses on pediatric genetic conditions

Health Professional Genetics Resources

Resource	URL	Content
Clinical Genetics: A Self Study for Health Care Providers	*http://www.vh.org/Providers/Textbooks/ClinicalGenetics/Contents.html*	Electronic textbook from Virtual Children's Hospital
Foundation for Genetic Education and Counseling	*http://www.fgec.org*	Genetics and common diseases, especially psychiatric disorders
Genetics and Your Practice	*http://www.marchofdimes.com/gyponline/index.bm2*	A practical "how-to" site on clinical genetics from the March of Dimes
Genetics in Clinical Practice: A Team Approach	*http://www.iml.dartmouth.edu/education/cme/Genetics/*	Virtual Genetics Clinic
Genetics in Primary Care	*http://genes-r-us.uthscsa.edu/resources/genetics/primary_care.htm*	Training program curriculum materials
Genetics in Psychology	*http://www.apa.org/science/genetics/homepage.html*	The American Psychological Association's site about genetics
Genetics Program for Nursing Faculty	*http://www.gpnf.org*	Links to genetics resources of particular interest to nurses

Table 14-1 continued

Health Professional Genetics Resources

Resource	URL	Content
Genetics: Educational Information	*http://www.faseb.org/genetics/careers.htm*	Medical school courses in genetics, some with syllabi
Information for Genetics Professionals	*http://www.kumc.edu/gec/geneinfo.html*	Educational, clinical, and research resources
National Coalition for Health Professional Education in Genetics	*http://www.nchpeg.org/*	Core competencies in genetics and reviews of education programs

Consumer Resources

Resource	URL	Content
Building and Understanding Your Medical Family History	*http://www.jamesline.com/patientsandpublic/prevention/cancergeneticsworkbook/*	Information on collecting family health history and assessing cancer risk
The DNA Files	*http://www.dnafiles.org/*	A series of 14 one-hour public radio documentaries and related information
Dolan DNA Learning Center	*hhttp://vector.cshl.org*	A variety of educational resources, including an interactive DNA timeline
Foundations of Classical Genetics	*http://www.esp.org/foundations/genetics/classical*	Complete versions of classic genetics work written between 350 AD and 1932
Generational Health	*http://www.generationalhealth.com/*	Tool to help trace a family's medical history and provide information on common diseases
Genetic Science Learning Center	*http://gslc.genetics.utah.edu/*	Basic genetics, genetic disorders, genetics in society, and several thematic units

Consumer Resources

Resource	URL	Content
Genetics and Rare Diseases Information Center	*http://www.genome.gov/10000409*	Information service for the general public, including patients and their families, health care professionals, and biomedical researchers
Genetics Education Center	*http://www.kumc.edu/gec/*	Material for educators
Genetics Home Reference–National Library of Medicine	*http://ghr.nlm.nih.gov/*	Consumer information about genetic conditions and the genes responsible for those conditions
Hispanic Educational Genome Project	*http://caldera.calstatela.edu/hgp/Bienvenido.html*	English and Spanish versions of a high school curriculum
The Human Genome Project: Exploring Our Molecular Selves	*http://www.genome.gov/Pages/EducationKit/*	Video about Human Genome Project, timeline about genetics, talking glossary, classroom activities, 3-D animation of cell
MendelWeb	*http://www.mendelweb.org/*	Mendel's papers in English and German and related materials
National Society for Genetic Counselors-Family History	*http://www.nsgc.org/consumer/familytree/index.asp*	Information on collecting family health history
The New Genetics: A Resource for Students and Teachers	*http://www4.umdnj.edu/camlbweb/teachgen.html*	Links to genetic education resources
Understanding Gene Testing	*http://press2.nci.nih.gov/sciencebehind/genetesting/genetesting01.htm*	Primer on genetic testing

TABLE 14-1 continued

ELSI, Policy, and Legislation

Resource	URL	Content
American Academy of Pediatrics: Ethical Issues with Genetic Testing in Pediatrics	*http://pediatrics.aappublications.org/cgi/content/full/107/6/1451*	Recommendations on newborn screening and genetic testing in children
Bioethics Resources on the Web	*http://www.nih.gov/sigs/bioethics/*	Links to bioethics resources
bioethics.net	*http://www.bioethics.net/genetics/genetics.php*	Links to articles on bioethics and genetics
DNA Patent Database	*http://dnapatents.georgetown.edu*	Searchable database of U.S. DNA-based patents issued by the USPTO
The Council for Responsible Genetics	*http://www.gene-watch.org/*	Information on the social, ethical, and environmental implications of genetic technologies
Ethical, Legal, & Social Issues	*http://www.ornl.gov/hgmis/elsi/elsi.html* *http://www.genome.gov/10001754*	Information, articles, and links on a wide range of issues
Foundation for Genetic Medicine, Inc.	*http://www.geneticmedicine.org/*	Information on the science of genetic medicine, genetic and genomic research, and ethical, legal, and social dimensions and implications
Genetics & Ethics	*http://www.genethics.ca/index.html*	Information on the social, ethical, and policy issues associated with genetic and genomic knowledge and technology
Genetics and the Law	*http://www.genelaw.info/*	A searchable online clearninghouse of information on emerging legal developments in human genetics

ELSI, Policy, and Legislation

Resource	URL	Content
Genetics and Public Policy Center	*www.dnapolicy.org*	Information on genetic technologies and genetic policies for the public, media, and policymakers
Genome Technology and Reproduction: Values and Public Policy and The Communities of Color & Genetics Policy Project	*http://www.sph.umich.edu/genpolicy/*	Two subprojects combined to form a 5-year project designed to provide policy recommendations based on public perceptions and responses to the explosion of genetic information and technology
HumGen	*http://www.humgen.umontreal.ca/en/*	Access to a comprehensive international database on the legal, social, and ethical aspects of human genetics
NHGRI Policy & Legislation Database	*http://www.genome.gov/Legislative Database*	Searchable database to laws and policies related to genetic issues.
National Information Resources on Ethics & Human Genetics	*http://www.georgetown.edu/research/nrcbl/nirehg/index.htm*	Links to resources on ethics and human genetics
NCSL-Genetic Technologies Project	*http://www.ncsl.org/programs/health/genetics.htm*	Status of legislative actions and access to policy briefs on genetic issues of concern to state legislators

TABLE 14-1 continued

ELSI, Policy, and Legislation

Resource	URL	Content
The President's Council on Bioethics	*http://www.bioethics.gov/*	Information on current bioethical issues
Scope Note Series (Kennedy Institute of Ethics/Georgetown University)	*http://www.georgetown.edu/research/nrcbl/nirehg/scope.htm*	Information on various aspects of genetics and ethics
THOMAS Legislative Information on the Internet	*http://thomas.loc.gov/*	Searchable database of U.S. legislation
Your Genes Your Choices	*http://ehrweb.aaas.org/ehr/books/index.html*	Describes the Human Genome Project, the science behind it, and the ethical, legal, and social issues that are raised by the project

Genetics Professional Groups

Resource	URL	Content
American Board of Genetic Counseling	*http://www.faseb.org/genetics/abgc/abgcmenu.htm*	Information about certification of genetic counselors
American Board of Medical Genetics	*http://www.faseb.org/genetics/abmg/abmgmenu.htm*	Information about medical genetic training programs and certification of geneticists
American College of Medical about Genetics	*http://www.acmg.net/*	Information and policy statements medical genetics
American Society for Human Genetics	*http://www.ashg.org/*	Information about human genetics

Genetics Professional Groups

Resource	URL	Content
Genetics Society of America	*http://www.faseb.org/genetics/gsa/gsamenu.htm*	Information about genetics
International Society of Nurses in Genetics	*http://www.isong.org/*	Information about genetics in nursing
National Society of Genetic Counselors	*http://www.nsgc.org/*	Information about genetic counseling
Society for the Study of Inborn Errors of Metabolism	*http://www.ssiem.org.uk/*	Information about inborn errors of metabolism

U.S. Government Genetics Agencies

Resource	URL	Content
Centers for Disease Control and Prevention, Office of Genomics & Disease Prevention	*http://www.cdc.gov/genomics/default.htm*	Information about human genetic discoveries and how to use to improve health and prevent disease
Department of Energy	*hhttp://www.doegenomes.org/*	Multiple genetics educational resources
Genetic Modification Clinical Research Information System (GeMCRIS)	*http://www.gemcris.od.nih.gov/*	Access to an array of information about human gene transfer trials registered with the NIH
Health Resources and Services Administration–Genetics Services Branch	*http://www.mchb.hrsa.gov/*	Organization with mission to improve and expand access to quality health care for all
National Cancer Institute's CancerNet	*http://www.cancer.gov/cancerinfo/prevention-genetics-causes*	Authoritative information about cancer genetics

Table 14-1 continued

U.S. Government Genetics Agencies

Resource	URL	Content
National Human Genome Research Institute	*http://www.genome.gov*	Research, health, policy, ethics, education, and training information and resources
National Institute of Environmental Health Sciences–Environmental Genome Project	*http://www.niehs.nih.gov/envgenom/home.htm*	Project to improve understanding of human genetic susceptibility to environmental exposures
National Institutes of Health Obesity Research	*http://obesityresearch.nih.gov/*	Information about NIH-supported research that seeks to identify genetic, behavioral, and environmental causes of obesity
National Institutes of Health	*http://www.nih.gov/*	Research, health, policy, ethics, education, and training information and resources
National Institute of Nursing Research–Summer Genetics Institute	*http://fmp.cit.nih.gov/ninr/*	Summer Genetics Institute program that is designed to provide training in molecular genetics for use in research and clinical practice
Office of Rare Diseases, National Institutes of Health	*http://rarediseases.info.nih.gov/*	Information on thousands of rare disorders
Secretary's Advisory Committee on Genetic Testing	*http://www4.od.nih.gov/oba/SACGT.htm*	Policy issues regarding genetic testing (archival)
Secretary's Advisory Committee on on Genetics, Health, and Society	*http://www4.od.nih.gov/oba/sacghs.htm*	Policy issues regarding the impact of genetic technologies on society

Sites That Track or Report on What's New in Genetics

Resource	URL	Content
Genetics and Molecular Medicine (American Medical Association)	*http://www.ama-assn.org/ama/pub/category/1799.html*	Links to current articles and other resources
Genome News Network (The Center for the Advancement of Genomics)	*http://www.genomenewsnetwork.org/*	Original articles and links
Science News Presented by BIO: Biotechnology Industry Organization	*http://science.bio.org/genomics.news.html*	Links to current articles

Family History Tools

Resource	URL	Content
American Medical Association–Family History Tools	*http://www.ama-assn.org/ama/pub/category/2380.html*	Tools for gathering family history
Cyrillic	*www.cyrillicsoftware.com*	Pedigree drawing software
Progeny	*www.progeny2000.com*	Pedigree drawing software

Risk Assessment

Resource	URL	Content
Harvard Center for Cancer Prevention–Your Cancer Risk	*http://www.yourcancerrisk.harvard.edu/*	Personalized estimation of cancer risk and tips for prevention
National Cancer Institute–Breast Cancer Risk Assessment Tool	*http://bcra.nci.nih.gov/brc/*	Interactive tool to measure a woman's risk of invasive breast cancer

Table 14-1 continued

IRB Related Resources

Resource	URL	Content
My Very Own Medicine: What Must I Know? Information Policy for Pharmacogenetics. Public Health Genetics Unit, National Health Service, UK. D. Melzer *et al.* (2003).	*http://www.phgu.org.uk/about_phgu/pharmacogenetics.html*	General information and background, looking ahead to future needs, including guidance for IRBs
New York State Task Force on Life and Law: Genetic Testing and Screening in the Age of Genomic Medicine (2001).	*http://www.health.state.ny.us/nysdoh/taskfce/screening.htm*	Includes general and state-specific information in a bulleted report that is relatively easy to scan by topic
Office for Human Research Protections: Protecting Human Research Subjects	*http://www.hhs.gov/ohrp/*	Discusses many issues that continue to challenge IRBs and investigators (and policy makers) today
Pharmacogenetics: Ethical Issues. Nuffield Council on Bioethics (2003).	*http://www.nuffieldbioethics.org/publications/pp_0000000018.asp*	Includes a section discussing the use of pharmacogenetics in clinical trials

Search Engines

Resource	URL	Content
Centers for Disease Control: Genomics and Disease Prevention Information System	*http://www2a.cdc.gov/genomics/GDPQueryTool/frmQueryBasicPage.asp*	Provides access to information and resources for guiding public health research, policy, and practice on using genetic information to improve health and prevent disease. Includes core competencies for public health genetics.

Search Engines

Resource	URL	Content
Genetics Resources on the Web	*http://www.geneticsresources.org*	Search engine for information related to human genetics
Georgetown University: National Information Resource on Ethics & Human Genetics	*http://www.georgetown.edu/research/nrcbl/nirehg/index.htm*	Search engine for literature on specific issues
National Library of Medicine: Pub Med	*http://www.ncbi.nlm.nih.gov/PubMed*	Basic search engine for biomedical research, including research and commentary regarding clinical research ethics and regulations
National Newborn Screening and Genetics Resource Center: Genetic Education Materials Database	*http://www.gemdatabase.org/GEMDatabase/index.asp*	Search engine for clinical issues

Courtesy of Susan Vasquez BA, Special Assistant to the Deputy Director, Office of the Director, National Human Genome Research Institute, NIH.

(More detail about the process and definitions of terms can be found at ACCE Project on the CDC Web site). Over a 3-year period, five tests for different disorders will be evaluated from the standpoint of analytical validity, clinical validity, clinical utility, and related ethical–legal–social issues. The goal of this effort is to facilitate the appropriate transition of genetic tests from investigational settings to use in clinical and public health practice. This goal is consistent with preliminary recommendations of the Department of Health and Human Services Secretary's Advisory Committee on Genetic Testing.

New Information Technologies and Genomics Resources

As genomics permeates all aspects of health promotion and health maintenance, new information technologies will be needed to locate appropriate genomic resources and to coordinate and analyze genomic information specific to individuals and populations. Nurses must have skills to be able to navigate these new technologies, especially the Internet, where patients often turn for information. **Table 14-1** provides a detailed listing of Web sites with current educational, clinical, and client support resources.

Chapter Activities

1. Visit the International Society of Nurses in Genetics, Inc. Web site at *http://www.isong.org.* Check out the announcement related to its annual meeting. Would that meeting be of benefit to you because you would be meeting other nurses who could enhance your understanding of using genetics in your practice? If so, make plans to join ISONG!
2. Look at Table 14-1 and choose 2 or 3 Web sites to visit. Share with your colleagues what you found and whether or not it is a useful site for you and/or your patients to know about.

Resources

Lea, D. H. (2003). Public health genetics: The nursing role of the future. Presented at the 16th Annual International Conference, International Society of Nurses in Genetics. November 1–4, 2003, Los Angeles, CA.

Lewis, J. (2003). Testimony found at: *http://www4.od.nih.gov/oba/SACGHS/meetings/March2004/PublicComment030104.pdf,* Accessed 9/15/04.

Suggested Readings

Bennett, R. (1999). *The practical guide to the genetic family history.* New York, NY: Wiley-Liss, Inc.

Collins, F., Green, E., Guttmacher, A., & Guyer, M. (2003). A vision for the future of genomics research. *Nature, 422,* 835–847.

Greendale, K. & Pyeritz, R. (2003). Empowering primary care health professionals in medical genetics: How soon? How fast? How far? *American Journal of Medical Genetics, 106,* 223–232.

Guttmacher, A., Jenkins, J., & Uhlmann, W. (2001). Genomic medicine: Who will practice it? A call to open arms. *American Journal of Medical Genetics, 106,* 216–222.

Guttmacher, A., Collins, F., & Drazen, J. (Eds.). (2004). *Genomic Medicine: Articles from the NEJM.* Maryland: The Johns Hopkins University Press.

HRSA. (2000). *Report of the Expert Panel on Genetics and Nursing: Implications for Education and Practice.* Washington, DC: HRSA.

International Society of Nurses in Genetics, Inc. (1998). *Scope and standards of genetics clinical nursing practice.* Washington, DC: American Nurses Association.

Lea, D. H., & Monsen, R. B. (2003). Preparing nurses for a 21st century role in genomics-based health care. *Nursing Education Perspectives, 24,* 75–80.

Leshner, A. (2004). Science at the leading edge. *Science, 303,* 729.

Tranin, A., Masny, A., & Jenkins, J. (2003). *Genetics in Oncology Practice: Cancer Risk Assessment.* Pittsburgh, PA.: Oncology Nursing Press.

Joseph D. McInerney, MA, MS
Executive Director
jdmcinerney@nchpeg.org www.nchpeg.org

2360 W. Joppa Road, Suite 320 Lutherville, MD 21093 Tel: 410-583-0600 Fax: 410-583-0520

January 2001

Dear Colleagues:

On behalf of the National Coalition for Health Professional Education in Genetics (NCHPEG), we are pleased to provide these *Core Competencies in Genetics Essential for All Health-Care Professionals.* This document, approved by NCHPEG's steering committee, is the product of extensive deliberation by a working group of specialists who have broad experience in genetics and the health professions (see page 389 for a list of working-group members). We are grateful to Dr. Jean Jenkins, of the National Cancer Institute, for her dedicated and capable leadership of the working group.

As research in genetics accelerates, bringing new insights into health and disease, all health-care professionals must be aware of the implications of genetics for education and practice. The core competencies are the first step in what will be a comprehensive effort by NCHPEG to help the health professions integrate genetics into all areas of health care.

We welcome your feedback on this document, and we hope you will draw upon the collective expertise of our member organizations as you consider the most appropriate mechanisms for the implementation of these core competencies in your own discipline.

Sincerely,

Francis S. Collins, MD, PhD   Michael J. Scotti, Jr., MD   Gladys White, PhD, RN
NCHPEG Executive Committee

The National Coalition for Health Professional Education in Genetics (NCHPEG) endorsed these core competencies on February 14, 2000. NCHPEG is an interdisciplinary group comprising leaders from more than 100 diverse health professional organizations, consumer and voluntary groups, government agencies, private industries, managed-care organizations, and genetics professional societies. NCHPEG is a national effort to promote health-professional education and access to information about advances in human genetics, to improve the nation's health.

If you have any questions about this document or would like information about NCHPEG, please contact:

Joseph D. McInerney, Executive Director
2360 W. Joppa Road, Suite 320
Lutherville, MD 21093
Phone: 410-583-0600/Fax: 410-583-0520
jdmcinerney@nchpeg.org
www.nchpeg.org

Third printing, October 2002

Purpose

The impetus for developing the *ideal* competencies related to genetics was to encourage health care providers to integrate genetics knowledge, skills, and attitudes into routine health care to provide effective care to individuals and families.

The Core Competency and Curriculum Working Group of NCHPEG recommends that all health professionals possess the core competencies in genetics, as identified in this report, to enable them to integrate genetics effectively and responsibly into their current practice. Competency in these areas represents the minimum knowledge, skills, and attitudes necessary for health professionals from all disciplines (medicine, nursing, allied health, public health, dentistry, psychology, social work, etc.) to provide patient care that involves awareness of genetic issues and concerns.

Each health care professional should at a minimum be able to:

- Appreciate limitations of his or her genetic expertise.
- Understand the social and psychological implications of genetic services.
- Know how and when to make a referral to a genetics professional.

Background

During the last decade, the evolution of scientific discoveries from the study of genetics has provided information with potential for tremendous influence on health care. Understanding the role genetics plays in health and disease provides the means to integrate such information into diagnosis, prevention, and treatment of many common diseases and to improve the health of society. Genetic discoveries are already making their way into mainstream health care (Collins, 1999). Patients are beginning to ask providers about genetic services. Primary care professionals face economic, institutional, and professional opportunities and challenges in managing persons at risk for inherited conditions (Touchette et al., 1997). As outlined by the Institute of Medicine Report on the Future of Public Health (IOM, 1988), public health agencies will have an increasing role in assessing the health needs of populations, working with the private sector in ensuring the quality of genetic tests (SACGT, 2000) and services, and evaluating the impact of interventions on medical, behavioral, and psychosocial outcomes. Ultimately, health care providers, regardless of specialty area, role, or practice setting, will face questions about implications of genetics for their patients. The fast pace of genetic advances and the paucity of professional training in genetics leave many providers without up-to-date answers for their patients.

Implementation

It is essential that persons and groups responsible for continuing education, curriculum development, licensing, certification, and accreditation bodies for all health care disciplines adopt these recommendations and integrate genetics content into ongoing education. The competencies provide direction for curriculum content that can be used in the design of seminars, workshops, and academic preparation. There is a need for commitment on the part of all educators to incorporate genetic information into all levels of professional education. Enhanced genetics competency will help us to meet the changing demands of the health care system and promote human benefit as a result of discoveries in genetics and genetic medicine. **Although this list may appear challenging, it is important to prepare for the reality of tomorrow and not only for the needs of today.**

This document is a work in progress, because it is likely that the knowledge produced by the Human Genome Project and related activities will create an ongoing need to assess and revise expectations. Although the list is extensive, NCHPEG believes that the recommendations provide a useful tool for organizing the teaching of basic genetics in many educational settings and can be modified for a particular discipline.

Those health professionals involved in the *direct* provision of genetics services may require additional training to achieve an appropriately higher level of competence. Indeed, there are a number of examples of specific recommendations for training of professionals who require specialized knowledge of genetics (AAFP, 1999; ASCO, 1997; APHMG, 1998; ASHG, 1995; Fine et al., 1996; Hayflick & Eiff, 1998; Jenkins et al., 2001; Reynolds & Benkendorf, 1999; Stephenson, 1998; Taylor-Brown & Johnson, 1998).

Recommendations

Note: Throughout this document, the term "clients" includes individuals and their sociological and biological families.

Knowledge

All health professionals should understand:

1.1 basic human genetics terminology
1.2 the basic patterns of biological inheritance and variation, both within families and within populations
1.3 how identification of disease-associated genetic variations facilitates development of prevention, diagnosis, and treatment options
1.4 the importance of family history (minimum three generations) in assessing predisposition to disease
1.5 the role of genetic factors in maintaining health and preventing disease
1.6 the difference between clinical diagnosis of disease and identification of genetic predisposition to disease (genetic variation is not strictly correlated with disease manifestation)
1.7 the role of behavioral, social, and environmental factors (lifestyle, socioeconomic factors, pollutants, etc.) to modify or influence genetics in the manifestation of disease
1.8 the influence of ethnoculture and economics in the prevalence and diagnosis of genetic disease
1.9 the influence of ethnicity, culture, related health beliefs, and economics in the client's ability to use genetic information and services

1.10 the potential physical and/or psychosocial benefits, limitations, and risks of genetic information for individuals, family members, and communities
1.11 the range of genetic approaches to treatment of disease (prevention, pharmacogenomics/prescription of drugs to match individual genetic profiles, gene-based drugs, gene therapy)
1.12 the resources available to assist clients seeking genetic information or services, including the types of genetics professionals available and their diverse responsibilities
1.13 the components of the genetic-counseling process and the indications for referral to genetic specialists
1.14 the indications for genetic testing and/or gene-based interventions
1.15 the ethical, legal, and social issues related to genetic testing and recording of genetic information (e.g., privacy, the potential for genetic discrimination in health insurance and employment)
1.16 the history of misuse of human genetic information (eugenics)
1.17 one's own professional role in the referral to genetics services, or provision, follow-up, and quality review of genetic services

Skills

All health professionals should be able to:

2.1 gather genetic family-history information, including an appropriate multigenerational family history
2.2 identify clients who would benefit from genetic services
2.3 explain basic concepts of probability and disease susceptibility, and the influence of genetic factors in maintenance of health and development of disease
2.4 seek assistance from and refer to appropriate genetics experts and peer support resources
2.5 obtain credible, current information about genetics, for self, clients, and colleagues
2.6 effectively use new information technologies to obtain current information about genetics
2.7 educate others about client-focused policy issues
2.8 participate in professional and public education about genetics

Skills 2.9–2.17 delineate the components of the genetic-counseling process and are not expected of all health care professionals. However, health professionals should be able to facilitate the genetic-counseling process and prepare clients and families for what to expect, communicate relevant information to the genetics team, and follow up with the client after genetics services have been provided. For those health professionals who choose to provide genetic-counseling services to their clients, all components of the process, as delineated in 2.9–2.17 should be performed.

- 2.9 educate clients about availability of genetic testing and/or treatment for conditions seen frequently in practice
- 2.10 provide appropriate information about the potential risks, benefits, and limitations of genetic testing
- 2.11 provide clients with an appropriate informed-consent process to facilitate decision making related to genetic testing
- 2.12 provide, and encourage use of, culturally appropriate, user-friendly materials media to convey information about genetic concepts
- 2.13 educate clients about the range of emotional effects they and/or family members may experience as a result of receiving genetic information
- 2.14 explain potential physical and psychosocial benefits and limitations of gene-based therapeutics for clients
- 2.15 discuss costs of genetic services, benefits, and potential risks of using health insurance for payment of genetic services; potential risks of discrimination
- 2.16 safeguard privacy and confidentiality of genetic information of clients to the extent possible
- 2.17 inform clients of potential limitations to maintaining privacy and confidentiality of genetic information

Attitudes

All health professionals should:

- 3.1 recognize philosophical, theological, cultural, and ethical perspectives influencing use of genetic information and services
- 3.2 appreciate the sensitivity of genetic information and the need for privacy and confidentiality
- 3.3 recognize the importance of delivering genetic education and counseling fairly, accurately, and without coercion or personal bias
- 3.4 appreciate the importance of sensitivity in tailoring information and services to clients' culture, knowledge, and language level
- 3.5 seek coordination and collaboration with interdisciplinary team of health professionals
- 3.6 speak out on issues that undermine clients' rights to informed decision making and voluntary action
- 3.7 recognize the limitations of their own genetics expertise
- 3.8 demonstrate willingness to update genetics knowledge at frequent intervals
- 3.9 recognize when personal values and biases with regard to ethical, social, cultural, religious, and ethnic issues may affect or interfere with care provided to clients
- 3.10 support client-focused policies

References

American Family Physician. *AAFP Core Educational Guidelines;* 1999. Available at: http://www.aafp.org/afp/990700ap/core.html.

American Society of Clinical Oncologists. *Cancer Genetics Curriculum;* 1997. Available at: www.asco.org.

American Society of Human Genetics Information and Education Committee. ASHG Report from the ASHG Information and Education Committee: Medical School Curriculum in Genetics. *American Journal of Human Genetics.* 1995; *56,*535–537.

Association of Professors of Human and Medical Genetics. Clinical objectives in medical genetics for undergraduate medical students. *Genetics in Medicine.* 1998; *1,* 54–55.

Collins F. Shattuck lecture: Medical and social consequences of the Human Genome Project. *The New England Journal of Medicine.* 1999; *341,* 28–37.

Fine B., Baker D., Fiddler M., ABGC Consensus Development Consortium. Practice-based competencies for accreditation of and training in graduate programs in genetic counseling. *Journal of Genetic Counseling.* 1996; *5,* 113–121.

Hayflick S., & Eiff, M. Role of primary care providers in the delivery of genetic services. *Community Genetics.* 1998; *1,* 18–22.

Institute of Medicine. *The future of public health.* Washington, DC: National Academy Press; 1988.

Jenkins J., Dimond E., & Steinberg, S. Preparing for the future through genetics nursing education. *Journal of Nursing Scholarship.* 2001; *33*(2), 191–195.

Reynolds P., & Benkendorf, J. Genes and generalists: Why we need professionals with added competencies. *Culture and Medicine.* 1999; *171,* 375–379.

Secretary's Advisory Committee on Genetic Testing (SACGT). *Enhancing the oversight of genetic tests;* 2000. Available at: http://www4.od.nih.gov/oba/sacgt.htm.

Stephenson J. Group drafts core curriculum for 'What docs need to know about genetics. *JAMA.* 1998; *279,* 735–736.

Taylor-Brown S., & Johnson, A. *Genetics Practice;* 1998. Update available at: www.naswdc.org.

Touchette N., Holtzman N.A., Davis, J.G., & Feetham, S. (Eds.). *Toward the 21st Century: Incorporating genetics into primary health care.* Cold Spring Harbor, NY: Cold Spring Harbor Laboratory Press, 1997.

Acknowledgments

Core Competencies Working Group of the NCHPEG, January 2001

Note: Organization representing

Jean Jenkins, PhD, RN, FAAN—*Oncology Nursing Society*
Charles G. Atkins, PhD—*American Osteopathic Association*
Michelle A. Beauchesne, DNSc, RN, PNP—*National Organization of Nurse Practitioner Faculties*
Judith Benkendorf, MS, CGC—*American Board of Genetic Counseling*
Miriam Blitzer, PhD—*American Society of Human Genetics*

Karina Boehm, MPH—*National Institute of Dental and Craniofacial Research*
Richard E. Braun, MD—*American Academy of Insurance Medicine*
Jessica G. Davis, MD—*American College of Medical Genetics*
Sue K. Donaldson, PhD, RN, FAAN—*American Academy of Nursing*
Louis J. Elsas, MD, FFACMG—*Association of Professors of Human and Medical Genetics*
Suzanne Feetham, PhD, RN, FAAN—*Society of Pediatric Nurses*
Elizabeth Gettig, MS—*Centers for Disease Control*
Alan Guttmacher, MD—*National Human Genome Research Institute*
Sheldon Horowitz, MD—*American Board of Medical Specialities*
Ann Johnson, PhD, MSW—*National Association of Social Workers*
Ronald Katz, DMD, PhD—*American Association of Dental Reserach & American Association of Dental Schools*
Carole Kenner, DNS, RNC, FAAN—*National Association of Neonatal Nurses*
Penny Kyler, MA, OTR/L, FAOTA—*American Occupational Therapy Association*
E. Virginia Lapham, PhD—*Human Genome Education Model Project*
Michelle A. Lloyd-Puryear, MD, PhD—*Health Resources and Services Administration*
Kenneth Miller, PhD—*American Association of Colleges of Pharmacy*
Andrea Farkas Patenaude, PhD—*American Psychological Association*
P. Preston Reynolds, MD, PhD, FACP—*Society of General Internal Medicine*
Robert Smith, PhD—*American Cancer Society*
Susan Taylor-Brown, PhD—*Council on Social Work Education*
Barry H. Thompson, MD, MS—*Department of Defense and Armed Forces Institute of Pathology*

Executive Committee of NCHPEG, January 2001

Francis S. Collins, MD, PhD—*National Human Genome Research Institute*
Rita Black Monsen, DSN, MPH, RN—*American Nurses Association*
Michael J. Scotti, Jr., MD—*American Medical Association*

NCHPEG Member Organizations October 2002

A

Aetna U.S. Healthcare
Agency for Healthcare Research and Quality
American Academy of Family Physicians
American Academy of Insurance Medicine
American Academy of Nursing
American Academy of Oral Medicine

American Academy of Pediatric Dentistry
American Academy of Pediatrics
American Academy of Physician Assistants
American Association for Dental Research
American Association for Marriage and Family Therapy
American Association for Respiratory Care
American Association of Colleges of Nursing
American Association of Colleges of Osteopathic Medicine
American Association of Colleges of Pharmacy
American Association of Critical-Care Nurses
American Association of Occupational Health Nurses
American Association of Orthodontists
American Association on Mental Retardation
American Board of Genetic Counseling
American Board of Medical Specialties
American Cancer Society
American Cleft Palate-Craniofacial Association
American College of Clinical Pharmacy
American College of Medical Genetics
American College of Obstetricians and Gynecologists
American College of Physicians
American College of Preventive Medicine
American Dental Education Association
American Dental Hygienists' Association
American Dietetic Association
American Medical Association
American Medical Informatics Association
American Nurses Association
American Occupational Therapy Association, Inc.
American Osteopathic Association
American Pediatric Society
American Physical Therapy Association
American Psychological Association
American Society for Clinical Laboratory Science
American Society for Parenteral and Enteral Nutrition
American Society of Clinical Oncology
American Society of Health-System Pharmacists
American Society of Human Genetics
Armed Forces Institute of Pathology
Association of American Medical Colleges
Association of Asian Pacific Community Health

Association of Family Practice Residency Directors
Association of Genetic Counseling Program Directors
Association of Genetic Technologists
Association of Professors of Human or Medical Genetics
Association of Professors of Medicine
Association of Schools of Allied Health Professions
Association of Schools of Public Health
Association of State and Territorial Health Officials
Association of Teachers of Preventive Medicine
Association of Women's Health, Obstetric Neonatal Nurses

B
Biotechnology Industry Organization

C
Centers for Disease Control and Prevention
Chicago Center for Jewish Genetic Disorders
Council of Medical Specialty Societies
Council on Social Work Education

D
Dana Alliance for Brain Initiatives
Department of Veterans Affairs

F
Federation of Special Care Dentistry
Foundation for Blood Research

G
Genetic Alliance
GlaxoSmithKline

H
health Resources and Services Administration
Hoffmann-LaRoche, Inc.
Howard Hughes Medical Institute
Human Genome Education Model Project

I
International Patient Advocacy Association
International Society of Nurses in Genetics

J
Joint Commission on Accreditation of Healthcare Organizations

M
March of Dimes

N
National Association of Catholic Chaplains
National Association of Neonatal Nurses
National Association of Pediatric Nurse Practitioners
National Association of School Nurses
National Association of Social Workers
National Board of Medical Examiners
National Cancer Institute
National Center for Genome Resources
National Heart, Lung, and Blood Institute
National Human Genome Research Institute
National Institute of Child Health & Human Development
National Institute of Dental and Craniofacial Research
National Institute of Diabetes and Digestive and Kidney Diseases
National Institute of Environmental Health Sciences
National Institute of General Medical Sciences
National Institute of Mental Health
National Institute of Neurological Disorders & Stroke
National Institute of Nursing Research
National Institute on Aging
National Institute on Alcohol Abuse & Alcoholism
National Institute on Deafness and Other Communication Disorders
National Institute on Drug Abuse
National Marfan Foundation
National Medical Association
National Organization for Rare Disorders
National Organization of Nurse Practitioner Faculties
National Society of Genetic Counselors
Nursing Organization Liaison Forum

O
Office of Rare Diseases
Office on Women's Health
Oncology Nursing Society

P
Pharmaceutical Research and Manufacturers of America

R
RSi Communications Group

S
Scoliosis Association, Inc.
Sigma Theta Tau International
Society for Academic Continuing Medical Education
Society for Inherited Metabolic Disorders
Society of General Internal Medicine
Society of Gynecologic Oncologists
Society of Pediatric Nurses
Society of Teachers of Family Medicine

T
Tourette Syndrome Association, Inc.

U
Uniformed Services University of the Health Sciences

INDEX

A

abacavir, 128
aborting pregnancy, 97, 276–277
  stem cell research, 281–282
absorption, drug, 117
ACCE Project, 199
Acceptance of Health Status outcome, 194, 235
access to pharmacogenetics, 136
accuracy of genetic testing, 77, 91, 96, 137–139
acetylators, 128, 130
ACMG (American College of Medical Genetics), 176, 177
acute lymphoblastic leukemia (ALL), 124. *See also* cancer
ADRs (adverse drug reactions), 116, 118, 128
adults, referring for genetics evaluation, 163–164
advance practice nurses (APNs), 189, 314, 359–360
adverse drug reactions. *See* ADRs
advocacy groups. *See* support groups and counseling
AFP screening tests, 70
African Americans, 249–250, 251. *See also* ethnicity and race
aging, treatments and, 127–128
AIDS. *See* HIV treatment
ALL (acute lymphoblastic leukemia), 124. *See also* cancer
alpha-fetoprotein, 95
Alzheimer's disease, 128
American College of Medical Genetics (ACMG), 176, 177
American Nurses Association, 366
American Society of Human Genetics (ASHG), 175–176
amniocentesis, 70
Amsterdam Criteria, 26
ANA (American Nurses Association), 366
analytical sensitivity and specificity, 77
antidepressants, 121–122
antipsychotic drugs, 121–122
antisocial behaviors, 324
Anxiety Control outcome, 194, 235
APN (advance practice nurse), 189, 314, 359–360
APNG credential, 359–360
apolopoprotein E (APOE 4), 128
ART (assisted reproductive technologies), 97–99
ASHG (American Society of Human Genetics), 175–176, 177
assessments, nurse
  deciding whether to test, 218, 223, 226–227, 233
  family history. *See* family history
  referrals for testing, 156–164
  resources for additional information, 377
  of risk. *See* risk estimation
  transcultural nursing, 262–263
assisted reproductive technologies (ART), 97–99
Association of Women's Health and Neonatal Nurses (AWHONN), 366
association studies, 294
associations, professional, 175–177, 363–366, 380, 390–394
aunts, information about. *See* family history
autosomal inheritance, 12–14
awareness, patient. *See* education, patient; informed consent
AWHONN (Association of Women's Health and Neonatal Nurses), 366

B

behavioral change, effecting, 54
behavioral genetics, 324
bioethics. *See* ethical issues
biological therapy, 300
biology of human genome, 6–16
  basic terminology, 7–9
  current questions and research, 16
  DNA, RNA, and protein synthesis, 9–11
  genetic variation and inheritance patterns, 12–16
bipolar disorder. *See* mental health disorders
birth control pills, 130
birth defects, teratogenic agents and, 158–160
Black Americans. *See* African Americans
bladder cancer, 129–130
BRCA1 and BRCA2 mutations, 221–222

breaking the news. *See* communication with patients; information, impact of
breast cancer, 125–127, 220–223. *See also* cancer
brothers, information about. *See* family history

C

CAD (coronary artery disease), 127, 129
cancer, 23–25
  cigarette smoking and, 129–130
  CML (chronic myeloid leukemia), 348–350
  drug response, 63
  families, impact on, 232
  family history questionnaire, 28–37
  hereditary breast–ovarian cancer, 220
  HNPCC (hereditary non-polyposis colorectal cancer), 25–27, 50–52, 330–332
  nursing roles. *See also* cancer
  ONS (Oncology Nursing Society), 196, 263, 348, 365
  personal and health history questionnaire, 38–47
  research, 300
  surveillance and health outcome modifiers, 50–52
  treating, 124–127
Cancer Genetics Editorial Board, 190
cardiology, 127
caregivers. *See* family and relatives; support groups and counseling
carrier testing. *See* preventative testing
case coordinators, 198, 351
Catholicism. *See* religion and spirituality
causation, addressing, 171
CDC (Centers for Disease Control), 177, 200, 366
cellular reactions to drugs (pharmacodynamics), 120
certification of nurses, 358–360
CF (cystic fibrosis), 2–5, 7, 16–20, 158, 190
  genetic testing, 18
  newborn and neonatal screening, 91
  preventative and prenatal testing, 74, 94–95
children, 225, 344. *See also* newborn screening; prenatal screening
  information about. *See* family history
  informing parents of genetic condition, 167–173
chromosomes, 7
  analyzing, 70. *See also* genetic testing
  disorders, 12
  Philadelphia (Ph) chromosome, 125, 349
chronic disease, nursing roles and, 347–350. *See also* cancer
chronic myeloid leukemia (CML), 348–350
cigarette smoking, 129–130
classic PKU, 85
CLIA (Clinical Laboratory Improvement Amendments), 80
clients. *See entries at patient*
clinical care, genomics in, 1–6, 23–56
  benefits. *See* health benefits of genomics
  biology of human genome, 6–16
    basic terminology, 7–9
    current questions and research, 16
    DNA, RNA, and protein synthesis, 9–11
    genetic variation and inheritance patterns, 12–16
  carrier testing. *See* preventative testing
  ethics. *See* ethical issues
  family history. *See also* family history
    importance of, 16–20
    tools for collecting, 27–52, 54
  newborn screening. *See* newborn screening
  pharmacogenetics applications, 120–129. *See also* pharmacogenetics
  policies for, establishing, 259–261
  public information resources, 55
  referring patients for genetic testing, 147–166, 234
    indications for, 153–154
    informing parents of child's genetic condition, 167–173
    nursing assessments, 156–164
    nursing framework for, 154–156
    peer support, finding, 164–165
  research. *See* genetics research
  resources for, 66. *See also* resources for additional information
  variations as "red flags", 25–27
clinical genetic testing, 62, 79, 296–298
  ethical and social considerations, 301–311
    barriers to participation, 301–302
  nursing roles and, 313–315
  OI (Osteogenesis Imperfecta), 289
  patience, 320
  protocols for clinical trials, 299–301
  resources for, 315
Clinical Laboratory Improvement Amendments (CLIA), 80
clinical specificity and sensitivity, 77
clinical utility of genetic testing, 77

cloning, 280–281
CML (chronic myeloid leukemia), 348–350
codeine, 123
coding regions (exons), DNA, 10
coding SNPs (cSNPs), 118
cohort research studies, 295
colon cancer, 330–332
colorectal cancer (CRC). *See* HNPCC
communication with patients, 65, 205–207, 336
  clinical trials, participation in, 301–302
  familial roles and relationships, 236–237. *See also* family and relatives
  helping to make decisions, 83, 99–103, 192, 194, 319, 341
    choosing to share information, 229–230
    clinical trials, participation in, 301–306
    nursing role in, 304
    pharmacogenetics, 137–139
    when to recommend testing. *See* referrals for testing
    whether to test, 218, 223, 226–227, 233
  informed consent, 83, 99–103, 303–306
    genotyping and pharmacogenetics, 132–135
    nursing role in, 304
    prenatal screening, 89
  informing parents of child's genetic condition, 167–173
  obtaining family history, 377. *See also* family history
  prenatal screening and diagnosis, 99–103
  providing information to clients, 65, 167–173, 186–187
  referrals for testing, 147–166, 234
    indications for, 153–154
    nursing assessments, 156–164
    nursing framework for, 154–156
    peer support, finding, 164–165
  relationships with community, 313
  risk, conveying information about, 65
community identity, 256–257
community research, 312–313
competencies for informed decisions, 100
competencies for nurses, 351–354, 383–388
conception, testing before. *See* preventative testing
confidentiality. *See* discrimination; information, use of
congenital anomalies, 162
consent. *See* informed consent
consultations. *See* genetic testing; referrals for testing
consumer resources, 370–371. *See also* resources for additional information
coordinating nurses, 198, 351
coping, 194, 218, 235
  profound change, 317–319
  psychologists, role of, 241–245
coronary artery disease, 127, 129
costs of genetic testing, 78, 80, 261. *See also* risk estimation
  DNA banking, 81
  PGD (preimplantation genetic diagnosis), 99
  pharmacogenetics, 136
counseling. *See* genetic testing
counseling, religious, 273
counselors (genetic), defined, 183, 189. *See also* genetics professionals; support groups and counseling
  psychologists, 241–245
cousins, information about. *See* family history
CRC (colorectal cancer), 25–27. *See also* cancer
credentialing and quality assurance, 353–354, 358–360, 383–388
cryopreservation of embryos, 98
cultural identity, 188, 252–253. *See also* ethnicity and race
  socioeconomics. *See* socioeconomic issues
CYP (cytochrome P450) system, 119, 121, 297
cystic fibrosis (CF), 2–5, 7, 16–20, 158, 190
  genetic testing, 18
  newborn and neonatal screening, 91
  preventative and prenatal testing, 74, 94–95
cytoplasm, 10

D

D9N polymorphism, 129
data management in clinical trials, 306
data storage, 53. *See also* information, use of
databanks for genetic research, 62, 335
daughters, information about. *See* family history
deafness, screening for, 92–93
decision making, 83, 99–103, 192, 194, 319, 341
  choosing to share information, 229–230
  clinical trials, participation in, 301–306
  consent. *See* informed consent
  nursing role in, 304
  pharmacogenetics, 137–139
  when to refer patients for genetic testing, 147–166, 234
    indications for, 153–154

decision making (*continued*)
informing parents of child's genetic condition, 167–173
nursing assessments, 156–164
nursing framework for, 154–156
peer support, finding, 164–165
whether to test, 218, 223, 226–227, 233
declining participation in clinical trials, 302
deCODE genetics project, 295–296
definition of self, 154–155, 227–229, 253–255
depressive disorders
association studies, 295
behavioral genetics, 324
treating, 111–116, 121–123
detection of disorders. *See* genetic testing; prevention and early detection
DGS (DiGeorge syndrome), 339–340
DiA (dimeric inhibin), 95
diagnosis, establishing, 186
diagnostic testing, 75, 299
common tests, 76
DiGeorge syndrome (DGS), 339–340
disclosure of information. *See* information, use of
discrimination, 53, 82, 83, 104. *See also* ethical issues; ethnicity and race; information, use of
clinical trials, participation in, 303
community research, 312–313
genetic engineering, 278–279
genotyping and pharmacogenetics, 132
pharmacogenetics, 141
research availability and direction, 311
diseases. *See also specific disorder by name*
CDC (Centers for Disease Control), 177, 200, 366
chronic, nursing roles and, 347–350. *See also* cancer
drug response differences, 63, 118–119. *See also* pharmacogenetics
genetic discrimination and, 82
genetic signatures, 66
monitoring for. *See* surveillance and health outcome modifiers
population-based genetics studies, 295, 342–344
disposition, calculating. *See* risk estimation
distribution, drug, 117
diversity, cultural. *See* socioeconomic issues
DNA
basics of, 7–11
cloning, 280–281
RNA and protein synthesis, 9–11
testing. *See* genetic testing
DNA banking, 80–81
DNA probes, 78
D9N polymorphism, 129
documentation of clinical trials, 306
documentation of information, 53. *See also* information, use of
dose-dependent ADRs, 118. *See also* ADRs
double helix structure of DNA, 7–9
Down syndrome, 12
drug response and efficacy. *See* variable drug response
drugs. *See* pharmaceuticals; pharmacogenetics; pharmacogenomics; pharmacokinetics
duty to warn, 53, 184

E

early detection. *See* prevention and early detection
economic issues and genetic testing. *See* costs; socioeconomic issues
ECT (electroconvulsive treatment), 113
education, nurse, 351–354, 383–388
credentialing nurses, 353–354, 358–360, 383–388
education, patient, 184, 192, 205, 322. *See also* socioeconomic issues
clinical trials, participation in, 301–302
consumer resources, list of, 370–371. *See also* resources for additional information
genotyping and pharmacogenetics, 131–132
helping to make decisions, 83, 99–103, 192, 194, 319, 341
choosing to share information, 229–230
clinical trials, participation in, 301–306
nursing role in, 304
pharmacogenetics, 137–139
when to recommend testing. *See* referrals for testing
whether to test, 218, 223, 226–227, 233
informed consent, 83, 99–103, 303–306
genotyping and pharmacogenetics, 132–135
nursing role in, 304
prenatal screening, 89
Internet-based genetic services, 137, 197–199. *See also* resources for additional information
maternal multiple-marker screening results, 96
newborn screening and, 90–93
pharmacogenetics, 139–140
privacy and confidentiality rights, 309–310
providing information to clients, 65, 167–173, 186–187

referrals for testing, 147–166, 234
indications for, 153–154
nursing assessments, 156–164
nursing framework for, 154–156
peer support, finding, 164–165
reproductive choices, 93–94. *See also* preventative testing
reproductive technologies, 99
support groups and counseling, 164–165, 187, 193, 368–369
clinical trials, 303
community identity, 256–257
community relationships, 313
familial roles and relationships, 230–232
supportive care trials, 299
effecting behavioral change, 54
electroconvulsive treatment (ECT), 113
elimination, drug, 117
ELSI branch, NHGRI, 259–260, 372–374
embryo cryopreservation, 98
embryo transfer (ET), 98
embryos, testing. *See* preimplantation genetic diagnosis
emotional impact, 287–292
concerns about privacy. *See* information, use of
coping, 194, 218, 235
psychologists, role of, 241–245
decision making. *See* decision making
fear, 290–292, 322
hope, 194, 235, 287–290, 321
patience, 320
profound change, 317–319, 317–319
employment discrimination. *See* discrimination
EMs (extensive metabolizers), 119, 123
endpoints of clinical studies, 300
engineering, genetic, 278–279
environmental agents, 65
behavioral genetics, 324
community research, 312–313
GEI (gene-environment interactions), 129–130, 135–136, 138
prenatal exposures, 158–160
equality of care. *See* socioeconomic issues
equity of testing availability. *See* socioeconomic issues
estimating risk. *See* risk estimation
estriol (UE3), 95
ET (embryo transfer), 98
ethical issues, 258–259, 277–278. *See also* socioeconomic issues
cloning, 280–281
community research, 312–313
discrimination. *See* discrimination
DNA banking, 81, 83
ethnicity and race, 258
family history, 52–54
future, concerns for, 284
genetic engineering, 278–279
identity, patient's sense of, 154–155, 227–229, 253–255
oversight of genetic testing, 80
pharmacogenetics, 136–141
policy. *See* policy
pregnancy termination, 97, 276–277
religion and, 273–274
research and clinical trials, 301–311
resources for additional information, 372–374
stem cell research, 281–282
tandem mass spectrometry (MS-MS), 92
use of information. *See* information, use of
warning family members of risk, 53, 184
ethnicity and race, 48, 248–253. *See also* discrimination; family history
carrier (preconception) testing, 93–94
counseling, considerations for, 188
ethical concerns, 258
genetic assessment and, 157–160
identity, patient's sense of, 154–155, 227–229, 253–255
patient's awareness of possibilities, 104–106
pharmacogenetics and drug response, 117, 141. *See also* variable drug response
research and, 257–258, 311
risks of genetic testing, 104
TMPT levels, 124
transcultural nursing, 262–264
tuberculosis treatment, 128
exons (coding regions), DNA, 10
extensive metabolizers (EMs), 119
codeine, 123

F

Factor V Leiden mutation, 128, 130, 336–338
false positives in genetic testing, 77, 91, 96
familial consent. *See* informed consent
family and relatives
clinical trials and, 306–307
cultural context and. *See* socioeconomic issues
duty to warn, 53, 184
emotions of. *See* emotional impact

family and relatives (*continued*)
genetic causation, addressing, 171
informing parents of child's genetic condition, 167–173
inheritance. *See* inheritance
nursing interventions and, 234–237
pediatric testing, 344
roles of, 230–232, 255–256
sharing information with, 229–230
support from. *See* support groups and counseling
family history, 23–25. *See also* ethnicity and race; inheritance
assessing risk from, 156–164, 185
pharmacogenetic risk, 131
behavioral change, effecting, 54
dangers regarding, 103–104
ethics, 52–54. *See also* information, use of
future of, 205
importance of, 16–20
parentage testing, 80–81
tools and questionnaires, 27–52, 377
cancer genetics family history, 28–37
personal and health history, 38–47
Family History Public Health Initiative, 54
Family Systems Theory, 231
father, information about. *See* family history
FBR (Foundation for Blood), 366
fear, 290–292, 322
fetal development, teratogenic exposure and, 158–160
FMH. *See* family history
Foundation for Blood (FBR), 366
Fragile X syndrome, 80–81, 149–152
future, 317–329, 335, 354
family history, taking, 205
genetic counseling and education, 194–200
impact of. *See* information, impact of
need for genetics professionals, 203–207
nursing roles, 335–351
religious viewpoints, 284
research direction, 311, 353–354
transcultural nursing, 262–264

G

gastroenterology, 127
GCN (Genetics Clinical Nurse) credential, 359
GEI (gene-environment interactions), 129–130, 135–136, 138
gene mutation scanning, 78–79
gene sequencing, 78
genes, basics of, 7
genes, testing. *See* genetic testing
Genetic Alliance, 177
genetic assessments. *See* nursing assessments
genetic association studies, 294
genetic causation, addressing, 171
genetic counseling. *See* genetic testing
genetic counselors, defined, 183, 189, 233. *See also* genetics professionals; support groups and counseling
genetic discrimination. *See* discrimination
genetic engineering, 278–279
genetic information. *See* information, impact of; information, use of
genetic inheritance. *See* inheritance
genetic mapping, 62, 294, 335
genetic mutations, 12
NAT2 mutation, 130
screening for, 78–79
genetic profiles, treatment and, 120–129
genetic signatures, 66
genetic testing, 61–63, 69–107
accuracy of, 77, 137–139
cultural and economic barriers, 104–106. *See also* costs
decision making, 83, 99–103, 192, 194, 319, 341
choosing to share information, 229–230
clinical trials, participation in, 301–306
consent. *See* informed consent
nursing role in, 304
pharmacogenetics, 137–139
when to recommend testing. *See* referrals for testing
whether to test, 218, 223, 226–227, 233
definitions of, 75–77, 139–141
discrimination and stigmatization. *See* discrimination
embryo preimplantation, 75, 97–99, 275
ethics of. *See* ethical issues
future of, 317–329, 335, 354
family history, taking, 205
genetic counseling and education, 194–200
impact of. *See* information, impact of
need for genetics professionals, 203–207
nursing roles, 335–351
religious viewpoints, 284
research direction, 311, 353–354
transcultural nursing, 262–264

impact of knowledge, social. *See* information, impact of
improving techniques of, 293
indications for, 75, 153–154. *See also* referrals for testing; risk estimation
information provided by, 80–83
Internet-based services, 137, 197–199. *See also* resources for additional information
oversight of, 80
pharmacogenics and genotyping, 120–131. *See also* pharmacogenetics
psychiatry, 121
understanding meaning of, 139–141
population-based genetics studies, 295, 342–344
privacy of results. *See* information, use of
psychologists, role of, 241–245
referring patients, when appropriate, 147–166, 234
indications for, 153–154
informing parents of child's genetic condition, 167–173
nursing assessments, 156–164
nursing framework for, 154–156
peer support, finding, 164–165
research testing, 61, 80
community research, 312–313
risks of testing, 103–104
as routine, 196
for specific disorders
cancer, 25. *See also* cancer
CF (cystic fibrosis), 18
PKU screening, 69, 71–73, 85–88
Rett's syndrome, 270
SCD (sickle cell disease), 180
specific types of, 78. *See also* newborn screening; prenatal screening; preventative testing; prevention and early detection
terminology and concepts, 77
genetic variation, 12–16. *See also* inheritance
Genetics and Health Workforce Research Center, 195–197
genetics cohort research studies, 295
genetics professionals, 175–200, 321–322
future models, 194–200
growing need for, 203–207
nursing roles. *See* roles of nurses
participation in clinical trials, 302
professional organizations, 175–177, 363–366, 380
psychologists, 241–245
referrals to, 147–166, 234
indications for, 153–154
informing parents of child's genetic condition, 167–173
nursing assessments, 156–164
nursing framework for, 154–156
peer support, finding, 164–165
relationships with community, 313
resources for additional information, 369–370, 374–375
role of, 183–188
specialists, 188–191
transcultural nursing, 262–264
genetics research. *See* research
genetics resources. *See* resources for additional information
genome. *See* human genome
genomics in clinical care, 23–56
benefits. *See* health benefits of genomics
biology of human genome, 6–16
basic terminology, 7–9
current questions and research, 16
DNA, RNA, and protein synthesis, 9–11
genetic variation and inheritance patterns, 12–16
ethics. *See* ethical issues
family history. *See also* family history
importance of, 16–20
tools for collecting, 27–52, 54
introduction and growth of, 1–6
pharmacogenetics applications, 120–129. *See also* pharmacogenetics
policies for, establishing, 259–261
public information resources, 55
referring patients for genetic testing, 147–166, 234
indications for, 153–154
informing parents of child's genetic condition, 167–173
nursing assessments, 156–164
nursing framework for, 154–156
peer support, finding, 164–165
research. *See* genetics research
resources for, 66. *See also* resources for additional information
types of genetic testing, 78. *See also* newborn screening; prenatal screening; preventative testing; prevention and early detection
variations as "red flags", 25–27

genomics in health care, 333–360
nurse education, 351–354, 383–388
credentialing and quality assurance, 353–354, 358–360, 383–388
nursing roles, 335–351
research advances, 334–335, 353–354
gerontology, 127–128
Gleevec, 125, 350
global genomics, 260–261
GM-CSF protein, 64
GNCC (Genetic Nursing Credentialing Commission), 359–360
God, perspectives on. *See* religion and spirituality
government genetics agencies, 375–377
grandparents, information about. *See* family history

H

half-siblings, information about. *See* family history
haplotypes, 335
HapMap project, 62, 335
hCG (human chorionic gonadotropin), 95
HD (Huntington Disease), 92, 209–217, 219
families, impact on, 232
presymptomatic testing, 221, 224–226
health benefits of genomics, 59–67
conveying information to patients, 65. *See also* communication with patients
drug response and efficacy, 63, 118–119. *See also* pharmacogenetics
genetic testing. *See* genetic testing
health modifiers, 51–52, 64–65. *See also* prevention and early detection
genetic counseling, 191–194, 235
medical surveillance, 50, 65–66
pharmacogenetic testing and GEI, 135–136
prevention and early detection, 60–66. *See also* prevention and early detection
resources for, 66
therapeutic approaches. *See* therapeutic approaches
health care systems, 194
discrimination. *See* discrimination
pharmacogenetics, 136–137
public health, 199–200, 322
community research, 312–313
education and. *See* education, patient
Family History Public Health Initiative, 54
infectious diseases, treating, 128
newborn screening and, 88–89
population-based genetics studies, 295, 342–344
resources for genomic-based care, 66
role of genomics in, 333–360
nurse education, 351–354, 383–388
nursing roles, 335–351
research advances, 334–335, 353–354
health history questionnaire, 38–47. *See also* family history
health insurance, 104. *See also* information, use of
health modifiers, 51–52, 64–65. *See also* prevention and early detection
genetic counseling, 191–194, 235
Health Seeking Behavior outcome, 194, 235
health teaching. *See* education, patient
hearing screenings, 92–93
heart disease, 127
Herceptin, 125
hereditary breast–ovarian cancer, 221–223
hereditary disposition, calculating. *See* risk estimation
hereditary hemochromatosis, 300
hereditary non-polyposis colorectal cancer. *See* HNPCC
heritage. *See* ethnicity and race
HERZ overexpression, 125
HH (hereditary hemochromatosis), 300
history, patient. *See* family history
HIV treatment, 128
HNPCC (hereditary non-polyposis colorectal cancer), 25–27, 50–52, 330–332
hope, 194, 235, 287–290, 321
profound change, 317–319
HPA (hyperphenylalaninemia), 85
human genome
structure and function, 6–16
basic terminology, 7–9
current questions and research, 16
DNA, RNA, and protein synthesis, 9–11
genetic variation and inheritance patterns, 12–16
variations in, 25–27
Huntington Disease (HD), 92, 209–217, 219
families, impact on, 232
hyperphenylalaninemia (HPA), 85
hypersensitivity to abacavir, 128

I

ICSI (intracytoplasmic sperm injection), 98
identity, patient's sense of, 154–155, 227–229, 253–255

clinical trials and, 306–307
community identity, 256–257
concept of "normal", 274–275
profound change, 317–319
religious viewpoints, 276–277
imatinib mesylate (Gleevec), 125, 350
immigration status and genetic testing, 105–106
immunotherapy, 300
IMs (intermedia metabolizers), 119
in vitro fertilization (IVF), 98, 275. *See also* preimplantation genetic diagnosis
indications for genetics referrals, 153–154
infections during pregnancy, 159
infectious diseases, treating, 128
inflammatory bowel disease, treating, 127
information, impact of, 209–237, 330–332. *See also* future; information, use of
choosing to share information, 229–230
coping strategies, 218
decision to test, 218, 223, 226–227
emotional impact, 287–292
coping, 194, 218, 235, 241–245
decision making. *See* decision making
fear, 290–292, 322
hope, 194, 235, 287–290, 321
patience, 320
profound change, 317–319
familial roles and relationships, 230–232
nursing implications and roles, 232–237
profound change, 317–319
psychologists, role of, 241–245
redefinition of self, 227–229
relationships with community, 313
information technologies, 380
information, use of, 5, 52–54, 82, 83, 134–135. *See also* genetic testing; information, impact of
choosing to share, 229–230
clinical trials, 303, 307–311
discrimination. *See* discrimination
genotyping results, 131
nursing roles and, 309–310
position statement (ISONG), 308–310
principles of genetic counseling, 184
providing information to clients, 65, 167–173, 186–187
informed consent, 83, 99–103, 303–306. *See also* decision making
genotyping and pharmacogenetics, 132–135
nursing role in, 304
prenatal screening, 89
inheritance, 12–16
22q11 deletion syndrome, 339
bipolar disorder, 112
causation issues, addressing, 171
CF (cystic fibrosis), 18
drug response variability. *See* variable drug response
Factor V Leiden mutation, 337
family, perception of, 255–256. *See also* family and relatives
Fragile X syndrome, 151
HD (Huntington Disease), 220
PKU (phenylketonuria), 86
Rett's syndrome, 269
SCD (sickle cell disease), 179
X-SCID, 346
Institutional Review Board (IRB), 301, 378
insurance (medical), 104. *See also* information, use of
discrimination. *See* discrimination
integrating genomics into health care, 333–360
nurse education, 351–354, 383–388
credentialing and quality assurance, 353–354, 358–360, 383–388
nursing roles, 335–351
research advances, 334–335, 353–354
intergenic SNPs, 118
interindividual variability in drug response. *See* variable drug response
intermedia metabolizers (IMs), 119
International Society of Nurses in Genetics. *See* ISONG
Internet, genetic services and, 137, 197–199. *See also* resources for additional information
interpretation of risk. *See* risk estimation
intervention strategies, 135–136, 191
on families, 234–237
on tested individuals, 232–235
intracytoplasmic sperm injection (ISCI), 98
introns (noncoding regions), DNA, 10, 16
SNPs in, 118
investigational genetic testing, 62. *See also* genetic testing
IRB (Institutional Review Board), 301, 378
Islamic opinion. *See* religion and spirituality
ISONG (International Society of Nurses in Genetics, Inc.), 100–102, 176, 177, 195, 364–365
educating nurses, 352
informed consent (position statement), 304
nurse credentialing, 353–354, 359–360, 383–388

ISONG (International Society of Nurses in Genetics, Inc.) (*continued*)
privacy and confidentiality (position statement), 308–310
IVF (in vitro fertilization), 98, 275. *See also* preimplantation genetic diagnosis

J

Judaism. *See* religion and spirituality

K

kinship. *See* family and relatives

L

laws government genomics. *See* policy
leucopenia, 127
leukemia, 124. *See also* cancer
licensing policies, impact of, 259
licensure of nurses, 358–360
life, where begins, 276–277
likelihood of risk. *See* risk estimation
linkage analysis research, 294
Long QT Syndrome, 323
lung disease, 300

M

making decisions. *See* decision making
managed health care systems, 194
discrimination. *See* discrimination
pharmacogenetics, 136–137
public health, 199–200, 322
community research, 312–313
education and. *See* education, patient
Family History Public Health Initiative, 54
infectious diseases, treating, 128
newborn screening and, 88–89
population-based genetics studies, 295, 342–344
resources for genomic-based care, 66
role of genomics in, 333–360. *See also* clinical care, genomics in
nurse education, 351–354, 383–388
nursing roles, 335–351
research advances, 334–335, 353–354
managing data of clinical trials, 306
mapping genes, 62, 294
HapMap project, 335
March of Dimes recommendations for newborn screening, 84, 90
marketing services on Internet. *See* Internet, genetic services and
maternal exposures, assessing, 158–160
medical genetic evaluation, 187
medical geneticists, 190
medical history. *See* family history
medical insurance, 104. *See also* information, use of
discrimination. *See* discrimination
medical records and disclosure, 53. *See also* information, use of
medical surveillance. *See* surveillance and health outcome modifiers
medically underserved communities. *See* socioeconomic issues
medications. *See* pharmaceuticals
melanoma. *See* cancer
Mendelian inheritance patterns, 12
mental health disorders
association studies, 295
behavioral genetics, 324
treating, 111–116, 121–123
messenger RNA (mRNA), 10
metabolism, drug, 117, 119–120
acetylator status, 128, 130
aging and gerontology, 127–128
codeine, 123
CYP2D6 and Tamoxifen trials, 297
psychiatric medications, 121–122
microarray technology, 125–127
minority communities. *See* socioeconomic issues
minors. *See* children
mitochondria, 10–11
modifiers to health outcomes. *See* health modifiers
monitoring clinical trials, 306
monitoring patient health. *See* surveillance and health outcome modifiers
mood disorders
association studies, 295
behavioral genetics, 324
treating, 111–116, 121–123
mood stabilizers, 113
morality perspectives. *See* religion and spirituality
Mormons. *See* religion and spirituality
mother, information about. *See* family history
mRNA (messenger RNA), 10
MS-MS (tandem mass spectrometry), 91–92
mtDNA, 10–11
multifactorial inheritance, 12
multiple-marker maternal serum screening, 95–96
mutations, genetic, 12

NAT2 mutation, 130
screening for, 78–79

N

NAT2 mutation, 130
National Bioethics Advisory Commission (NBAC), 312–313
National Cancer Institute (NCI), 331
National Human Genome Research Institute (NHGRI), 247, 259–260
National Society of Genetics Counselors (NSGC), 176, 177
NBAC (National Bioethics Advisory Commission), 312–313
NCHPEG (National Coalition for Health Professional Development in Genetics), 352, 383, 390–394
NCI (National Cancer Institute), 331
negative predictive value, defined, 77
negative results, risks of, 226, 228
neonatal screening. *See* newborn screening
nephews, information about. *See* family history
neural tube defects, screening for, 70, 95
newborn screening, 70–73, 75, 81, 84–90, 341, 366
common tests, 76
family education and support, 90–93
indications for genetics referrals, 160–163
PKU screening, 69, 71–73
detailed information about, 85–88
potential risks of, 103–104
Newborn Screening Task Force, 90–91
NHGRI (National Human Genome Research Institute), 247, 259–260
nieces, information about. *See* family history
non-dose-dependent ADRs, 118. *See also* ADRs
non-PKU hyperphenylalaninemia, 85
noncoding regions (introns), DNA, 10, 16
SNPs in, 118
nonresponsiveness to drug treatment. *See* variable drug response
"normal," concept of, 274–275
NSGC (National Society of Genetics Counselors), 176, 177
nucleotides, 118
nurse coordinators, 198, 351
nurse education, 351–354, 383–388
credentialing, 353–354, 358–360, 383–388
nurses, role of, 191–194
case coordinators, 198
chronic disease, 347–350
future models, 196–197, 335–351
genetics professionals, 183–188
in genetics research, 313–315
genotyping and, 130–131
informed decisions and consent, 100–102, 304
informing parents of child's genetic condition, 167–173
patience, 320
privacy and confidentiality, 309–310
profound change, 317–319
religion and, 283
transcultural nursing, 262–264
nursing assessments
deciding whether to test, 218, 223, 226–227, 233
family history. *See* family history
referrals for testing, 156–164
resources for additional information, 377
of risk. *See* risk estimation
transcultural nursing, 262–263
nursing interventions, 135–136, 191
on families, 234–237
on tested individuals, 232–235
nursing outcomes, 193, 235. *See also* health modifiers; treatment
nursing resources. *See* resources for additional information

O

OI (Osteogenesis Imperfecta), 288–289
oncology. *See* cancer
ONS (Oncology Nursing Society), 196, 263, 348, 365
oral contraceptives, 130
Osteogenesis Imperfecta (OI), 288–289
outcome modification, 191–194. *See also* health modifiers; treatment
genetic counseling, 235
ovarian cancer, 221–223
oversight of genetic testing, 80

P

PAH (phenylalanine hydroxylase) deficiency, 85
pain management, 123, 299
parentage testing, 80–81, 103
parental consent. *See* informed consent

parents. *See* family and relatives; family history
patent policies, impact of, 259
paternity testing, 80–81, 103
patience, 320
patient competencies for informed decisions, 100
patient education, 184, 192, 205, 322. *See also* socioeconomic issues
  clinical trials, participation in, 301–302
  consumer resources, list of, 370–371. *See also* resources for additional information
  genotyping and pharmacogenetics, 131–132
  helping to make decisions, 83, 99–103, 192, 194, 319, 341
    choosing to share information, 229–230
    clinical trials, participation in, 301–306
    nursing role in, 304
    pharmacogenetics, 137–139
    whether to test, 218, 223, 226–227, 233
  informed consent, 83, 99–103, 303–306
    genotyping and pharmacogenetics, 132–135
    nursing role in, 304
    prenatal screening, 89
  informing parents of child's genetic condition, 167–173
  Internet-based genetic services, 137, 197–199. *See also* resources for additional information
  maternal multiple-marker screening results, 96
  newborn screening and, 90–93
  pharmacogenetics, 139–140
  privacy and confidentiality rights, 309–310
  providing information to clients, 65, 167–173, 186–187
  referrals for testing, 147–166, 234
    indications for, 153–154
    informing parents of child's genetic condition, 167–173
    nursing assessments, 156–164
    nursing framework for, 154–156
    peer support, finding, 164–165
  reproductive choices, 93–94. *See also* preventative testing
  reproductive technologies, 99
  support groups and counseling, 164–165, 187, 193, 368–369
    clinical trials, 303
    community identity, 256–257
    community relationships, 313
    familial roles and relationships, 230–232
    supportive care trials, 299
patient emotions. *See* emotional impact
patient history, 38–47. *See also* family history
patient identity, sense of, 154–155, 227–229, 253–255
patient safety. *See* accuracy of genetic testing; risk estimation
patients, communication with, 65, 205–207, 336
  clinical trials, participation in, 301–302
  familial roles and relationships, 236–237. *See also* family and relatives
  helping to make decisions, 83, 99–103, 192, 194, 319, 341
    choosing to share information, 229–230
    clinical trials, participation in, 301–306
    nursing role in, 304
    pharmacogenetics, 137–139
    whether to test, 218, 223, 226–227, 233
  informed consent, 83, 99–103, 303–306
    genotyping and pharmacogenetics, 132–135
    nursing role in, 304
    prenatal screening, 89
  informing parents of child's genetic condition, 167–173
  obtaining family history, 377. *See also* family history
  prenatal screening and diagnosis, 99–103
  providing information to clients, 167–173, 186–187, 186–187
  referrals for testing, 147–166, 234
    indications for, 153–154
    informing parents of child's genetic condition, 167–173
    nursing assessments, 156–164
    nursing framework for, 154–156
    peer support, finding, 164–165
  relationships with community, 313
  risk, conveying information about, 65
PD (pharmacodynamics), 120
PDQ (Physician Data Query) databases, 190
pediatric testing, 344. *See also* newborn screening; prenatal screening
pedigree construction, 48–49. *See also* family history
peer support. *See* support groups and counseling
perigenic SNPs, 118
personal health history, 38–47. *See also* family history; information, impact of
persuasion, 321
PGD (preimplantation genetic diagnosis), 75, 97–99, 275
pharmaceuticals. *See also* pharmacogenetics

cellular reactions to drugs, 120
drug response information, 63, 118–119
environmental agents, 65
behavioral genetics, 324
community research, 312–313
GEI (gene-environment interactions), 129–130, 135–136, 138
prenatal exposures, 158–160
pharmacodynamics, 120
pharmacogenetics, 116–119, 141–142
clinical applications, 120–129
CYP2D6 and Tamoxifen trials, 297
dangers of reliability, 137–139
ethical and societal issues, 136–141
GEI (gene-environment interactions), 129–130
nursing implications and roles, 130–136
research and informed consent, 133, 298–301, 353–354
pharmacogenomics, 117, 141–142, 353–354
pharmacokinetics, 119–120
phases of clinical trials, 298
PhD scientists, 190
phenotyping, 121
phenylketonuria. *See* PKU screening
Philadelphia (Ph) chromosome, 125, 349
philosophical perspectives. *See* religion and spirituality
physical assessment, genetic testing referral from, 162, 164
PK (pharmacokinetics), 119–120
PKU screening, 69, 71–73
detailed information about, 85–88
PMs (poor metabolizers), 119, 120
antidepressants, 122
codeine, 123
policy, 259–261, 320
nurse credentialing and quality assurance, 353–354, 358–360, 383–388
religious considerations, 282–283
resources for additional information, 372–374
polymorphisms, 116, 118, 334–335
D9N and cigarette smoking, 129
poor metabolizers (PMs), 119, 120
antidepressants, 122
codeine, 123
population-based genetics studies, 295, 342–344
position statements, 365–366
informed decisions and consent, 100–102, 304
privacy and confidentiality, 308–310
religious, 273
positive predictive value, defined, 77
post-birth assessment. *See* newborn screening
potential risks. *See* risk estimation
preconception testing. *See* preventative testing
predicting genetic disorders and conditions. *See* genetic testing
predictive (presymptomatic) testing, 75, 81, 323. *See also* preventative testing
common tests, 76
HD (Huntington Disease), 221, 224–226
hereditary breast–ovarian cancer, 221–222
predictive values, defined, 77
pregnancy, exposures during, 158–160
pregnancy termination, 97, 276–277
stem cell research, 281–282
pregnancy, testing before. *See* preventative testing
pregnancy, testing during. *See* prenatal screening
preimplantation genetic diagnosis (PGD), 75, 97–99, 275
prenatal screening and diagnosis, 69–70, 74, 75, 81, 95–103, 341
common tests, 76
cystic fibrosis. *See* CF (cystic fibrosis)
family history assessment, 157–160
maternal exposures, 158–160
nursing care during, 99–103
PGD (preimplantation genetic diagnosis), 75, 97–99
potential risks of, 103–104
religious viewpoints, 276–277
presymptomatic testing. *See* predictive testing
preventative (carrier) testing, 70, 74, 81, 93–95. *See also* predictive testing
indications for, 75
potential risks of, 103–104
prevention and early detection, 51, 60–66, 323. *See also* risk estimation
behavioral change, effecting, 54
carrier (preconception) testing. *See* preventative testing
clinical trials, 299
drug response and efficacy, 63, 118–119. *See also* pharmacogenetics
genetic counseling and, 191–194
genetic testing. *See* genetic testing
health modifiers, 51–52, 64–65. *See also* prevention and early detection
genetic counseling, 191–194, 235
medical surveillance, 50, 65–66
pharmacogenetic testing and GEI, 135–136

prevention and early detection (*continued*)
newborn screening. *See* newborn screening
prenatal screening. *See* prenatal screening
presymptomatic testing. *See* predictive testing
for specific disorders
22q11 deletion syndrome, 340
Fragile X syndrome, 151
SCD (sickle cell disease), 153–154
therapeutic approaches, 63–64. *See also* genomics in clinical care
monitoring pharmacogenetic responses, 135–136
primary care genetics, 194
primary care physicians, 196
priorities for genetic research, 328–329
privacy. *See* discrimination; information, use of
probes, DNA, 78
professional organizations, 175–177, 363–366, 380, 390–394
professionals. *See* genetics professionals
profiles, treatment and, 120–129
profiling, racial, 254. *See also* ethnicity and race
protein synthesis, basics of, 9–11
prototype family history tools, 54
psychiatric illness
association studies, 295
behavioral genetics, 324
treating, 111–116, 121–123
psychological impact. *See* information, impact of
public education. *See* education, patient
public health, 199–200, 322
community research, 312–313
education and. *See* education, patient
Family History Public Health Initiative, 54
infectious diseases, treating, 128
newborn screening and, 88–89
population-based genetics studies, 295, 342–344
public information resources, 55

Q

quality of assurance of nursing care, 358–360
quality of life trials, 299
questionnaires for family history, 27–52, 377
behavioral change, effecting, 54
cancer genetics family history, 28–37
personal and health history, 38–47

R

race. *See* ethnicity and race
reactions to drugs. *See* ADRs
receptor-target affinity, 117, 120
records of information, 53. *See also* information, use of
recurrence. *See* inheritance
"red flags," discovering, 25–27
redefinition of self, 154–155, 227–229, 253–255
referrals for testing, 147–166, 234
indications for, 153–154
informing parents of child's genetic condition, 167–173
nursing assessments, 156–164
nursing framework for, 154–156
peer support, finding, 164–165
Regional Genetics Program, 182
registered nurses, 343
regulation. *See* policy
relatives. *See* family and relatives; family history
reliability. *See* accuracy of genetic testing
relief from uncertainty, 223, 226–227
religion and spirituality, 267–284. *See also* ethnicity and race
cloning, 280–281
future, concerns for, 284
genetic engineering, 278–279
identity as human, 277
"normal," concept of, 274–275
nursing implications and roles, 283
policies, establishing, 282–283
prenatal screening, 276–277
stem cell research, 281–282
reproductive choices. *See* preventative testing
reproductive technologies, 97–99
cloning, 280
research, 61, 80, 292
clinical trials, 296–298
cloning, 280
community research, 312–313
emotional impact, 287–292
concerns about privacy. *See* information, use of
coping, 194, 218, 235, 241–245, 317–319
decision making. *See* decision making
fear, 290–292, 322
hope, 194, 235, 287–290, 321
patience, 320
profound change, 317–319
ethics of. *See* ethical issues
ethnicity and, 257–258, 311
future of, 317–329, 335, 354
family history, taking, 205

genetic counseling and education, 194–200
impact of. *See* information, impact of
need for genetics professionals, 203–207
nursing roles, 335–351
religious viewpoints, 284
research direction, 311, 353–354
transcultural nursing, 262–264
to improve testing techniques, 293
integrating genomics into health care, 334–335
pharmacogenetics and pharmacogenomics, 133, 353–354
policies for, establishing, 259–261
priorities for, 328–329
resources for, 66, 367
stem cell research, 281–282
residual disease, monitoring for, 65–66. *See also* surveillance and health outcome modifiers
resources for additional information, 66, 208, 363–380. *See also specific resource topic*
clinical trials, 315
costs of genetic testing, 78, 80
ethics, 278
pharmacogenetics research, 140
professional and resource organizations, 177
public information, 55
Rett's syndrome, 271
subspecialty genetics resources, 190
support groups, 155
transcultural nursing, 263
Rett's syndrome, 267–271
revealing test results. *See* information, impact of
ribosomes, 10
rights to privacy. *See* information, use of
Risk Detection outcome, 194
risk estimation, 50, 186. *See also* nursing assessments; prevention and early detection
assessing from family history, 156–164, 185
pharmacogenetics, 131
behavioral change, effecting, 54
conveying information to patients, 65
genetic research, 303
hereditary breast–ovarian cancer, 223
identity, patient's sense of, 154–155, 227–229, 253–255
informed consent, 83, 99–103, 303–306. *See also* decision making
genotyping and pharmacogenetics, 132–135
nursing role in, 304
prenatal screening, 89
predictive (presymptomatic) testing, 75, 81, 323
common tests, 76
HD (Huntington Disease), 221, 224–226
hereditary breast–ovarian cancer, 221–222
preventative (carrier) testing, 70, 74, 81, 93–95
indications for, 75
potential risks of, 103–104
resources for additional information, 377
risks of genetic testing, 103–104
RNA (ribonucleic acid), 10–11
roles of nurses, 191–194
case coordinators, 198
chronic disease, 347–350
future models, 196–197, 335–351
genetics professionals, 183–188
in genetics research, 313–315
genotyping and, 130–131
informed decisions and consent, 100–102, 304
informing parents of child's genetic condition, 167–173
patience, 320
privacy and confidentiality, 309–310
profound change, 317–319
religion and, 283
transcultural nursing, 262–264
when to refer patients for genetic testing, 147–166, 234
indications for, 153–154
nursing assessments, 156–164
nursing framework for, 154–156
peer support, finding, 164–165
Roman Catholicism. *See* religion and spirituality
rRNA (ribosomal RNA), 10

S

SACGHS (Secretary's Advisory Committee of Genetics, Health, and Society), 259–260
safeguarding privacy. *See* information, use of
safety, patient. *See* accuracy of genetic testing; risk estimation
SCDs (sickle cell diseases), 84, 89, 104, 177–182, 251
SCID (severe combined immunodeficiency), 318–319, 345–347
screening, genetic. *See* genetic testing
screening trials, 299
search engines, list of, 378–379
selective serotonin reuptake inhibitors. *See* SSRI
self, patient's sense of, 154–155, 227–229, 253–255

sensitivity of genetic testing, 77
sequence analysis, 78
services, genetic. *See* genetic testing
severe combined immunodeficiency (SCID), 318–319, 345–347
sharing information. *See* information, use of
siblings, information about. *See* family history
sickle cell diseases. *See* SCDs
Sigma Theta Tau International, 366
signatures, genetic, 66
single-gene disorders, 12
single-nucleotide polymorphisms. *See* SNPs
sisters, information about. *See* family history
slow acetylators, 128, 130
smoking tobacco, 129–130
SNPs (single-nucleotide polymorphisms), 118, 334–335
social impact. *See* information, impact of
societal issues
  community identity, 256–257
  ethics. *See* ethical issues
  family. *See* family and relatives
  genomics policies, 259–261, 320
    nurse credentialing and quality assurance, 353–354, 358–360, 383–388
    religious considerations, 282–283
    resources for additional information, 372–374
  impact of knowledge. *See* information, impact of
  public health. *See* public health
  spirituality. *See* religion and spirituality
  transcultural nursing, 262–264
societies, professional, 175–177, 363–366, 380, 390–394
socioeconomic issues, 104–106. *See also* costs; ethnicity and race
  barriers to clinical trial participation, 301–302
  community identity, 256–257
  costs. *See* costs of genetic testing
  counseling, considerations for, 188
  equality of care, 261
  ethics. *See* ethical issues
  family, role of. *See* family and relatives
  identity, patient's sense of, 154–155, 227–229, 253–255
  impact of knowledge. *See* information, impact of
  pharmacogenetics, 136
  research among diverse populations, 311
  transcultural nursing, 262–264
sons, information about. *See* family history
"soul," concept of, 276–277
specialists, genetic, 188–191
specificity of genetic testing, 77
spirituality. *See* religion and spirituality
SSRI (selective serotonin reuptake inhibitors), 122
standardized pedigree symbols, list of, 49
stem cell research, 281–282
stigmatization. *See* discrimination
storage of information, 53. *See also* information, use of
structure of DNA, 7–9
subspecialty genetics resources, 190
support groups and counseling, 164–165, 187, 193, 368–369
  clinical trials, 303
  community identity, 256–257
  community relationships, 313
  familial roles and relationships, 230–232
  supportive care trials, 299
surveillance and health outcome modifiers, 50, 65–66
  pharmacogenetic testing and GEI, 135–136
susceptibility of risk. *See* risk estimation
susceptibility testing. *See* predictive testing
symptom alleviation, 299
syphilis study, 251

T

talking to patients. *See* communication with patients
tandem mass spectrometry (MS-MS), 91–92
teaching. *See* education, patient
teratogenic agents, exposure to, 158–160
terminating pregnancy, 97, 276–277
  stem cell research, 281–282
testing. *See* genetic testing
theological perspectives. *See* religion and spirituality
therapeutic agents. *See* pharmaceuticals
therapeutic approaches, 63–64. *See also* genomics in clinical care
  monitoring pharmacogenetic responses, 135–136
therapeutic cloning, 280
therapy. *See* treatment
thrombophilia, 128, 130
TMPT (thiopurine S nethyltransferase), 124, 127
tobacco smoking, 129–130
training nurses, 351–354, 383–388
transcultural nursing, 262–264
transfer RNA (tRNA), 10

treatment
22q11 deletion syndrome, 340
CML (chronic myeloid leukemia), 350
Factor V Leiden mutation, 338
Fragile X syndrome, 151
infectious diseases, 128
nursing outcomes, 193, 235. *See also* health modifiers; treatment
pharmacogenetics. *See* pharmacogenetics
research. *See* clinical genetic testing
SCD (sickle cell disease), 180
X-SCID, 346
tricyclic antidepressants, 121–122
tRNA (transfer RNA), 10
trust in genetic testing. *See* accuracy of genetic testing
tumors. *See* cancer
Tuskegee syphilis study, 251
22q11 deletion syndrome, 339–340

U

UE3 (estriol), 95
ultrarapid metabolizers (UMs), 119, 120, 122
ultrasound screening, 95
uncertainty, relief from, 223, 226–227
uncles, information about. *See* family history
U.S. government genetics agencies, 375–377
utility of genetic testing, 77

V

vaccine development, 63–64
validity of genetic testing, 77, 137–139
variable drug response, 63, 118–119. *See also* pharmacogenetics
variant PKU, 85
VCFS (velo-cardio-facial syndrome), 339–340
venous thrombosis, 128, 130

W

W128X mutation, 74. *See also* CF (cystic fibrosis)
warning family members of risk, 53, 184
Web-based resources. *See* resources for additional information
WHO (World Health Organization), 261
workforce, nursing, 363
world health. *See* public health

X

X-linked recessive inheritance, 12, 15
X-SCID (X-linked severe combined immunodeficiency), 346–347

CPSIA information can be obtained at www.ICGtesting.com
Printed in the USA
BVOW021308230413

318908BV00002B/4/P